AF675299

Silver and Gold Compounds as Antibiotics

Silver and Gold Compounds as Antibiotics

Guest Editors

Sotiris K Hadjikakou
Christina N. Banti

MDPI

Basel • Beijing • Wuhan • Barcelona • Belgrade • Novi Sad • Cluj • Manchester

Guest Editors

Sotiris K Hadjikakou
Department of Chemistry
Univeristy of Ioannina
Ioannina
Greece

Christina N. Banti
Department of Chemistry
University of Ioannina
Ioannina
Greece

Editorial Office
MDPI AG
Grosspeteranlage 5
4052 Basel, Switzerland

This is a reprint of the Special Issue, published open access by the journal *Antibiotics* (ISSN 2079-6382), freely accessible at: https://www.mdpi.com/journal/antibiotics/special_issues/silver_antibiotics.

For citation purposes, cite each article independently as indicated on the article page online and as indicated below:

Lastname, A.A.; Lastname, B.B. Article Title. *Journal Name* **Year**, *Volume Number*, Page Range.

ISBN 978-3-7258-4008-3 (Hbk)
ISBN 978-3-7258-4007-6 (PDF)
https://doi.org/10.3390/books978-3-7258-4007-6

© 2025 by the authors. Articles in this book are Open Access and distributed under the Creative Commons Attribution (CC BY) license. The book as a whole is distributed by MDPI under the terms and conditions of the Creative Commons Attribution-NonCommercial-NoDerivs (CC BY-NC-ND) license (https://creativecommons.org/licenses/by-nc-nd/4.0/).

Contents

About the Editors

Sotiris K Hadjikakou

Sotiris K. Hadjikakou is Professor of Inorganic Chemistry at the Department of Chemistry, University of Ioannina, Greece, and Director of the Laboratory of Biological Inorganic Chemistry. He also serves as Director of the International Graduate and PhD Programs in Biological Inorganic Chemistry. He earned his B.Sc. and Ph.D. with honors from the Department of Chemistry at Aristotle University of Thessaloniki. His postdoctoral research was carried out in bio-organometallic chemistry at the University of Dortmund, Germany, and he was a visiting researcher at the University of Essex, UK.

His research lies in the field of biological inorganic chemistry, with emphasis on the development of metal-based anticancer and antimicrobial therapeutic agents. His work further involves drug delivery systems and the design of antimicrobial medical devices.

Professor Hadjikakou has authored a high number of peer-reviewed publications and is the holder of patents. He has coordinated several research programs. He has supervised many postdoctoral researchers, PhD candidates, and master's students. He has also given invited lectures at scientific conferences and academic institutions.

He is currently Editor of *Inorganic Chemistry Communications* and serves on the editorial boards of the *International Journal of Molecular Sciences; Antibiotics; Frontiers in Chemical Biology;* and *Phosphorus, Sulfur, and Silicon and the Related Elements.* He has served as Guest Editor for multiple Special Issues and is an active reviewer for many scientific publishers.

He is a member of professional societies, and his participation in scientific networks includes COST Actions related to antimicrobial resistance and multidrug-resistant tumors. He has served on advisory committees of major international conferences and was Chair of the 16th International Symposium on Applied Bioinorganic Chemistry and the 20th Hellenic Symposium on Medicinal Chemistry.

Christina N. Banti

Dr. Banti received her bachelor's degree in Biological Applications and Technologies, University of Ioannina, Greece, in 2009. She was awarded an M.Sc. in Bioinorganic Chemistry, University of Ioannina, and obtained her PhD from the Department of Physiology, School of Medicine, University of Ioannina, Greece. She is now an adjunct lecturer in the Department of Chemistry, University of Ioannina. She is a member of the Editorial Board in the *Journal of Pharmaceutical Sciences* (Elsevier), a member of the Editorial Advisory Board in *Cell Biochemistry and Function* (Wiley), and Associate Editor in *Frontiers in Chemical Biology*, Section Molecular Sciences (Frontiers). Her research interests are mainly focused on biological inorganic chemistry and particularly on the evaluation of anticancer and antibacterial properties of metal complexes by studying their interaction with intracellular molecules.

antibiotics

Editorial

Silver and Gold Compounds as Antibiotics

Christina N. Banti and Sotiris K. Hadjikakou *

Laboratory of Biological Inorganic Chemistry, University of Ioannina, 45110 Ioannina, Greece; cbanti@uoi.gr
* Correspondence: shadjika@uoi.gr

This Special Issue entitled "Silver and Gold Compounds as Antibiotics" covers a selection of recent research and review articles focused on biological inorganic chemistry. This Special Issue offers a comprehensive summary of the latest advancements and emerging trends in the rapidly evolving field of the role of silver and gold compounds as antibiotics. The insights provided in this Special Issue will enhance readers' understanding of antimicrobial metal compounds.

Infectious diseases remain a significant global health threat, being among the most difficult challenges in modern medicine [1]. With the growing resistance of bacteria to conventional antibiotics, the need to discover and develop new antimicrobial agents has become increasingly urgent [1]. Silver and gold salts have been used for centuries to combat microbial infections, and the development of new silver and gold compounds is now crucial in overcoming the limitations of existing antibiotics and preventing the emergence of multidrug-resistant bacteria [2,3].

This Special Issue brings together a wide range of articles examining the potential of silver and gold compounds in antimicrobial applications. It is composed by eleven articles or reviews, including the following:

- Judita Puišo et al. in "Antimicrobial Properties of Newly Developed Silver-Enriched Red Onion–Polymer Composites" show that biogenic silver nanoparticles, produced in a cyanobacterial culture in various sizes, effectively combat a range of local fruit pathogens, highlighting their potential as versatile, eco-friendly biocontrol agents that support food security and sustainability [4].
- Munirah F. Aldayel et al. in "Differential Antimicrobial Effect of Three-Sized Biogenic Silver Nanoparticles as Broad-Spectrum Antibacterial Agents against Plant Pathogens" report the synthesis and characterization of seven different silver nanoparticles (AgNPs) from various fungi isolated in Brazil, demonstrating their broad-spectrum antifungal activity against both pathogenic yeasts and agricultural phytopathogens, with potential for diverse applications due to their effective and non-specific action [5].
- Luiz Gustavo Ribeiro et al. in "Antifungal Activity of Mycogenic Silver Nanoparticles on Clinical Yeasts and Phytopathogens" demonstrate the development of antimicrobial 3D-printed collagen scaffolds with in situ synthesized silver nanoparticles (AgNPs) using UV irradiation, revealing that the method effectively controls the size and shape of AgNPs, enhances the thermal stability and swelling capacity of the scaffolds, and demonstrates high bactericidal activity against both Gram-negative and Gram-positive bacteria [6].
- Sofia Municoy et al. in "Development of 3D-Printed Collagen Scaffolds with In-Situ Synthesis of Silver Nanoparticles" investigate the antifungal and antiamoebic effects of silver nanorings (AgNRs) and compare them with other silver nanoparticles, revealing that AgNRs demonstrate notable antimicrobial activity against both fungi and amoebae, offering a promising alternative to traditional antifungal and antiamoebic therapies in the face of increasing drug resistance [7].
- Sara González-Fernández et al. in "A Promising Antifungal and Antiamoebic Effect of Silver Nanorings, a Novel Type of AgNP" investigate how varying the size of

Citation: Banti, C.N.; Hadjikakou, S.K. Silver and Gold Compounds as Antibiotics. *Antibiotics* **2024**, *13*, 850. https://doi.org/10.3390/antibiotics13090850

Received: 15 August 2024
Accepted: 29 August 2024
Published: 5 September 2024

Copyright: © 2024 by the authors. Licensee MDPI, Basel, Switzerland. This article is an open access article distributed under the terms and conditions of the Creative Commons Attribution (CC BY) license (https://creativecommons.org/licenses/by/4.0/).

silver nanoparticles biosynthesized by Cyanothece-like cyanobacteria affects their antimicrobial efficacy, revealing that smaller nanoparticles are more effective against pathogenic bacteria, including MRSA and *Streptococcus* sp., with optimal size control achieved by adjusting precursor concentrations [8].

- Nermin A. El Semary et al. in "Multidrug-Resistant Bacterial Pathogens and Public Health: The Antimicrobial Effect of Cyanobacterial-Biosynthesized Silver Nanoparticles" demonstrate that cyanobacteria can biosynthesize silver nanoparticles that are active against pathogenic bacteria, and the size of silver nanoparticles can be controlled [9].
- Erika Alejandra Jardón-Romero et al. in "Antimicrobial Activity of Biogenic Silver Nanoparticles from *Syzygium aromaticum* against the Five Most Common Microorganisms in the Oral Cavity" found that biogenic silver nanoparticles synthesized using extracts of *Syzygium aromaticum* through green synthesis demonstrated effective antimicrobial activity against a range of microorganisms in oral cavities [10].
- Elvira Ivonne Murillo-Rábago et al. in "Optimized Synthesis of Small and Stable Silver Nanoparticles Using Intracellular and Extracellular Components of Fungi: An Alternative for Bacterial Inhibition" used fungal extracts to create small-sized, stable silver nanoparticles, demonstrating their efficacy as antibacterial agents with potential clinical applications [11].
- Amal Adnan Ashour et al. in "Comparison and Advanced Antimicrobial Strategies of Silver and Copper Nanodrug-Loaded Glass Ionomer Cement against Dental Caries Microbes" show that adding metronidazole and copper nanoparticles extracted from *Thymus vulgaris* to glass ionomer cement increases the cement's antimicrobial efficacy against bacteria, while maintaining compressive strength and possibly enhancing dental restorations [12].
- Yolice Patricia Moreno Ruiz et al. in "Advanced Hydrogels Combined with Silver and Gold Nanoparticles against Antimicrobial Resistance" reviewed the potential of hydrogels combined with metallic nanoparticles as a promising approach to combat multidrug-resistant bacteria, emphasizing the antibacterial and antibiofilm properties, delivery mechanisms, and antimicrobial resistance [13].
- Michele Fiore et al. in "Is Silver the New Gold? A Systematic Review of the Preclinical Evidence of Its Use in Bone Substitutes as Antiseptic" reviewed the preclinical data of silver ions or silver nanoparticles in bone substitutes as antiseptic to bone infection treatments [14].

The contributions in this Special Issue highlight the significant potential of silver and gold compounds in combating bacterial infections. However, bringing these laboratory findings into clinical use will require collaboration among researchers, healthcare professionals, and policymakers. Ongoing research into innovative synthesis methods, deeper insights into their mechanisms of action, and comprehensive assessments of their safety and effectiveness are essential for progressing these compounds toward clinical applications.

We sincerely thank all the authors, reviewers, and the editorial team for their commitment and effort in making this Special Issue possible. We also express our gratitude to our readers for their interest and involvement in this important topic.

In light of the challenges posed by bacterial resistance and the ongoing search for new antibiotics, the research in this Special Issue provides a beacon of hope. The innovative strategies and encouraging outcomes showcased here emphasize the potential of silver and gold compounds as valuable tools in our antimicrobial arsenal. We encourage you to explore these articles, immerse yourself in the intriguing realm of antimicrobial metal compounds, and join us in the fight against bacterial infections to improve public health.

Conflicts of Interest: The authors declare no conflicts of interest.

References

1. Fischbach, M.A.; Walsh, C.T. Antibiotics for Emerging Pathogens. *Science* **2009**, *325*, 1089–1093. [CrossRef] [PubMed]
2. Ketikidis, I.; Banti, C.N.; Kourkoumelis, N.; Tsiafoulis, C.G.; Papachristodoulou, C.; Kalampounias, A.G.; Hadjikakou, S.K. Conjugation of Penicillin-G with Silver(I) Ions Expands Its Antimicrobial Activity against Gram Negative Bacteria. *Antibiotics* **2020**, *9*, 25. [CrossRef] [PubMed]
3. Büssing, R.; Bublitz, A.; Karge, B.; Brönstrup, M.; Strowig, T.; Ott, I. An organometallic hybrid antibiotic of metronidazole with a Gold(I) N-Heterocyclic Carbene overcomes metronidazole resistance in Clostridioides difficile. *J. Biol. Inorg. Chem.* **2024**, *29*, 511–518. [CrossRef] [PubMed]
4. Puišo, J.; Žvirgždas, J.; Paškevičius, A.; Arslonova, S.; Adlienė, D. Antimicrobial Properties of Newly Developed Silver-Enriched Red Onion–Polymer Composites. *Antibiotis* **2024**, *13*, 441. [CrossRef] [PubMed]
5. Aldayel, M.F.; El Semary, N.; Adams, D.G. Differential Antimicrobial Effect of Three-Sized Biogenic Silver Nanoparticles as Broad-Spectrum Antibacterial Agents against Plant Pathogens. *Antibiotics* **2023**, *12*, 1114. [CrossRef] [PubMed]
6. Ribeiro, L.G.; Roque, G.S.C.; Conrado, R.; De Souza, A.O. Antifungal Activity of Mycogenic Silver Nanoparticles on Clini-cal Yeasts and Phytopathogens. *Antibiotics* **2023**, *12*, 91. [CrossRef] [PubMed]
7. Municoy, S.; Antezana, P.E.; Bellino, M.G.; Desimone, M.F. Development of 3D-Printed Collagen Scaffolds with In-Situ Synthesis of Silver Nanoparticles. *Antibiotics* **2023**, *12*, 16. [CrossRef] [PubMed]
8. González-Fernández, S.; Lozano-Iturbe, V.; Menéndez, M.F.; Ordiales, H.; Fernández-Vega, I.; Merayo, J.; Vazquez, F.; Quirós, L.M.; Martín, C. A Promising Antifungal and Antiamoebic Effect of Silver Nanorings, a Novel Type of AgNP. *Antibiotics* **2022**, *11*, 1054. [CrossRef] [PubMed]
9. El Semary, N.A.; Bakir, E.M. Multidrug-Resistant Bacterial Pathogens and Public Health: The Antimicrobial Effect of Cy-anobacterial-Biosynthesized Silver Nanoparticles. *Antibiotics* **2022**, *11*, 1003. [CrossRef] [PubMed]
10. Jardón-Romero, E.A.; Lara-Carrillo, E.; González-Pedroza, M.G.; Sánchez-Mendieta, V.; Salmerón-Valdés, E.N.; Toral-Rizo, V.H.; Olea-Mejía, O.F.; López-González, S.; Morales-Luckie, R.A. Antimicrobial Activity of Biogenic Silver Nanopar-ticles from Syzygium aromaticum against the Five Most Common Microorganisms in the Oral Cavity. *Antibiotics* **2022**, *11*, 834. [CrossRef] [PubMed]
11. Murillo-Rábago, E.I.; Vilchis-Nestor, A.R.; Juarez-Moreno, K.; Garcia-Marin, L.E.; Quester, K.; Castro-Longoria, E. Opti-mized Synthesis of Small and Stable Silver Nanoparticles Using Intracellular and Extracellular Components of Fungi: An Alternative for Bacterial Inhibition. *Antibiotics* **2022**, *11*, 800. [CrossRef] [PubMed]
12. Ashour, A.A.; Felemban, M.F.; Felemban, N.H.; Enan, E.T.; Basha, S.; Hassan, M.M.; Gad El-Rab, S.M.F. Comparison and Advanced Antimicrobial Strategies of Silver and Copper Nanodrug-Loaded Glass Ionomer Cement against Dental Caries Microbes. *Antibiotics* **2022**, *11*, 756. [CrossRef] [PubMed]
13. Moreno Ruiz, Y.P.; de Almeida Campos, L.A.; Alves Agreles, M.A.; Galembeck, A.; Macário Ferro Cavalcanti, I. Advanced Hydrogels Combined with Silver and Gold Nanoparticles against Antimicrobial Resistance. *Antibiotic* **2023**, *12*, 104. [CrossRef] [PubMed]
14. Fiore, M.; Bruschi, A.; Giannini, C.; Morante, L.; Rondinella, C.; Filippini, M.; Sambri, A.; De Paolis, M. Is Silver the New Gold? A Systematic Review of the Preclinical Evidence of Its Use in Bone Substitutes as Antiseptic. *Antibiotics* **2022**, *11*, 995. [CrossRef] [PubMed]

Disclaimer/Publisher's Note: The statements, opinions and data contained in all publications are solely those of the individual author(s) and contributor(s) and not of MDPI and/or the editor(s). MDPI and/or the editor(s) disclaim responsibility for any injury to people or property resulting from any ideas, methods, instructions or products referred to in the content.

antibiotics

MDPI

Review

Is Silver the New Gold? A Systematic Review of the Preclinical Evidence of Its Use in Bone Substitutes as Antiseptic

Michele Fiore *, Alessandro Bruschi, Claudio Giannini, Lorenzo Morante, Claudia Rondinella, Matteo Filippini, Andrea Sambri and Massimiliano De Paolis

Orthopaedic and Traumatology Unit, IRRCS Azienda Ospedaliera-Universitaria di Bologna, 40138 Bologna, Italy; alessandro.bruschi@ior.it (A.B.); claudio.giannini@ior.it (C.G.); lorenzo.morante@ior.it (L.M.); claudia.rondinella@ior.it (C.R.); matteo.filippini@ior.it (M.F.); andrea.sambri2@unibo.it (A.S.); massimiliano.depaolis@aosp.bo.it (M.D.P.)
* Correspondence: michelefiore.md@gmail.com

Abstract: Antibiotic-laden bone substitutes represent a viable option in the treatment of bone and joint infections with bone defects. In particular, the addition of silver ions or silver nanoparticles to bone substitutes to achieve local antiseptic activity could represent a further contribution, also helping to prevent bacterial resistance to antibiotics. An in-depth search of the main scientific databases was performed regarding the use of silver compounds for bone substitution. The available evidence is still limited to the preclinical level: 22 laboratory studies, 2 animal models, and 3 studies, with both in vitro and in vivo analysis, were found on the topic. Numerous biomaterials have been evaluated. In vitro studies confirmed that silver in bone substitutes retains the antibacterial activity already demonstrated in coatings materials. Cytotoxicity was generally found to be low and only related to silver concentrations higher than those sufficient to achieve antibacterial activity. Instead, there are only a few in vivo studies, which appear to confirm antibacterial efficacy, although there is insufficient evidence on the pharmacokinetics and safety profile of the compounds investigated. In conclusion, research on bone substitutes doped with silver is in its early stages, but the preliminary findings seem promising.

Keywords: bone and joint infections; orthopaedic; odontology; bone substitutes; silver compounds; silver ions; silver nanoparticles

Citation: Fiore, M.; Bruschi, A.; Giannini, C.; Morante, L.; Rondinella, C.; Filippini, M.; Sambri, A.; De Paolis, M. Is Silver the New Gold? A Systematic Review of the Preclinical Evidence of Its Use in Bone Substitutes as Antiseptic. *Antibiotics* **2022**, *11*, 995. https://doi.org/10.3390/antibiotics11080995

Academic Editors: Sotiris K Hadjikakou, Christina N. Banti and Anthony William Coleman

Received: 3 July 2022
Accepted: 22 July 2022
Published: 24 July 2022

Publisher's Note: MDPI stays neutral with regard to jurisdictional claims in published maps and institutional affiliations.

Copyright: © 2022 by the authors. Licensee MDPI, Basel, Switzerland. This article is an open access article distributed under the terms and conditions of the Creative Commons Attribution (CC BY) license (https://creativecommons.org/licenses/by/4.0/).

1. Introduction

Bone and joint infections (BJIs) represent an extremely heterogeneous group of diseases, including implant-associated infections (both in the fields of orthopaedics and odontology), septic arthritis, and osteomyelitis [1]. The increase in BJIs' incidence shown in recent years, mainly due to the increase in joint replacements and the use of orthopaedic hardware, currently represents a growing social and economic issue for health systems [1,2]. Indeed, implant-related infection rate is reported to be 5% for primary cases, 6% for revision cases and increases to 43% for previously infected cases [3]. Osteomyelitis, defined as a bone inflammation caused by infection, may be the common endpoint of BJIs. Bacterial infections complicate the bone-healing process following fractures or surgical treatment, often resulting in significant bone loss [3]. Once occurred, bone infections are very challenging to treat, due to the difficulty of achieving a suitable antibiotic concentration in the affected area that may permit bacteria eradication [4]. Hence, the treatment of BJIs generally requires wide debridement with removal of all infected bone and soft tissues, irrigation, and, subsequently, dead-space filling [5,6]. Bone defects wider than 2 cm or circumferential losses involving more than 50% of bone are defined as critical-size bone defects (CSDs). CSDs usually progress to healing failure, even after optimal fixation [7]. In order to restore the continuity of the bone loss resulting from the surgical treatment, autologous, allogenic and artificial bone can be implanted [6,8]. Autograft substitutes are still considered the gold standard for bone repair and regeneration due to their osteogenic nature combined with no immunological side effects of the graft [7,9]; however, they are associated with donor site morbidity (hematomas, infection, and neurovascular injury) and longer operative time [10]. On the other hand, when using allografts, the principal concerns are related to mechanical resistance, limited osteoconduction and risk of infections [8]. Therefore, greater attention has increasingly been given to bone graft substitutes. They have been defined as "a synthetic or biologically organic combinations which can be inserted for the treatment of a bone defect instead of autogenous or allogenous bone" [10]. Theoretically, the bone substitutes mimic bone graft, combining advantages of natural and synthetic biomaterials [8,9] and supporting local bone healing [7]. The ideal bone substitute should be biocompatible, osteoconductive, osteoinductive, resorbable, thermally nonconductive, sterilizable, and available at a reasonable cost [10]. Bone substitutes can be broadly categorized into ceramics (nonresorbable and biodegradable), hydroxyapatite, β-tri-calcium phosphate, calcium sulfate, calcium carbonate, silicate ("Bioglass"), magnesium composites and calcium phosphate cements [7]. Current advances have been made with the development of tissue-engineered products, incorporating growth factors and stem cells [10].

Despite the development in biomaterials' properties, risk of infection remains a major issue after implantation. Thus, antibacterial properties should be considered during the development and choice of a bone substitute, as well as biocompatibility and physic-chemical features [11,12]. This is in order to prevent infections of the bone substitutes, or to optimise their use in case of BJIs. In fact, the result of bacterial adhesion to implants or grafts usually progresses with complete removal [12]. Moreover, to prevent infection recurrence, systemic or local antibiotics should be administrated after surgery; nevertheless, inappropriate and excessive use of antibiotics, in addition to systemic side effects for the patient, increases the risk of the emergence of multidrug-resistant bacteria [6]. The addition of antimicrobial nanomaterials, such as silver, zinc, copper, carbon nanotube, graphene oxide, molybdenum disulfide and titanium oxide, into the biomaterials has significantly shown to inhibit microbial infection, determining also a lower tendency to develop bacterial resistance [9,13,14].

In particular, silver's efficacy and safety has been reported in several in vitro and animal studies [15]. Silver ions (Ag) and silver nanoparticles (AgNPs) have garnered prominent consideration in recent years due to their broad spectrum of antibacterial properties, low bacterial resistance, and relatively low cytotoxicity [13]. For these reasons, the use of silver gained interest in the clinical practice with applications such as wound healing and cardiac and orthopaedic implant coating. Particularly in the orthopaedic field, the use of silver has proved to be effective in the treatment of megaprosthesis infections [15].

Although only preclinical studies about the use of silver combined with bone substitutes have been published, this issue may have important clinical implications for the prevention and treatment of BJIs. The aim of this review is to provide an overview of the evidence currently available in the literature.

2. Results

A total of 271 studies were found through the electronic search and 8 studies were added after cross-referenced research on the bibliography of the examined full-text articles. After a preliminary analysis, a total of 27 studies were included in this systematic review [3,4,6,8,9,11–13,16–34]. To date, there are no clinical studies on the use of bone substitutes containing adjuvant silver. Twenty-two laboratory studies [4,6,8,9,11–13,16–20,22–32,34], two animal models [21,33] and three studies in which both in vitro and in vivo analysis were performed [3,6,31] were found on the topic.

A wide range of biomaterials were evaluated as possible carriers of silver in bone. The tests used to appraise the in vitro or in vivo antibacterial activity of the compounds in the various studies included: cultures from bone samples, agar diffusion, halo test, agar dilution, broth microdilution, spread plate method, bacterial count through scanning electron microscope, confocal laser scanning microscopy, epi-fluorescence microscopy, histopathological examinations, RT-PCR bacterial DNA measurement, and radiological examination. All in vitro studies reported a partial or total inhibition of the bacterial growth. All in vivo studies that directly investigated the antibacterial effect of silver compounds confirmed their efficacy in the treatment of osteomyelitis. The data on cell and tissue toxicity are inconsistent; however, in all the studies, the antibacterial activity of the compounds tested was reported at nontoxic concentrations. Extended data from the included studies are reported in Table 1.

Table 1. Data from included studies.

Study	Type of Study	Material Tested	Antimicrobial Activity Evaluations	Bacteria Tested	Reported Results	Toxicity
Afzal, 2012 [12]	In vitro	Hydroxyapatite–silver (Ag-HA) and carbon nanotube–silver (CNT-Ag) composites	Bacterial count through SEM	*Escherichia coli* *Staphylococcus epidermidis*	Partial response.	N/A
Bee, 2020 [13]	In vitro	Antibacterial silver-nanoparticle-decorated hydroxyapatite (HAp/AgNP)	Agar diffusion	*Staphylococcus aureus*	Zone of inhibition of bacterial growth.	N/A
Bostancıoğlu, 2015 [11]	In vitro	Silver-doped calcium-phosphate-based inorganic powder (ABT)	Agar diffusion Agar dilution	*Escherichia coli* *Pseudomonas aeruginosa* *Staphylococcus aureus*	Partial response or total response depending on dilution and concentration.	Concentration-dependent cytotoxicity on V79 379A and HUVEC lines. ABT is noncytotoxic and bears good biocompatibility even at 1000 μg mL^{-1} of ABT with the highest content of silver.
Correia, 2016 [8]	In vitro	Tricalcium phosphate (TCP)/sodium alginate scaffold doped with AgNP	Agar diffusion	*Staphylococcus aureus*	Halo of 0.820 cm with 1 cm scaffold.	No cytotoxicity on osteoblast cells.
Dalavi, 2020 [9]	In vitro	Alginate-nanohydroxyapatite doped with chitooligosaccharide-coated silver nanoparticles (COS-Ag-Alg-HA)	Broth microdilution	*Staphylococcus aureus*	Total response at higher concentration than 77.2% using 3 mg/mL of microsphere.	No cytotoxicity on human osteosarcoma osteoblast-like MG-63 cells.
Deng, 2017 [16]	In vitro	PEEK doped with Ag + nanoparticles	Agar diffusion	*Staphylococcus aureus* *Escherichia coli*	Halo of 14 mm of inhibition for both the bacteria with 0.9 mm scaffold.	Initial low proliferation rate of human osteosarcoma osteoblast-like MG-63 cells.
Gong, 2017 [17]	In vitro	Silver-doped hydroxyapatite (Ag-HA) + Bio-Oss	RT-PCR bacterial DNA measurement	*Porphyromonas gingivalis* *Fusobacterium nucleatum*	Partial response, with decreasing of bacterial DNA at 2 h, 4 h, and 24 h compared to control group in which no inhibition was seen.	AgHA showed obvious cytotoxicity against periodontal fibroblasts and rat bone-marrow stromal cells, with relative survival rates of <80%. Bio-Oss only showed survival rates exceeding 95% of periodontal.
Jacquart, 2013 [18]	In vitro	Calcium carbonate–calcium phosphate bone cement doped with silver (Ag-$CaCO_3$-CaP)	Broth microdilution	*Staphylococcus aureus* *Escherichia coli*	Complete response.	No cytotoxicity on human bone marrow stroma cells.
Jegatheeswaran, 2015 [19]	In vitro	Polyethylene-glycol/hydroxyapatite doped with silver (Ag-HAp-PEG)	Epi-fluorescence microscopy	*Escherichia coli*	Partial response with increasing bacteria death in analyses at 6 and 12 h.	N/A
Jiang, 2016 [20]	In vitro	Hydroxyapatite/polyurethane composite scaffolds doped with silver phosphate particles (Ag_3PO_4-n-HA/PU)	Agar diffusion	*Staphylococcus aureus* *Escherichia coli*	The bacteriostatic rate resulted time and weight percentage of Ag incorporated depending.	Scaffolds with no more than 5 wt% appear to have no cytotoxicity on human osteosarcoma osteoblast-like MG-63 cells. Higher concentration (>5%) would weaken cytocompatibility.
Kose, 2020 [21]	In vivo (rabbit)	Calcium phosphate (CP) with silver ions	Radiological examination Bacterial cultures from bone samples Histopathological examinations	*Staphylococcus aureus*	No MRSA was found at cultures, no X-ray signs of osteomyelitis and no sign of chronic inflammation in histological analysis, compared to the control groups.	No inflammatory reactions.
Sampath Kumar, 2015 [22]	In vitro	Calcium-deficient hydroxyapatite (CDHA) carrier of doxycycline and Ag+ ions	MIC/MBC studies and time-kill assay	*Staphylococcus aureus* *Escherichia coli*	When compared with doxycycline, the antibiotic release provided the initial high antibacterial activity, while the sustained ion release provided a long-term antibacterial activity.	No cytotoxicity on L6 myoblast cells.
Lim, 2014 [23]	In vitro	Silver and silicon-containing apatite (Ag,Si-HA)	Bacterial count through SEM	*Staphylococcus aureus* *Escherichia coli*	No bacteria growth compared to negative control: complete response.	MSCs treated with Ag,Si-HA showed an initial low proliferation rate compared to controls, and faster proliferation after day 3.
Nam, 2017 [24]	In vitro	Portland cement doped with silver nanoparticles (SNPC)	Agar diffusion	*Streptococcus mutans* *Streptococcus sobrinus*	1.0% wt of SNPC has no antibacterial effect; 3.0 wt% SNPC inhibited *S. sorbinus* by 1.9 ± 0.5 mm, while no inhibition halos were shown for *S. mutans* at the same dose. SNPC of 5.0 wt% significantly inhibited S. *sorbinus* (halo diameter 4.2 ± 0.3 mm) and *S. mutans* (halo diameter 2.2 ± 0.4 mm).	N/A

Table 1. *Cont.*

Study	Type of Study	Material Tested	Antimicrobial Activity Evaluations	Bacteria Tested	Reported Results	Toxicity
Paterson, 2020 [4]	In vitro	Polycaprolactone scaffolds with silver-doped hydroxyapatite (Ag-nHA)	Agar diffusion	*Staphylococcus aureus* *Escherichia coli*	The scaffold reduced the viable bacteria count to undetectable levels by 48 h for *E. coli* and 96 h for *S. aureus*: complete response.	Silver-doped nHA to enhance MSC differentiation down an osteogenic path. Scaffolds containing 10 mol.% silver may be toxic for MSCs.
Sethmann, 2018 [25]	In vitro	Phosphatized Calcium Carbonate biomineral (PCCB) doped with Ag + silver ions	Agar diffusion	*Pseudomonas aeruginosa* *Staphylococcus aureus*	Samples treated with an AgNO3 solution with 10 mmol/L showed nearly the same antibacterial performance as samples treated with 100 mmol/L. Halo of 1.1–1.2 mm for Gram- and 3 mm for Gram+.	N/A
Shimabukuro, 2021 [6]	In vitro + in vivo (rabbit)	Silver phosphate in carbonate apatite (Ag_3PO_4-CO_3Ap)	Agar diffusion immunofluorescence	*Staphylococcus epidermidis*	Antibacterial effect if concentration of Ag3Po4 is more than 0.1 wt %. Complete response.	Ag_3PO_4 content of 0.1–0.95 wt % may show antibacterial properties without cytotoxicity. Higher concentrations showed increasing toxicity for MC3T3-E1 cells. Ag_3PO_4 content of 0.1–0.3 wt % in the samples did not affect bone formation in vivo.
Sonamuthu, 2018 [26]	In vitro	Fluorinate-hydroxyapatite/polyvinyl alcohol doped with silver nanoparticles (AgNp-fHA)	Agar diffusion CLSM Broth microdilution	*Staphylococcus aureus* *Escherichia coli*	Antibacterial activity is time- and concentration-dependent. More effect on Gram + due to the different composition of membrane; complete response G+ and G- partial response in CLSM.	No cytotoxicity on human osteosarcoma osteoblast-like MG-63 cells.
Sowmya-Srinavasan, 2013 [27]	In vitro	Bioactive alpha- and beta-chitin hydrogel/nanobioactive glass ceramic doped with silver	Agar diffusion	*Staphylococcus aureus* *Escherichia coli*	Antibacterial activity of Ag dose dependent, similar effect between G+ and G-, but less effective than gentamicin alone.	No cytotoxicity on human primary osteoblasts and human periodontal ligament cells.
Verné, 2009 [29] + Miola, 2009 [28]	In vitro	SiO-CaO-NaO-AlO doped with silver (Ag-SCNA)	Agar diffusion Broth microdilution	*Staphylococcus aureus* *Escherichia coli*	Same antimicrobial activity against G+ and G-, halo of 2 mm.	No cytotoxicity on fibroblasts. Slightly lower proliferation rate compared to control cells.
Vollmer, 2016 [30]	In vitro	Calcium phosphate (CaP) doped with silver	Agar diffusion Bacterial count through SEM	*Escherichia coli*	Antimicrobial activity with halo in agar diffusion (no dimensions reported) and characteristics of poor health of bacteria at SEM compared to control.	No cytotoxicity on human osteoblasts.
Weng, 2020 [31]	In vitro + in vivo (rabbit)	Loaded nano-hydroxyapatite-reduced graphene oxide doped with Ag nanoparticles (AgNp-AHRG)	Agar diffusion Kirby–Bauer diffusion WBC count CRP Radiological examination	*Staphylococcus aureus*	Antibacterial activity in vitro and the halo zone is dependent on the concentration of Ag. In vivo, it significantly reduced the levels of inflammatory markers, such as leukocytes and CRP, after implantation in the infected site. In subsequent observations, the healing of the bone in the implanted group was significantly improved compared to the untreated group.	Concentration-dependent cytotoxicity on bone marrow stromal cells. No cytotoxicity for 1% and 2% silver AgNp-AHRG scaffolds.
Wilcock, 2017 [32]	In vitro	Hydroxyapatite paste silver doped (Ag-nHA)	Agar diffusion	*Pseudomonas aeruginosa* *Staphylococcus aureus*	Antibacterial activity dependent on Ag concentration.	N/A
Yuan, 2016 [3]	In vitro + in vivo (rabbit)	Porous β-tricalcium phosphate with Ag nanoparticles (AgNp-βTCP)	Agar diffusion Bacterial count through SEM	*Staphylococcus aureus* *Escherichia coli*	Antibacterial activity dependent on concentration. Difference in activity between G+ and G- was not reported. At SEM, there is some bacteria visible, but no biofilm was seen.	No local and systemic toxicity.
Zhang, 2019 [33]	In vivo (rabbit)	Nano-hydroxyapatite/polyurethane composite scaffolds doped with silver phosphate particles (Ag/n-HA/PU)	WBC count Radiological examination Histopathological examinations	*Staphylococcus aureus*	Radiological healing of infection with no difference between 3% wt and 10% wt concentration as well as no difference in histological analysis for trabeculae formation.	Local toxicity for highest concentration of silver (Ag/n-HA/10PU).
Zhang, 2020 [34]	In vitro	Brushite/Ag3PO4-coated Mg-based scaffolds (Mg-DCPD-Ag)	Spread plate method Bacterial count through SEM	*Staphylococcus aureus* *Escherichia coli* *Staphylococcus epidermidis*	Antibacterial activity with complete response depending on concentration of Ag.	Cytotoxicity for highest concentration of silver (Mg-DCPD-0.46 Ag)

Abbreviations: SEM, scanning electron microscope; MRSA, methicillin-resistant Staphylococcus aureus; WBC, white blood cells; CLSM, confocal laser scanning microscopy; MSC, mesenchymal Stem Cell.

3. Discussion

Silver antimicrobial activity relies on several mechanisms. Principally, it stops cells' respiratory chain, affecting the cells' energy generation, due to its affinity to the sulfhydryl and thiol groups [35]. Additionally, silver leads to a release of potassium [36], binds DNA and RNA, disrupting the cells' translation and transcription processes [37], and produces intracellular reactive oxygen species [38]. Consequently, silver has the ability to eliminate a broad spectrum of pathogens that can be found at implant sites [39]. Surfaces or materials containing silver particles act by directly releasing ions into the water solution in which they lie. More recently, silver technology has focused on the use of nanoparticles. AgNPs seems to be more reactive, with a stronger antibiofilm potential than their bulk metal counterparts, partially due to the increased active surface area [40,41]. AgNPs are usually 1 nm to 20 nm in size. Because of their small dimensions, the surface area is taken advantage of, passing more into cell membranes, thereby contributing to augmented antimicrobial action [42]. Furthermore, antibacterial mechanisms of AgNPs have been hypothesized that do not depend on the release of ions but are related to the interaction between silver and other substrates. For example, interaction with some titanium alloys can lead to the production of an electron cloud around the surface of the compounds [35]. This cathodic reaction, which produces a proton depletion region, would appear to reduce the transmembrane proton electrochemical gradient and lead to bacterial death by interfering with ATP synthesis [35]. In addition, to date, there is limited evidence that AgNPs possess osteoconductive capabilities, promoting the proliferation of mesenchymal stem cells and their osteogenic differentiation in vitro, as well as enhancing bone fracture healing in animal models [43].

The clinical use of silver for antibacterial purposes in implantable devices has been mainly investigated as a coating material [15]. Therefore, there is already evidence regarding the efficacy and safety profile of this application. Studies largely agree on time and concentration influencing the bactericidal effect of silver in bone substitutes, with higher Ag concentration and longer exposure time associated with better antibacterial responses [3,22]. Accordingly, a cytotoxic effect of silver at excessively high concentrations is to be expected, by means of the same antibacterial mechanisms, to which, however, eukaryotic cells are less sensitive at low concentrations because of their extremely better antioxidant and DNA repair activity [44]. It has been reported that silver has toxic effects for humans at a high blood concentration of 300–1200 ppb [45]. Instead, silver concentrations below 200 ppb can be considered as normal [37]. Instead, regarding AgNPs' safety profile, various studies suggested that there is a large gap, at least by an order of magnitude, between the toxic and antibacterial doses of AgNPs [46,47]. All available clinical data on the pharmacokinetics of silver are related to its use in coating materials, especially regarding serum levels, which have never been found to exceed the threshold of toxicity. Nevertheless, several side effects have been related to silver in clinical studies, including diffuse argyria, kidney and liver damage, leukopenia, and peripheral neuropathies [36,48,49]. No embryotoxic side effects in humans are described [50].

In contrast to silver used in coatings, this review highlighted that there are currently no clinical data on the use of silver as an adjuvant in biomaterials for bone substitution. However, there are some in vivo studies using animal models and a substantial number of laboratory studies investigating the antibacterial efficacy of silver on a plethora of different biomaterials.

All in vitro studies included in this review supported that even when used as a biomaterial constituent, silver appears to preserve its antibacterial activity [3,4,6,8,9,11–13,16–20,22–32,34]. This evidence emerged regardless of the type of analysis used by individual studies to highlight the effect of silver. In particular, the impact of silver was found to be cross-cutting for both Gram+ and Gram- bacteria. Some studies hypothesised that due to the different composition of the membranes, silver-doped materials could have more antibacterial effect against Gram+ [14,25,26]. However, other studies did not report differential antibacterial effect, showing a complete response in both cases [3,27,29]. Unfortunately, no study included in this review conducted quantitative evaluations specifically aimed at investigating this aspect using different bacterial species clustered according to membrane characteristics. Three out of five studies with animal models directly investigated the in vivo antibacterial effect of silver compounds. All confirmed the efficacy of using biomaterials containing adjuvant silver in the treatment of S. aureus osteomyelitis [21,31,33]. In detail: (1) Zhang et al. investigated silver application in two hydroxyapatite compounds with different silver concentrations [33]; (2) calcium phosphate beads doped with silver ions were used in the study by Kose et al. [21]; (3) Weng et al. instead considered the use of silver nanoparticle (AgNP)-loaded nano-hydroxyapatite-reduced graphene oxide scaffolds [31]. In all in vivo studies, it was also confirmed that the presence of silver did not affect the osteoconductive and osteoinductive activity of the bone substitutes employed [3,6,21,31,33].

The ever-increasing resistance of human pathogens to antibiotics makes silver a convenient alternative to be harnessed as an antibacterial weapon. Indeed, bacterial resistance hardly arises in the presence of silver ions in the environment [51,52], while no relevant data describing bacterial resistance to AgNPs have been reported yet [53]. Therefore, it would be interesting to examine the antibacterial effect of silver compared with antibiotics in bone substitutes. However, only few conflicting data are available: Kose et al. showed a better result for silver compared to vancomycin in murine models [21]; on the other side, two studies showed a better antibacterial effect of bone substitutes doped with doxycycline or gentamicin alone [22,27].

Little can be stated about the safety profile of silver use in biomaterials. Indeed, it might be expected that the pharmacokinetics of silver used as a constituent material in bone substitutes would differ radically from its use in coatings. This could lead to the ineffective release of silver in vivo, exceeding toxicity thresholds, or even irregular or excessively phasic kinetics that alternate between these two possibilities. Several in vitro studies have evaluated the kinetics of silver release from biomaterials in water or in different solutions of simulated body fluid. However, on the one hand, the extreme heterogeneity of the materials tested, and on the other hand, the radical differences in environment compared to a real setting, fully restrict the deducible conclusions. In fact, only in vivo assessments may allow approximating the pharmacokinetics. Similar remarks can be made about cytotoxicity from in vitro studies, with results varying widely due to the broad spectrum of materials analysed. However, in all in vitro studies, the antibacterial activity of the compounds analysed was achieved at noncytotoxic concentrations. From this point of view, with regard to the in vivo studies, Zhang et al. found that the silver concentration of the bone tissue was found to be over 2 ppm in the n-HA/PU10 group with local toxic risk increasing when the silver concentration exceeds 1 ppm in tissue, as recommended by the safety guidelines [33]. Moreover, the same compound has shown the possibility of inducing liver toxicity [33]. In contrast, the studies by Kose et al. and Yuan et al. found no local and systemic toxicity in the animal models used, both investigating calcium phosphate derivates doped with silver ions or nanoparticles, respectively [3,21]. Shimabukuro et al. reported that the inflammatory effect of silver phosphate (Ag_3PO_4) in a bone substitute composed of carbonate apatite (CO_3Ap) was concentration-dependent [6]. Similarly, the study by Schneider et al., not included in this review as it did not directly investigate the antibacterial effect, showed no inflammatory reactions to cotton-wool-like silver-doped calcium phosphate nanocomposites in sheep [54]. Another study, not included in this review for the same reason, performed by Wnukiewicz et al., examined the soft-tissue reaction to corundum ceramic with colloidal silver in rabbits, finding no difference with the control group [55].

4. Materials and Methods

An in-depth search of the scientific research was performed according to 2020 PRISMA (preferred reporting items for systematic reviews and meta-analyses) guidelines [56]. The search algorithm according to these guidelines is shown in Figure 1.

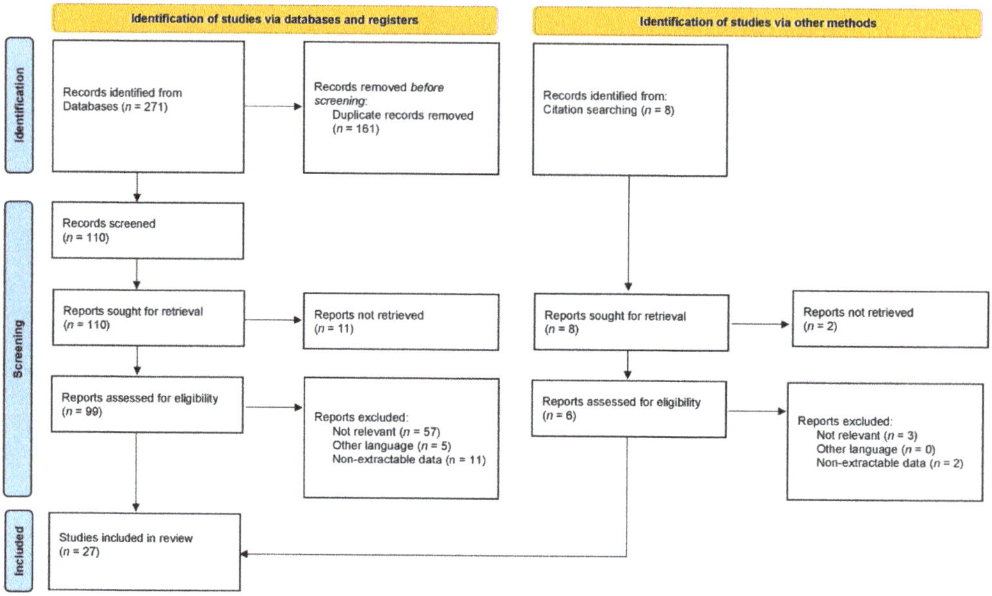

Figure 1. PRISMA 2020 flow diagram and the selection of studies.

A search regarding the existing evidence on the use of silver compounds and silver nanoparticles combined with biomaterials for bone substitution with no restriction on date of publication, up to the end of June 2022, was performed on the PubMed (https://pubmed.ncbi.nlm.nih.gov/ (accessed on 30 June 2022)), Scopus (https://www.scopus.com (accessed on 30 June 2022)), and Web of Science (www.webofscience.com (accessed on 30 June 2022)) databases. Various combinations of the following keywords were used: "silver compound", "silver nanoparticles", "bone substitutes", "bone biomaterials". The inclusion criteria were as follows: original research reporting preclinical results on in vitro testing and in vivo animal models of the antibacterial activity, pharmacokinetics and pharmacodynamics of silver combined with biomaterials used for bone substitution. Only studies in English were retained. Articles that were considered relevant during the electronic search were retrieved in full-text, and a cross-referencing hand-search of their bibliography was performed, in order to find further related articles. Reviews and meta-analysis were also analysed, in order to broaden the search for studies that might have been missed through the electronic search.

A formal assessment of the quality of the articles was not conducted, as there is no clear evidence of validated tools for the evaluation of preclinical laboratory studies. Only descriptive statistics were used for this study as the type of data provided.

The following data were independently extracted by all the investigators and summarized in Table 1: study type, material tested, type of antimicrobial activity evaluations, microorganisms tested, and findings about toxicity.

5. Conclusions

The introduction of bone substitutes doped with ionic silver or silver nanoparticles into clinical practice would provide a valuable further contribution to the management of challenging diseases such as osteomyelitis and peri-prosthetic or implant-related infections, as well as to the prevention of bacterial resistance to antibiotics. Numerous materials have already been evaluated for this purpose, but the available evidence is still limited to the preclinical level. In vitro studies have confirmed that when silver is added to bone substitutes, it retains the antibacterial activity already demonstrated in coatings materials. The antibacterial effect against Gram+ might be higher than against Gram-. However, conclusive data are lacking as well as it is unclear whether silver could provide greater efficacy than antibiotic loading. The cytotoxicity of silver compounds has generally been shown to be low and only related to concentrations of silver significantly higher than those sufficient to achieve antibacterial activity. On the other hand, there are only a few in vivo studies which appear to confirm antibacterial efficacy, although there is insufficient evidence on the pharmacokinetics and safety profile of the biomaterials investigated. In conclusion, research on bone substitutes doped with silver is in its early stages but the preliminary findings seem promising.

Author Contributions: Conceptualization, M.F. (Michele Fiore), A.B. and M.D.P.; methodology, M.F. (Matteo Filippini); software, A.S. and L.M.; investigation, A.B. and C.G.; data curation, M.F. (Matteo Filippini) and C.R.; writing—original draft preparation, A.B., C.G. and C.R.; writing—review and editing, M.F. (Michele Fiore) and A.S.; supervision, M.D.P. All authors have read and agreed to the published version of the manuscript.

Funding: This research received no external funding.

Institutional Review Board Statement: Not applicable.

Informed Consent Statement: Not applicable.

Data Availability Statement: The data reported in this study are available in the literature.

Conflicts of Interest: The authors declare no conflict of interest.

References

1. Sambri, A.; Spinnato, P.; Tedeschi, S.; Zamparini, E.; Fiore, M.; Zucchini, R.; Giannini, C.; Caldari, E.; Crombe, A.; Viale, P.; et al. Bone and Joint Infections: The Role of Imaging in Tailoring Diagnosis to Improve Patients' Care. *J. Pers. Med.* **2021**, *11*, 1317. [CrossRef] [PubMed]
2. Mandracchia, V.J.; Sanders, S.M.; Jaeger, A.J.; Nickles, W.A. Management of osteomyelitis. *Clin. Podiatr. Med. Surg.* **2004**, *21*, 335–351. [CrossRef]
3. Yuan, J.; Wang, B.; Han, C.; Huang, X.; Xiao, H.; Lu, X.; Lu, J.; Zhang, D.; Xue, F.; Xie, Y. Nanosized-Ag-doped porous beta-tricalcium phosphate for biological applications. *Mater. Sci. Eng. C Mater. Biol. Appl.* **2020**, *114*, 111037. [CrossRef] [PubMed]
4. Paterson, T.E.; Shi, R.; Tian, J.; Harrison, C.J.; De Sousa Mendes, M.; Hatton, P.V.; Li, Z.; Ortega, I. Electrospun Scaffolds Containing Silver-Doped Hydroxyapatite with Antimicrobial Properties for Applications in Orthopedic and Dental Bone Surgery. *J. Funct. Biomater.* **2020**, *11*, 58. [CrossRef] [PubMed]
5. Sambri, A.; Fiore, M.; Tedeschi, S.; De Paolis, M. The Need for Multidisciplinarity in Modern Medicine: An Insight into Orthopaedic Infections. *Microorganisms* **2022**, *10*, 756. [CrossRef]
6. Shimabukuro, M.; Hayashi, K.; Kishida, R.; Tsuchiya, A.; Ishikawa, K. No-Observed-Effect Level of Silver Phosphate in Carbonate Apatite Artificial Bone on Initial Bone Regeneration. *ACS Infect. Dis.* **2022**, *8*, 159–169. [CrossRef]
7. Busch, A.; Wegner, A.; Haversath, M.; Jager, M. Bone Substitutes in Orthopaedic Surgery: Current Status and Future Perspectives. *Z. Orthop. Unfallchir.* **2021**, *159*, 304–313. [CrossRef]
8. Correia, T.R.; Figueira, D.R.; de Sá, K.D.; Miguel, S.P.; Fradique, R.G.; Mendonca, A.G.; Correia, I.J. 3D Printed scaffolds with bactericidal activity aimed for bone tissue regeneration. *Int. J. Biol. Macromol.* **2016**, *93*, 1432–1445. [CrossRef]
9. Dalavi, P.A.; Prabhu, A.; Shastry, R.P.; Venkatesan, J. Microspheres containing biosynthesized silver nanoparticles with alginate-nano hydroxyapatite for biomedical applications. *J. Biomater. Sci. Polym. Ed.* **2020**, *31*, 2025–2043. [CrossRef]
10. Campana, V.; Milano, G.; Pagano, E.; Barba, M.; Cicione, C.; Salonna, G.; Lattanzi, W.; Logroscino, G. Bone substitutes in orthopaedic surgery: From basic science to clinical practice. *J. Mater. Sci. Mater. Med.* **2014**, *25*, 2445–2461. [CrossRef]
11. Bostancioglu, R.B.; Peksen, C.; Genc, H.; Gurbuz, M.; Karel, F.B.; Koparal, A.S.; Dogan, A.; Kose, N.; Koparal, A.T. Analyses of the modulatory effects of antibacterial silver doped calcium phosphate-based ceramic nano-powder on proliferation, survival, and angiogenic capacity of different mammalian cells in vitro. *Biomed. Mater.* **2015**, *10*, 045024. [CrossRef]
12. Afzal, M.A.; Kalmodia, S.; Kesarwani, P.; Basu, B.; Balani, K. Bactericidal effect of silver-reinforced carbon nanotube and hydroxyapatite composites. *J. Biomater. Appl.* **2013**, *27*, 967–978. [CrossRef] [PubMed]
13. Bee, S.L.; Bustami, Y.; Ul-Hamid, A.; Lim, K.; Abdul Hamid, Z.A. Synthesis of silver nanoparticle-decorated hydroxyapatite nanocomposite with combined bioactivity and antibacterial properties. *J. Mater. Sci. Mater. Med.* **2021**, *32*, 106. [CrossRef] [PubMed]
14. Kolmas, J.; Groszyk, E.; Kwiatkowska-Rozycka, D. Substituted hydroxyapatites with antibacterial properties. *BioMed Res. Int.* **2014**, *2014*, 178123. [CrossRef] [PubMed]
15. Fiore, M.; Sambri, A.; Zucchini, R.; Giannini, C.; Donati, D.M.; De Paolis, M. Silver-coated megaprosthesis in prevention and treatment of peri-prosthetic infections: A systematic review and meta-analysis about efficacy and toxicity in primary and revision surgery. *Eur. J. Orthop. Surg. Traumatol.* **2021**, *31*, 201–220. [CrossRef]

16. Deng, L.; Deng, Y.; Xie, K. AgNPs-decorated 3D printed PEEK implant for infection control and bone repair. *Colloids Surf. B Biointerfaces* **2017**, *160*, 483–492. [CrossRef]
17. Gong, J.; Yang, L.; He, Q.; Jiao, T. In vitro evaluation of the biological compatibility and antibacterial activity of a bone substitute material consisting of silver-doped hydroxyapatite and Bio-Oss((R)). *J. Biomed. Mater. Res. B Appl. Biomater.* **2018**, *106*, 410–420. [CrossRef]
18. Jacquart, S.; Siadous, R.; Henocq-Pigasse, C.; Bareille, R.; Roques, C.; Rey, C.; Combes, C. Composition and properties of silver-containing calcium carbonate-calcium phosphate bone cement. *J. Mater. Sci. Mater. Med.* **2013**, *24*, 2665–2675. [CrossRef]
19. Jegatheeswaran, S.; Sundrarajan, M. PEGylation of novel hydroxyapatite/PEG/Ag nanocomposite particles to improve its antibacterial efficacy. *Mater. Sci. Eng. C Mater. Biol. Appl.* **2015**, *51*, 174–181. [CrossRef]
20. Jiang, J.; Li, L.; Li, K.; Li, G.; You, F.; Zuo, Y.; Li, Y.; Li, J. Antibacterial nanohydroxyapatite/polyurethane composite scaffolds with silver phosphate particles for bone regeneration. *J. Biomater. Sci. Polym. Ed.* **2016**, *27*, 1584–1598. [CrossRef]
21. Kose, N.; Asfuroglu, Z.M.; Kose, A.; Sahinturk, V.; Gurbuz, M.; Dogan, A. Silver ion-doped calcium phosphate-based bone-graft substitute eliminates chronic osteomyelitis: An experimental study in animals. *J. Orthop. Res.* **2021**, *39*, 1390–1401. [CrossRef] [PubMed]
22. Sampath Kumar, T.S.; Madhumathi, K.; Rubaiya, Y.; Doble, M. Dual mode antibacterial activity of ion substituted calcium phosphate nanocarriers for bone infections. *Front. Bioeng. Biotechnol.* **2015**, *3*, 59. [CrossRef] [PubMed]
23. Lim, P.N.; Shi, Z.; Neoh, K.G.; Ho, B.; Tay, B.Y.; Thian, E.S. The effects of silver, silicon-containing apatite towards bacteria and cell responses. *Biomed. Mater.* **2014**, *9*, 015010. [CrossRef]
24. Nam, K.Y. Characterization and antimicrobial efficacy of Portland cement impregnated with silver nanoparticles. *J. Adv. Prosthodont.* **2017**, *9*, 217–223. [CrossRef]
25. Sethmann, I.; Volkel, S.; Pfeifer, F.; Kleebe, H.J. Development of Phosphatized Calcium Carbonate Biominerals as Bioactive Bone Graft Substitute Materials, Part II: Functionalization with Antibacterial Silver Ions. *J. Funct. Biomater.* **2018**, *9*, 69. [CrossRef] [PubMed]
26. Sonamuthu, J.; Samayanan, S.; Jeyaraman, A.R.; Murugesan, B.; Krishnan, B.; Mahalingam, S. Influences of ionic liquid and temperature on the tailorable surface morphology of F-apatite nanocomposites for enhancing biological abilities for orthopedic implantation. *Mater. Sci. Eng. C Mater. Biol. Appl.* **2018**, *84*, 99–107. [CrossRef] [PubMed]
27. Srinivasan, S.; Kumar, P.T.; Nair, S.V.; Nair, S.V.; Chennazhi, K.P.; Jayakumar, R. Antibacterial and bioactive alpha- and beta-chitin hydrogel/nanobioactive glass ceramic/nano silver composite scaffolds for periodontal regeneration. *J. Biomed. Nanotechnol.* **2013**, *9*, 1803–1816. [CrossRef] [PubMed]
28. Miola, M.; Ferraris, S.; Di Nunzio, S.; Robotti, P.F.; Bianchi, G.; Fucale, G.; Maina, G.; Cannas, M.; Gatti, S.; Masse, A.; et al. Surface silver-doping of biocompatible glasses to induce antibacterial properties. Part II: Plasma sprayed glass-coatings. *J. Mater. Sci. Mater. Med.* **2009**, *20*, 741–749. [CrossRef]
29. Verne, E.; Miola, M.; Vitale Brovarone, C.; Cannas, M.; Gatti, S.; Fucale, G.; Maina, G.; Masse, A.; Di Nunzio, S. Surface silver-doping of biocompatible glass to induce antibacterial properties. Part I: Massive glass. *J. Mater. Sci. Mater. Med.* **2009**, *20*, 733–740. [CrossRef]
30. Vollmer, N.L.; Spear, J.R.; Ayers, R.A. Antimicrobial activity and biologic potential of silver-substituted calcium phosphate constructs produced with self-propagating high-temperature synthesis. *J. Mater. Sci. Mater. Med.* **2016**, *27*, 104. [CrossRef]
31. Weng, W.; Li, X.; Nie, W.; Liu, H.; Liu, S.; Huang, J.; Zhou, Q.; He, J.; Su, J.; Dong, Z.; et al. One-Step Preparation of an AgNP-nHA@RGO Three-Dimensional Porous Scaffold and Its Application in Infected Bone Defect Treatment. *Int. J. Nanomed.* **2020**, *15*, 5027–5042. [CrossRef] [PubMed]
32. Wilcock, C.J.; Stafford, G.P.; Miller, C.A.; Ryabenkova, Y.; Fatima, M.; Gentile, P.; Mobus, G.; Hatton, P.V. Preparation and Antibacterial Properties of Silver-Doped Nanoscale Hydroxyapatite Pastes for Bone Repair and Augmentation. *J. Biomed. Nanotechnol.* **2017**, *13*, 1168–1176. [CrossRef] [PubMed]
33. Zhang, D.; Liu, W.; Wu, X.D.; He, X.; Lin, X.; Wang, H.; Li, J.; Jiang, J.; Huang, W. Efficacy of novel nano-hydroxyapatite/polyurethane composite scaffolds with silver phosphate particles in chronic osteomyelitis. *J. Mater. Sci. Mater. Med.* **2019**, *30*, 59. [CrossRef]
34. Zhang, L.; Jia, G.; Tang, M.; Chen, C.; Niu, J.; Huang, H.; Kang, B.; Pei, J.; Zeng, H.; Yuan, G. Simultaneous enhancement of anti-corrosion, biocompatibility, and antimicrobial activities by hierarchically-structured brushite/Ag3PO4-coated Mg-based scaffolds. *Mater. Sci. Eng. C Mater. Biol. Appl.* **2020**, *111*, 110779. [CrossRef]
35. Cao, H.; Liu, X.; Meng, F.; Chu, P.K. Biological actions of silver nanoparticles embedded in titanium controlled by micro-galvanic effects. *Biomaterials* **2011**, *32*, 693–705. [CrossRef] [PubMed]
36. Tweden, K.S.; Cameron, J.D.; Razzouk, A.J.; Holmberg, W.R.; Kelly, S.J. Biocompatibility of silver-modified polyester for antimicrobial protection of prosthetic valves. *J. Heart Valve Dis.* **1997**, *6*, 553–561. [PubMed]
37. Wan, A.T.; Conyers, R.A.; Coombs, C.J.; Masterton, J.P. Determination of silver in blood, urine, and tissues of volunteers and burn patients. *Clin. Chem.* **1991**, *37*, 1683–1687. [CrossRef]
38. Russell, A.D.; Hugo, W.B. Antimicrobial activity and action of silver. *Prog. Med. Chem.* **1994**, *31*, 351–370. [CrossRef]
39. Hetrick, E.M.; Schoenfisch, M.H. Reducing implant-related infections: Active release strategies. *Chem. Soc. Rev.* **2006**, *35*, 780–789. [CrossRef]

40. Palanisamy, N.K.; Ferina, N.; Amirulhusni, A.N.; Mohd-Zain, Z.; Hussaini, J.; Ping, L.J.; Durairaj, R. Antibiofilm properties of chemically synthesized silver nanoparticles found against Pseudomonas aeruginosa. *J. Nanobiotechnol.* **2014**, *12*, 2. [CrossRef]
41. Afkhami, F.; Ahmadi, P.; Chiniforush, N.; Sooratgar, A. Correction to: Effect of different activations of silver nanoparticle irrigants on the elimination of Enterococcus faecalis. *Clin. Oral Investig.* **2022**, *26*, 1103. [CrossRef] [PubMed]
42. Yin, I.X.; Zhang, J.; Zhao, I.S.; Mei, M.L.; Li, Q.; Chu, C.H. The Antibacterial Mechanism of Silver Nanoparticles and Its Application in Dentistry. *Int. J. Nanomed.* **2020**, *15*, 2555–2562. [CrossRef] [PubMed]
43. Zhang, R.; Lee, P.; Lui, V.C.; Chen, Y.; Liu, X.; Lok, C.N.; To, M.; Yeung, K.W.; Wong, K.K. Silver nanoparticles promote osteogenesis of mesenchymal stem cells and improve bone fracture healing in osteogenesis mechanism mouse model. *Nanomedicine* **2015**, *11*, 1949–1959. [CrossRef] [PubMed]
44. AshaRani, P.V.; Low Kah Mun, G.; Hande, M.P.; Valiyaveettil, S. Cytotoxicity and genotoxicity of silver nanoparticles in human cells. *ACS Nano* **2009**, *3*, 279–290. [CrossRef]
45. Clement, J.L.; Jarrett, P.S. Antibacterial silver. *Met.-Based Drugs* **1994**, *1*, 467–482. [CrossRef]
46. Pauksch, L.; Hartmann, S.; Rohnke, M.; Szalay, G.; Alt, V.; Schnettler, R.; Lips, K.S. Biocompatibility of silver nanoparticles and silver ions in primary human mesenchymal stem cells and osteoblasts. *Acta Biomater.* **2014**, *10*, 439–449. [CrossRef]
47. Necula, B.S.; van Leeuwen, J.P.; Fratila-Apachitei, L.E.; Zaat, S.A.; Apachitei, I.; Duszczyk, J. In vitro cytotoxicity evaluation of porous TiO(2)-Ag antibacterial coatings for human fetal osteoblasts. *Acta Biomater.* **2012**, *8*, 4191–4197. [CrossRef]
48. Brutel de la Riviere, A.; Dossche, K.M.; Birnbaum, D.E.; Hacker, R. First clinical experience with a mechanical valve with silver coating. *J. Heart Valve Dis.* **2000**, *9*, 123–129.
49. Lansdown, A.B. Critical observations on the neurotoxicity of silver. *Crit. Rev. Toxicol.* **2007**, *37*, 237–250. [CrossRef]
50. Shavlovski, M.M.; Chebotar, N.A.; Konopistseva, L.A.; Zakharova, E.T.; Kachourin, A.M.; Vassiliev, V.B.; Gaitskhoki, V.S. Embryotoxicity of silver ions is diminished by ceruloplasmin—Further evidence for its role in the transport of copper. *Biometals* **1995**, *8*, 122–128. [CrossRef]
51. Percival, S.L.; Bowler, P.G.; Russell, D. Bacterial resistance to silver in wound care. *J. Hosp. Infect.* **2005**, *60*, 1–7. [CrossRef] [PubMed]
52. Landsdown, A.B.; Williams, A. Bacterial resistance to silver in wound care and medical devices. *J. Wound Care* **2007**, *16*, 15–19. [CrossRef] [PubMed]
53. Gallo, J.; Panacek, A.; Prucek, R.; Kriegova, E.; Hradilova, S.; Hobza, M.; Holinka, M. Silver Nanocoating Technology in the Prevention of Prosthetic Joint Infection. *Materials* **2016**, *9*, 337. [CrossRef] [PubMed]
54. Schneider, O.D.; Mohn, D.; Fuhrer, R.; Klein, K.; Kampf, K.; Nuss, K.M.; Sidler, M.; Zlinszky, K.; von Rechenberg, B.; Stark, W.J. Biocompatibility and Bone Formation of Flexible, Cotton Wool-like PLGA/Calcium Phosphate Nanocomposites in Sheep. *Open Orthop. J.* **2011**, *5*, 63–71. [CrossRef] [PubMed]
55. Wnukiewicz, W.; Rutowski, R.; Zboromirska-Wnukiewicz, B.; Reichert, P.; Gosk, J. Evaluation of Soft Tissue Reaction to Corundum Ceramic Implants Infiltrated with Colloidal Silver. *Adv. Clin. Exp. Med.* **2016**, *25*, 129–133. [CrossRef]
56. Page, M.J.; McKenzie, J.E.; Bossuyt, P.M.; Boutron, I.; Hoffmann, T.C.; Mulrow, C.D.; Shamseer, L.; Tetzlaff, J.M.; Akl, E.A.; Brennan, S.E.; et al. The PRISMA 2020 statement: An updated guideline for reporting systematic reviews. *BMJ* **2021**, *372*, n71. [CrossRef]

antibiotics MDPI

Review

Advanced Hydrogels Combined with Silver and Gold Nanoparticles against Antimicrobial Resistance

Yolice Patricia Moreno Ruiz [1,2], Luís André de Almeida Campos [1,3], Maria Andressa Alves Agreles [3], André Galembeck [2] and Isabella Macário Ferro Cavalcanti [1,3,*]

[1] Laboratory of Microbiology and Immunology, Academic Center of Vitória (CAV), Federal University of Pernambuco (UFPE), Vitória de Santo Antão 55608-680, Pernambuco, Brazil
[2] Department of Fundamental Chemistry, Federal University of Pernambuco (UFPE), Av. Jorn. Aníbal Fernandes, Cidade Universitária, Recife 50740-560, Pernambuco, Brazil
[3] Institute Keizo Asami (iLIKA), Federal University of Pernambuco (UFPE), Av. Prof. Moraes Rego, 1235, Cidade Universitária, Recife 50670-901, Pernambuco, Brazil
* Correspondence: isabella.cavalcanti@ufpe.br; Tel.: +55-81-98648-2081

Abstract: The development of multidrug-resistant (MDR) microorganisms has increased dramatically in the last decade as a natural consequence of the misuse and overuse of antimicrobials. The World Health Organization (WHO) recognizes that this is one of the top ten global public health threats facing humanity today, demanding urgent multisectoral action. The UK government foresees that bacterial antimicrobial resistance (AMR) could kill 10 million people per year by 2050 worldwide. In this sense, metallic nanoparticles (NPs) have emerged as promising alternatives due to their outstanding antibacterial and antibiofilm properties. The efficient delivery of the NPs is also a matter of concern, and recent studies have demonstrated that hydrogels present an excellent ability to perform this task. The porous hydrogel structure with a high-water retention capability is a convenient host for the incorporation of the metallic nanoparticles, providing an efficient path to deliver the NPs properly reducing bacterial infections caused by MDR pathogenic microorganisms. This article reviews the most recent investigations on the characteristics, applications, advantages, and limitations of hydrogels combined with metallic NPs for treating MDR bacteria. The mechanisms of action and the antibiofilm activity of the NPs incorporated into hydrogels are also described. Finally, this contribution intends to fill some gaps in nanomedicine and serve as a guide for the development of advanced medical products.

Keywords: antibiotic resistance; hydrogel; bacterial; antibiofilm; nanotechnology

Citation: Moreno Ruiz, Y.P.; de Almeida Campos, L.A.; Alves Agreles, M.A.; Galembeck, A.; Macário Ferro Cavalcanti, I. Advanced Hydrogels Combined with Silver and Gold Nanoparticles against Antimicrobial Resistance. *Antibiotics* **2023**, *12*, 104. https://doi.org/10.3390/antibiotics12010104

Academic Editors: Christina N. Banti and Sotiris K Hadjikakou

Received: 24 November 2022
Revised: 21 December 2022
Accepted: 4 January 2023
Published: 6 January 2023

Copyright: © 2023 by the authors. Licensee MDPI, Basel, Switzerland. This article is an open access article distributed under the terms and conditions of the Creative Commons Attribution (CC BY) license (https://creativecommons.org/licenses/by/4.0/).

1. Introduction

Antimicrobial drug resistance is considered one of the greatest threats to global public health. Multidrug resistance (MDR) is of particular concern, especially among 'ESKAPE' organisms: *Enterococcus faecium*, *Staphylococcus aureus*, *Klebsiella pneumoniae*, emphAcinetobacter baumannii, *Pseudomonas aeruginosa*, and *Enterobacter* spp., as they are responsible for many nosocomial severe infections [1]. Multi-resistant bacteria such as *P. aeruginosa*, *A. baumannii*, and *Enterobacteriaceae* have been declared priority one pathogens in a "List of Priority Pathogens for Research and Development of New Antibiotics" [2]. According to the World Health Organization (WHO), antibiotic resistance is one of the biggest public health problems, and around 79% of bacteria have developed resistance to one or more antibiotics. Approximately 700,000 people worldwide die from drug-resistant bacterial infections; by 2050, this number is estimated to reach 10 million [3,4]. In the US only, the healthcare cost related to antibiotic resistance is approximately $20 billion per year [4–6].

Antibiotic resistance comes from changes in the structure of bacteria due to changes in the genetic material or through the acquisition of genetic material from external sources, such as viruses, other bacteria, and the environment [7]. Moreover, it is known that about 40 to 80% of bacteria can form biofilms [8]. Biofilms are sets of sessile microorganisms attached to a substrate, or to each other, enclosed in a self-produced polymeric matrix. When embedded in the matrix, bacterial cells exhibit an altered rate of growth and gene transcription. Additionally, they produce specific proteins that help them to become more resistant, and they present high gene exchange, raising the recombination rate among the strains [9,10]. According to the National Institute of Health, biofilms are responsible for up to 80% of all microbial infections in humans, including cases of endocarditis, cystic fibrosis, non-healing chronic wounds, meningitis, kidney infections, and infections related to implantable devices such as urinary catheters, prosthetic joints, and heart valves [11].

Both innate and acquired host immune responses are activated during a biofilm infection. However, neither of these immune responses can eradicate the pathogen in the biofilm due to the polymeric matrix, which acts as a structural barrier. In addition, sessile bacteria are less responsive to traditional antibiotic therapy because they are 500 to 5000 times more resistant to drugs than planktonic cells. Thus, new strategies to inhibit biofilm formation and to eradicate already formed ones, are mandatory [10,12].

Amongst the pathogens with MDR to antibiotics, these bacteria can be highlighted: *S. aureus*, resistant to methicillin and vancomycin (MRSA and VRSA, respectively); *E. faecium*, resistant to vancomycin or fluoroquinolone; *E. coli*, resistant to polymyxin; and *Acinetobacter* spp., which is resistant to aminoglycosides [13,14]. Between 2018 and 2019, the United States Food and Drug Administration (US FDA) approved nine new antibiotics from 107 molecules, to fight against MDR bacteria [15].

Numerous studies have aimed to understand the phenotypic and genotypic evolution of antibiotic resistance [16]. Although some promising agents have been explored [17,18], there is an urgent need for new active antibiotic molecules, but usually, new antibiotics take a long time to be developed [4]. Thus, to reduce the problem of antibacterial resistance in a short period, will be challenging.

Before the discovery of penicillin, certain metals, oxides, or metallic salts were used to treat bacterial and fungal infections [19], but their use have declined. In recent years, inorganic materials, especially nanostructured systems, have proven to be effective against pathogenic microorganisms [20,21]. Nanoparticles of metals and metal oxides, such as silver (Ag), gold (Au), MgO, ZnO, and TiO_2, with antimicrobial activity, have been proposed as antibiotics (Table 1) [4].

Silver nanoparticles (Ag NPs) and gold nanoparticles (Au NPs) have been proposed as a new class of antibiotics. These NPs have shown a broad antibacterial activity against *E. coli* [22–24], *S. aureus* [25,26], *P. aeruginosa* [27,28], *Proteus vulgaris*, *S. aureus*, *Proteus mirabilis* [13], *Enterobacter cloacae* [29], and *Staphylococcus epidermidis* [28]. These metallic NPs are able to inhibit the growth of bacteria by inhibiting the formation of bacterial biofilms and/or destroying pathogenic microorganisms [30,31].

For in vivo application, the colloid of NPs requires a platform that acts as a carrier, providing stability to the NPs, regulating their controlled release at the local site of the bacterial infection. Among the supportive materials for metallic NPs, hydrogels are the most commonly investigated in nanomedicine. Hydrogels are tridimensional structures that can be functionalized due to the presence of distinct functional groups inside the network. Recently several antimicrobial agents have been incorporated into hydrogels such as antibiotics [32], nanoparticles [22–24], bacteriophages, antibacterial peptides [33], biological extracts [34,35], and antimicrobial enzymes [36]. Some of the prerequisites of hydrogels for health applications are: non-toxicity, sustainability, environmentally friendly [13], flexibility, elasticity [37], biocompatibility [38], immunogenic [22], biodegradability [39], resistance to severe conditions [40], good extensibility [41], and the ability to stimulate nutrients and metabolic exchange [22]. Hydrogels are capable of improving cellular internalization [29], absorbing wound exudates [42], expediting skin healing, stimulating the collagen proliferation [40], and exhibiting antibiofilm activity [23,37,39,43,44]. They may also be applied in medical devices such as venous or urinary catheters, artificial voice prosthesis, and prosthetic heart valves [45].

Table 1. Examples of metallic NPs used against resistant bacteria and their mechanism of bactericidal action.

NPs	Bacteria	Mechanism of Action	Ref.
Ag	*E. coli*, *B. subtilis*, and *S. aureus*	• Ag^+ ion liberation; • Cell membrane destruction and electron transport; • Bacterial DNA damage	[20,46]
Au	*P. aeruginosa* and *E. coli*	• Interaction with Au NPs • Mg^{2+} or Ca^{2+} ion sequestration to damage the cell membrane • Competition for the virus binding to the cell	[47–50]
ZnO	*E. coli*, *S. aureus*, and *Botrytis cinerea*	• Intracellular NP accumulation • Damage to the cell membrane • H_2O_2 production • Zn^{2+} ion liberation	[51–53]
TiO_2	*E. coli* and *Bacillus megaterium*	• Production of active oxygen species • Cellular membrane destruction • Generation of electron-hole pair by visible light excitation with low recombination rate	[54,55]
Cu	*E. coli* and *Bacillus subtilis*	• Cu^{2+} ion liberation • Cellular membrane damage • DNA alteration	[56]
MgO	*E. coli*, *S. aureus*, *Bacillus subtilis*, and *Bacillus megaterium*	• Cellular membrane damage • Alkalinization by MgO • Hydration • Active oxygen liberation	[57,58]

Ag NPs and Au NPs are loaded into polymer-based hydrogels with a porous structure, such as alginate [59], chitosan [27,60], gelatin [61], konjac glucomannan [23], hydroxypropyl methylcellulose [24], carboxymethyl cellulose [13], carboxymethyl chitosan [62], polyvinyl alcohol [37], carbopol [63,64], gelatin methacrylate [65], polyacrylamide [40], polyethyleneimine [66], and polyvinylpyrrolidone [40,63,67–69]. Occasionally, hydrogels exhibit poor mechanical properties, and other agents are required as additives in the manufacturing process, such as tannin acid [60], graphene [66,70–72], aluminosilicate nanotubes (NTs) [65], and metal-organic frameworks (MOFs) [69].

In this review, several significant aspects are presented, such as (i) biocompatible natural and synthetic polymers; (ii) synthesis strategies to produce antibacterial hydrogels; (iii) the physical, chemical, and biological properties of the hydrogels and the NPs; (iv) the synergism between the hydrogels and NPs characteristics; (v) NPs aspects that stand out in the antibacterial or antibiofilm efficiency; (vi) mechanisms of antibacterial action, of action for inhibiting biofilm, and for biofilm eradication.

2. Silver Nanoparticles (Ag NPs)

Ag NPs can damage the extracellular membrane of bacteria and their intracellular components, exhibiting a broad-spectrum antimicrobial effect [4]. Many Ag NP synthesis strategies have been developed to allow specific Ag NP surface properties, which, in turn, strongly depend on the characteristics of the reducing agent and the type of stabilizer used during their synthesis [4,73–79].

According to Sondi et al. [80], Ag NPs can cause harm to *E. coli* by forming pits in the cell wall, which could increase its permeability and affect the membrane vesicles.

Such damage has also been observed in other bacteria, such as *Scrub typhus*, *P. aeruginosa*, and *Vibrio cholerae* [81], which is attributed to the ability of Ag NPs to interact with some of its components, such as lipopolysaccharides (LPSs), and phosphatidylethanolamines (PEs) [82].

Mirzajani et al. [83] suggested that the ability of Ag NPs to harm the bacterial cell wall may result from their interaction with the peptidoglycan layer, since Ag NPs attack the β-1-4 bonds of N-acetylglucosamine and N-acetylmuramic acid of the glycan chain in the cell membrane of *S. aureus*. Additionally, Ag NPs may produce free radicals, such as reactive oxygen species (ROS), inside and outside the bacteria [84,85]. Elevated ROS levels are known to damage cell DNA, proteins, and enzymes, which could, in turn, interfere with the normal metabolism of bacteria [86]. It was found that Ag^+ ions released from Ag NPs can damage bacterial membrane function. In particular, the differences in Ag^+ concentration can induce a difference between the pH and the electrical potential inside and outside the membrane vesicles in *Vibro cholerae*, leading to the failure of membrane respiration and H^+ leakage [87].

The effect of Ag NPs on the bacterial membrane is related to their physicochemical properties, such as size, shape, surface area, surface charge, oxidation state, and surface chemistry. It has been reported that Ag NPs with small size and colloidal stability are preferred rather than those susceptible to aggregation [4,13,22,66,73,88]. The size of NPs is one of the most critical aspects determining their interaction with cells. Actually, Ag NP interaction is size-dependent [81,89]. Several works have shown that Ag NPs with a diameter of 3–10 nm, are the most effective in killing bacteria due to their preferential direct interaction with the bacterial membrane [52] and how fast the bacterial killing took place after their interaction [89].

The shape of Ag NPs can directly influence the available contact area needed to facilitate interactions of Ag NPs with the bacterial membrane. A comparative study using polyvinylpyrrolidone (PVP)-coated Ag NPs with different shapes suggested a strong correlation between the shape of the Ag NPs and their bactericidal properties. For example, Ag nanoplates (two-dimensional structure, 2D) showed the highest antimicrobial activity against *S. aureus* and *E. coli*, when compared to Ag nanorods (one-dimensional structure, 1D) and spherical Ag NPs (zero-dimensional structure, 0D). Sadeghi et al. [90] showed that Ag nanoplates exhibited the largest surface area, providing the most significant contact area to interact with the bacterial cell wall.

The Ag NP surface charge is also important. It was observed that positively charged Ag NPs using capping agents such as poly(amide amine) dendrimers (PAMAM) [91], poly(ethyleneimine) (PEI) [92], poly(ethylene glycol) (PEG), and polyvinylpyrrolidone (PVP) [67] facilitated the interaction between the particles and the negatively charged bacterial membrane [91]. Ag NPs with a negative surface charge have shown lower antimicrobial activity [93] due to the strong repulsion between the particles and the bacterial wall. This limits the interaction between Ag NPs and bacteria and considerably weakens their antimicrobial effect.

Ag NPs have been also combined with antibiotics (ampicillin, amoxicillin, chloramphenicol, erythromycin, among others) [94] by chelation of the active groups. The combination of Ag NPs with other materials, such as polycationic chitosan, has shown promising results by facilitating the attachment of Ag NPs to the negatively charged bacterial wall [95]. Mishra et al. [96] developed a multifunctional system of Ag NPs embedded in the chitosan-polyethylene glycol (CS-PEG) hydrogel. This implantable device inhibited biofilm formation and the released the drug payload at the same time. Chen et al. [97] prepared a chitosan sponge containing Ag NPs and used it as a tissue for wound healing. Both in vitro and in vivo composite tests showed excellent antibacterial activity against drug-resistant pathogenic bacteria.

Recently, researchers discovered that Ag nanoclusters (NCs) are effective for this type of application [98,99]. NCs are NPs whose sizes are smaller than 2 nm and contain "countable" Ag atoms as a nucleus protected by organic ligands [4]. Ag NCs have shown promising results for biomedical applications, such as bioimaging, biosensing, and antimicrobial agents [100,101]. These NPs have also been used to functionalize natural cellulose nanofibers [102], silk fibers [103], textiles [104], and natural or synthetic polymer-based hydrogels to exhibit antimicrobial activity. Although there are many studies of antimicrobial Ag NPs embedded into hydrogels as platforms for delivering metallic nanostructures as alternative to standard drugs; their mechanism of action has not been entirely elucidated. Nevertheless, all the above examples demonstrate this as a promising strategy in preventing and eradicating infections [7,105].

2.1. Antibacterial Activity of Ag NPs Loaded into Hydrogels

Ag NPs incorporated into hydrogels have shown antibacterial properties and the ability to control infections [37]. The NPs are incorporated into a hydrogel with a porous structure by in situ polymer synthesis or by adding the NP colloid to the polymer. Additionally, microwave radiation is another approach to produce NPs within hydrogels. The polymer-based hydrogels help to control the morphology and size of the nanostructures and participate as a stabilizing medium for nucleation sites to produce silver seeds [13,66]. Biocompatible polymers, such as chitosan [27], konjac glucomannan [23], carboxymethyl cellulose [13], carboxymethyl chitosan [62], polyvinyl alcohol [37], carbopol-934 [63], graphene [41,70–72], gelatin methacrylate [65], polyacrylamide [40], polyethyleneimine [66], and polyvinylpyrrolidone [40,63,67–69], have been used to fabricate antibacterial and antibiofilm materials.

These polymeric biomaterials have helped to treat and prevent infections caused by pathogenic bacteria and are capable of improving the healing and regeneration of the skin. For example, Ag/chitosan/hydrogel has been shown to help the healing process, reduced inflammation at skin wounds, and accelerated the re-epithelization rate to treat post-operative infection [22].

Chitosan (CS) is derived from chitin, the second most abundant biopolymer in nature, after cellulose. It has been used in the synthesis of hydrogels due to its biodegradability, biocompatibility, and antibacterial activity [22]. Chemical crosslinking, the addition of nanofillers, blending with other polymers, and using alkali–urea solutions, are some of the methods used to improve chitosan processability. To improve the water solubility of chitosan, quaternization method has been used, in which a quaternary ammonium moiety was introduced into the chitosan structure by chemical reactions, thus producing quaternate chitosan. Some studies have reported the use of chitosan as a matrix to incorporate Ag nanoparticles. Ag NPs were also synthesized in situ within oxidized dextran (ODex), adipic dihydrazide-grafted hyaluronic acid (HA-ADH), and quaternized chitosan (HACC), resulting in the Ag–ODex/HA-ADH/HACC hydrogel [27]. The Ag NPs had a particle distribution size of around 50–190 nm. The hydrogel displayed antibacterial properties against *E. coli* ATCC 8739, *S. aureus* ATCC 14458, and *P. aeruginosa* CMCCB10104, and the inhibition zones were 24, 24, and 27 mm, respectively. These results were associated with the hydrogel's positive charge due to the quaternate chitosan's cationic group, that favor the interaction with the negatively charged bacterial cell walls. This system reduced the wound area in rats up to 41.3% after 7 days, decreased inflammation, and improved re-epithelialization [27].

In a similar study, Xie et al. [22] prepared an Ag/chitosan hydrogel using an alkali–urea solution, LiOH (4.5% wt.)/KOH (7% wt.)/CH_4N_2O (8% wt.) by the freeze/thaw process, $AgNO_3$, and $Na_3C_6H_5O_7$. The silver concentration in the hydrogel increased, leading to spherical and ellipsoidal Ag NPs with a size distribution of 4.45 nm $\pm$ 0.37 nm to 9.22 $\pm$ 0.54 nm. The hydrogel composite had large tensile mechanical properties (15.95 $\pm$ 1.95 MPa). The antimicrobial activity was 99.86 $\pm$ 0.12% against *E. coli* and 99.94 $\pm$ 0.10% against *S. aureus* tested on rats for 14 days. The wound contraction was 70.5% on the 4th day and 99.75% on the 14th day. Thus, Ag NPs coated with chitosan accelerated the healing process. The authors determined that Ag NPs destroyed the bacterial cell wall due to interactions between the NPs and the lipid layer of the bacterial cell membrane. The

Ag NPs would merge with bacterial DNA damaging bacterial replication and impairing bacterial respiratory function.

Furthermore, carboxymethyl chitosan is a derivative of chitosan, non-toxic, and also capable of forming gels [62]. Carboxymethyl chitosan-based hydrogels have shown enhanced physicochemical, and biological properties, including antimicrobial, antioxidant, and antifungal activities. This hydrogel has been used in applications such as wound healing, drug-carrying, smart tissue, and biomedical nanodevices [106]. Additionally, it has been well explored in the cosmetic and food industry [62].

Ag/chitosan-carboxymethyl β-cyclodextrin hydrogel (CM-βCD) is an alternative approach to inhibit the growth of bacteria. It has been shown to display antibacterial activity against *E. coli* and *S. aureus* [25]. The interactions between Ag^+ ions and bacteria were improved through ions exchange between Ag^+ and H^+ from the carboxylic and amino groups within the Ag NPs-CM-βCD hydrogel. The inhibition zone increased when the concentration of CM-βCD was increased in the hydrogel [25].

Pandian et al. [37] fabricated a Ag/N, O-carboxymehtyl chitosan (N, O-CMC) hydrogel with self-healing properties. The ethylenediaminetetraacetic acid (EDTA, $C_{10}H_{16}N_2O_8$) and ferric ions (Fe^{3+}, $FeCl_3$, 2%) were used in the synthesis process to produce a self-healing hydrogel. The size distribution of Ag NPs was 25 ± 14 nm according to TEM images. The hydrogel displayed an antibacterial activity against ATCC and clinical strains of *E. coli* ATCC 25922, *S. aureus* ATCC 35556, MRSA ATCC 43300, *P. aeruginosa* ATCC 47085, and *K. pneumonia* ATCC 700603. The minimum inhibitory concentration (MIC) for *P. aeruginosa* was 48.5 mg/mL, 32.5 mg/mL for MRSA, and 32 mg/mL for *S. aureus*. The Ag NPs/N, O-CMC hydrogel was more efficient against *E. coli* and *K. pneumonia* with MIC values of 17.5 and 23.0 mg/mL, respectively. At the same time, the minimum bactericidal concentration (MBC) values were 55 and 71 mg/mL, respectively. The authors described that the interaction between Fe^{3+} (metal) and -COOH (ligand) was responsible for the self-healing property of the Ag NPs/N, O-CMC hydrogel [37].

The carboxymethyl chitosan (CMCS) has been mixed with oxidized konjac glucomannan (OKGM). The OKGM is a natural polysaccharide, soluble in water, that was shown to improve the microstructure and mechanical properties of chitosan [107,108], gelatin [109], and oxidized hyaluronic acid [110], acting as a macromolecular cross-linker [108]. The OKGM-based hydrogel exhibited self-healing characteristics in a recent study, where Ag NPs/OKGM/CMCS hydrogel demonstrated antibacterial properties against *E. coli* and *S. aureus* [23]. This hydrogel was tested on rats' skin. The hydrogel pore size distribution was in the range of 59.4 to 230 μm, increasing as the concentration of OKGM increased, but the swelling capacity decreased. Higher concentrations of polymers accelerated the gelation time from 600 to 57 s [23]. Similar to a previous study, Ag/konjac glucomannan hydrogel was tested against *S. aureus* and *E. coli* showing good antibacterial efficiency on rabbit skin infections [111].

Hydrogels based on carboxymethyl cellulose (CMC), polyvinyl alcohol (PVA), and $C_8H_{14}O_4$ (EDGE) has been prepared using microwave radiation as a carrier of Ag NPs (8–14 nm). The Ag release rate from this hydrogel was 85% over five days [13]. Ag^+ ions are bound to the hydrogel composite via electrostatic interactions. This Ag/hydrogel acted as a bactericide against pathogenic microorganisms of the urinary tract, such as *E. coli*, *K. pneumoniae*, *P. aeruginosa*, *P. vulgaris*, *S. aureus*, and *P. mirabilis* [13]. The Ag/hydrogels with 5 mg/mL of Ag presented a growth inhibition diameter of 16.6 mm against *E. coli*, 15.8 mm against *K. pneumoniae*, 15.6 mm against *P. aeruginosa*, and 15.2 mm against *P. vulgaris*.

Hydrogels based on carbopol-934 and *Aloe vera* supported Ag spherical NPs encapsulating quercetin (QCT) [63]. This system was designed to take advantage of (i) the properties of QCT as an anti-inflammatory and antioxidant; (ii) of carbopol-934, as a biodegradable and bioadhesive polymer with good tensile strength; (iii) *Aloe vera* that stimulates collagen production; and finally, (iv) of Ag NPs that have broad antimicrobial activity. The QCT-Ag/carbopol-*Aloe vera* hydrogel presented antibacterial activity against *S. aureus* MTCC

3160 and *E. coli* BL-21 with inhibition zone values of about 19.0 and 17.0 mm, respectively. Ag NPs improved the release rate of quercetin from the hydrogel for the treatment of wounds in diabetic patients.

Some studies have explored the incorporation of graphene into hydrogel structures due to its high thermal and electrical conductivity, and sizeable mechanical strength [66,70–72]. The graphene embedded in hydrogel reduced hydrogel breaking and reinforced its mechanical properties. The Ag/graphene composite hydrogel was prepared using acrylic acid and N,N′-methylene bisacrylamide ($C_7H_{10}N_2O_2$), with a mass ratio of 5:1 silver to graphene [41]. The Ag NPs of an average size of 39 nm were deposited onto the surface of graphene nanosheets. The Ag NPs/graphene hydrogel was evaluated against *E. coli* and *S. aureus* using the shaking flask method. The antimicrobial activity was enhanced as the Ag NP concentration increased. Larger nanoparticle sizes displayed better antimicrobial activity than smaller ones. The graphene promoted the incorporation of a higher number of NPs and avoided their aggregation onto its surface.

Another approach that has been explored is to combine chitosan and graphene to produce an antibacterial hydrogel with enhanced durability [71]. For instance, Nešović et al. prepared Ag/poly(vinyl alcohol)/chitosan/graphene hydrogels [70–72] by electrochemical synthesis of nanoparticles in a hydrogel network. The hydrogel displayed better mechanical characteristics, such as tensile strength and elastic modulus. The Ag NPs size distribution was from 6.38 to 10.00 nm depending on the chitosan content. The antimicrobial activity was evaluated against *E. coli* ATCC 25922 and *S. aureus* TL. The number of bacteria colonies decreased quickly in 15 min, when the $AgNO_3$ concentration was 0.25 mM and 0.5% wt. of chitosan, during the composite hydrogel preparation (0.25Ag/PVA/0.5CHI/Gr). Increasing the chitosan content resulted in a slower Ag release rate from the hydrogel. Nešović et al. [71] found that Ag NPs prevented adenosine 5′-triphosphate (ATP) formation within the microorganism.

Figure 1 summarizes the hydrogels embedded with Ag and Au NPs against multidrug-resistant bacteria.

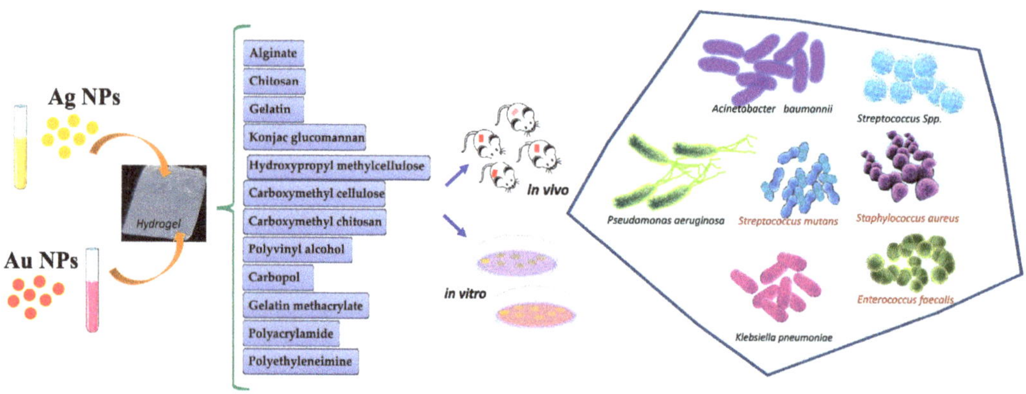

Figure 1. Silver and gold NPs loaded into hydrogels for antibacterial application.

A Ag-polyethyleneimine (PEI)-graphene oxide (GO) hydrogel was produced using Pluronic F127 gel [66]. In this case, Pluronic F-127 was used to create a sustained antimicrobial effect, presenting reverse thermal gelation properties. PEI decreased the aggregation of nanostructures within the hydrogel. The antimicrobial activity against *E. coli* was 99.86%, and 99.94% against *C. albicans*, using 10 μg/mL of the hydrogel. The Ag release rate from the hydrogel was 72% in 7 days. The authors proposed that the graphene oxide nanosheets damaged the bacterial cell wall due to the sharp edges leading to a faster disruption of the plasmatic membrane by the Ag NPs.

Furthermore, Ag NPs have been incorporated into nanotubes/polymer hydrogels to explore NP delivery. For instance, aluminosilicate nanotubes (NTs) are platforms with a great capacity to store and carry molecules and drugs. They also help to reduce the hydrogel degradation rate and can be loaded, as additives, into the hydrogels, such as gelatin methacrylate (GelMA), a biocompatible hydrogel with many biological characteristics [112]. For example, Ag NPs were loaded into aluminosilicate nanotubes (NTs) and then within a methacrylate gelatin [65] matrix, to produce an antibacterial hydrogel capable of improving bone regeneration. The hydrogel was prepared using photopolymerization by UV irradiation of 365 nm and 400 W. According to the inhibition zone results, the Ag/NTs/GelMA hydrogel showed higher antibacterial activity against *E. coli* ATCC 8739 than *S. aureus* ATCC 29213.

In addition, the morphology of Ag NPs is another relevant aspect that influences a hydrogel's antibacterial efficiency. Different NP shapes may present a distinct surface area to interact with bacterial membranes, leading to diverse antibacterial activity [40,113,114]. In this context, Ag NPs with different morphologies (spherical, triangular, and rod) were incorporated into polyacrylamide (PPA) and N-mehtylene bisacrylamide (MBA) hydrogels, named PAA-MBA [40]. The mechanical strength of the Ag NPs-PAA-MBA hydrogel (4 to 5 KPa) did not depend on Ag NP shape. Rod-shaped nanoparticles were poorly absorbed within the hydrogel network due to the formation of aggregates on the hydrogel surface. However, these NPs showed antibacterial activity. The hydrogel doped with spherical NPs of 12.7 ± 5.9 nm and triangular NPs of 37.1 ± 15.0 nm demonstrated high antimicrobial activity against *E. coli*.

Table 2 summarizes the hydrogels embedded with Ag NPs for antibacterial application.

2.2. Antibiofilm Activity of Hydrogels Loaded with Ag NPs

Taking the usefulness of non-invasive therapy into consideration, and the elimination of drug-resistant biofilms in oral infections and wound healing, hydrogels loaded with Ag NPs is an alternative method of infection management [115–117]. In this scenario, Haidari et al. [43] investigated the effectiveness of applied Ag NP hydrogels in mature *S. aureus* biofilms, both in vitro and in vivo. In vitro tests were performed by flow cytometry, where bacterial cells with compromised membranes were stained red by propidium iodide, whereas cells with intact membranes were stained green by SYTO9. The test showed that after treating the *S. aureus* biofilm with the Ag NP hydrogel, most of the cells were stained in a high red fluorescence intensity, associated with a substantially lower biofilm biomass, indicating severe disruption of the mature biofilm. For in vivo tests, an established *S. aureus* mouse model of a mature biofilm wound infection was utilized. The antibiofilm treatment started after biofilms had been fully established. IVIS bioluminescent imaging was used to track 10 days of Ag NP hydrogel treatment in real-time. The Ag NP hydrogel treatment gradually decreased the *S. aureus* biofilm starting on day 4.

Table 2. Ag NPs loaded into hydrogel for antibacterial application.

System	Materials	Ag NP Properties (Size and Surface Charge)	NP Synthesis Method	Bacteria	Target	Antibacterial Properties: Inhibition Zone (mm) and MIC Values	Ref.
Ag–ODex HA-ADH/HACC	Dextran, sodium hyaluronic, chitosan quaternary ammonium salt, and $AgNO_3$	50–190 nm	Chemical reduction, in situ, Schiff-base reaction to form hydrogel	*E. coli* ATCC8739, *S. aureus* ATCC14458, and *P. aeruginosa* CMCCB10104	In vitro; In vivo, rats	The Kirby–Bauer (KB) method. The inhibition zone was 24, 24, and 27 mm, respectively	[27]
Ag/CS	LiOH, KOH, CH_4N_2O, $AgNO_3$, and $Na_3C_6H_5O_7$	Spherical and ellipsoidal NPs; 4.45–9.22 nm	Chemical reduction with sodium citrate, in situ	*E. coli* and *S. aureus*	In vivo; rats	Antibacterial activity: 99.86% and 99.94%, respectively	[22]
Ag/CM- βCD	Chitosan, $NaBH_4$, $AgNO_3$, NaOH, cyclodextrin, CH_3CO_2H, and C_5H_8O	50 nm	Chemical reduction with $NaBH_4$, in situ	*E. coli* and *S. aureus*	In vitro	The inhibition zone increased when the CM-βCD concentration was increased in the hydrogel	[25]
Ag/N, O-carboxymehtyl chitosan (N, O-CMC)	Chitosan, $AgNO_3$, $C_{10}H_{16}N_2O_8$ (EDTA), $CaCl_2$, $FeCl_3$, and $C_2H_3ClO_2$	25 nm	Chemical reduction using $C_2H_3ClO_2$	*E. coli* ATCC25922, *S. aureus* ATCC35556, MRSA ATCC 43300, *P. aeruginosa* ATCC47085, and *K. pneumonia* ATCC700603	In vitro, L929 cells	MIC values: 48.5 mg/mL for *P. aeruginosa;* 32.0 mg/mL for *S. aureus* and MRSA; 17.5 mg/mL for *E. coli*, and 23.0 mg/mL for *K. pneumonia*	[37]
Ag/OKGM-CMCS	Oxidized konjac glucomannan (OKGM) and Carboxymethyl chitosan (CMCS)	60 nm	Schiff-base reaction	*S. aureus* and *E. coli*	In vitro, L929 cells; In vivo, rats	The Ag/hydrogel achieved high antimicrobial activity, but the inhibition zone values were not displayed	[23]
Ag/KGM	Eggs, konjac glucomannan, $AgNO_3$, and NaOH	9.5–30.2 nm	In situ	*S. aureus* and *E. coli*	In vitro; In L929 cells; in vivo, rabbits	Good antibacterial efficiency on rabbits' skin infections	[111]
Ag/CMC/PVA/EGDE	Carboxymethyl cellulose (CMC), polyvinyl alcohol (PVA), and ethylene glycol diglycidyl ether (EGDE)	8–14 nm	Microwave radiation	*E. coli*, *K. pneumoniae*, *P. aeruginosa*, *Proteus vulgaris*, *S. aureus*, and *Proteus mirabilis*	In vitro, patient urine	The inhibition zone: 16.6 mm for *E. coli*, 15.8 mm for *K. pneumoniae*, 15.6 mm for *P. aeruginosa* and 15.2 mm for *P. vulgaris*	[13]
QCT-Ag/Carbopol- *aloe vera*	Carbopol 934, $AgNO_3$, QCT, polyvinylpyrrolidone (PVP), *Aloe vera*, $C_3H_8O_3$, and $NaBH_4$	44.1 nm; ζ: −14.76 mV	Chemical reduction with $NaBH_4$	*S. aureus* MTCC 3160 and *E. coli* BL-21	In vitro, L929 cells; In vivo, mice skin	The inhibition zone: 17 mm for *E. coli* and 19 mm for *S. aureus*	[63]
Ag/graphene	$AgNO_3$, $C_7H_{10}N_2O_2$, $(NH_4)_2S_2O_8$, and NH_3H_2O	39 nm	Hummer's method	*E. coli* and *S. aureus*	In vitro, L929 cells; In vivo, rats	The disc diffusion method. Large Ag concentration led to great antibacterial activity using 5:1% wt. of Graphene	[41]
Ag/poly(vinyl alcohol)/chitosan/graphene	Graphene, chitosan, CH_3CO_2H, KNO_3, $AgNO_3$, and K_2HPO_4	6.38–10.00 nm	Electrochemical synthesis in situ using 90 V	*E. coli* ATCC 25922 and *S. aureus* TL	In vitro, MRC-5 and L929 cells;	The inhibition zone: 15.5 mm for *S. aureus* and 13.5 mm for *E. coli;* great antimicrobial activity with the 0.25Ag/PVA/0.5CHI/Gr	[70–72]

Table 2. *Cont.*

System	Materials	Ag NP Properties (Size and Surface Charge)	NP Synthesis Method	Bacteria	Target	Antibacterial Properties: Inhibition Zone (mm) and MIC Values	Ref.
Ag/PEI- graphene oxide	Pluronic F 127, graphene oxide, $C_8H_{17}N_3$.HCl, $AgNO_3$, NH_4OH, and NaCl	10 nm; ζ: 42.6 mV	Amidation reaction with $Ag(NH_3)_2OH$ by microwave reactor	*E. coli* and *C. albicans*	In vitro	*E. coli* (99.86%) and *C. albicans* (99.94%)	[66]
Ag/PAA-MBA	$K_2S_2O_8$, $NaBH_4$, PVP, C_3H_5NO, $C_6H_9Na_3O_9$, and $AgNO_3$	Spherical: 12.7 nm; triangular: 37.1 nm; hexagonal: 26.9 nm	Chemical reduction using $NaBH_4$	*E. coli* W3110	In vitro	The spherical and triangular shapes of the Ag NPs displayed better antibacterial activity than the rod-shaped NPs.	[40]
Ag/halloysite/ gelatin methacrylate	$AgNO_3$, $NaBH_4$, $(CH_3)_2SO$, and C_2H_4O	Ag NPs changed the microstructure and roughness of the hydrogel	In situ by photopolymerization using UV radiation (365 nm and 400 W)	*E. coli* ATCC 8739 and *S. aureus* ATCC 29213	In vitro; In vivo, crania of rats	The inhibition zone test showed that the hydrogel restrained the growth of the bacteria	[65]
Ag/KGM	Chitosan, carboxymethyl, β-cyclodextrin, etc.	50 nm	Chemical reduction	*S. aureus* and *E. coli*	In vitro	Inhibition zone: 22 and 19 mm, respectively	[25]

From 5 to 10 days after the infection, there was a statistically significant decrease in the concentration of bacterial cells, showing the high efficiency of the Ag NPs in eradicating established mature biofilms in wounds. This study demonstrated the use of an Ag NP hydrogel as a valid therapeutic approach for the effective and safe elimination of mature *S. aureus* biofilms in wounds.

Consistent with these results, Imran et al. [118], also reported the antibiofilm activity of a Ag NP-loaded hydrogel against *B. subtilis* and *E. coli*. It was revealed that the hydrogel showed a dose-dependent biofilm inhibition activity, with a minimum biofilm inhibition of approximately 27% when the Ag NPs were used at a concentration of 10 ppm and a maximum inhibition of 97% when the Ag NPs were used at a concentrations of 100 ppm. Additionally, the half maximal inhibitory concentration (IC_{50}) values obtained were 29.88 and 27.36 for *E. coli* and *B. subtilis*, respectively. Pandian et al. [37], in turn, evaluated the antibiofilm activity of in situ Ag NPs incorporated in an N, O-carboxymethyl chitosan self-healing hydrogel. After 48 h, a decrease of 68.86 ± 0.05%, 75.07 ± 0.02%, and 83.22 ± 0.01% was observed in *E. coli*-, *S. aureus*-, and *P. aeruginosa*-treated biofilms, respectively.

Alfuraydi et al. [119] described the preparation of novel cross-linked chitosan and PVA hydrogels impregnated with Ag NPs, as well as its activity against different strains of fungi, Gram-positive and Gram-negative bacteria. In their results, The minimal biofilm inhibition concentration (MBIC) for the chitosan hydrogels alone ranged from 15.63 to 125 µg/mL, differing from the MBIC values of the hydrogel containing Ag NPs at 1 and 3%, which ranged from 1.95 to 7.81 µg/mL. These data demonstrated how the dispersion of Ag NPs inside the matrix of the chitosan hydrogel significantly improved its ability to prevent the formation of biofilms.

Similarly, the antibiofilm action of the chitosan hydrogel containing Ag NPs was previously explored by Pérez-Díaz et al. [120]. In their work, the hydrogels demonstrated a great impact on the multi-species biofilm of oxacillin-resistant *S. aureus* (ORSA), achieving a 6 Log_{10} reduction at a Ag NP concentration of 100 ppm. The antibiofilm activity against *P. aeruginosa* was lower, with a Log_{10} decrease of 3.3 at a concentration of 1000 ppm. As stated in the study conducted by Arinah et al. [121], the different results on the tested drugs' antibiofilm activity could be attributed to structural variations in the bacterial membrane walls, which differ in Gram-negative or Gram-positive bacteria. In their work, the authors incorporated *Pleurotus ostreatus*-biosynthesized Ag NPs into a genipin-crosslinked gelatin hydrogel to investigate the antibiofilm properties against the biofilms of *S. aureus*, *P. aeruginosa*, *Bacillus* sp., and *E. coli*. Stronger biofilm inhibition of about 58 ± 4% was observed in Gram-negative strains. For Gram-positive bacteria, the percentage of inhibition was 55 ± 5% for *S. aureus* and 38 ± 1% for *Bacillus* spp.

Furthermore, many recent studies have reported antibacterial and antibiofilm activity improvement of drugs when they are associated with metallic nanoparticles, such as Ag NPs [122,123]. Thus, in the research conducted by Lopez-Carrizales et al. [124], chitosan hydrogel loaded with Ag NPs and the antibiotic ampicillin (AMP) were tested against resistant bacterial pathogens, evaluating its capacity to prevent the early formation of biofilms by the colony biofilm model. The biofilm produces thick, layered structures, and the counting of colony-forming units (CFU) was Log_{10}-transformed. The antibiofilm action of the hydrogel changed depending on the Ag NP and ampicillin concentrations and the tested strain. The biofilms of *A. baumannii*, *E. faecium*, and *S. epidermidis*, were significantly inhibited by the hydrogel with the lowest concentration of Ag NPs and ampicillin (25 ppm Ag NPs/50 ppm AMP), exhibiting Log_{10} reductions of 10 ± 0.01, 8.9 ± 0.02, and 7.8 ± 0.13, respectively. However, the *E. cloacae* biofilm was only inhibited by a higher antimicrobial dose (250 ppm Ag NPs/500 ppm AMP), resulting in a Log_{10} reduction of 9.9 ± 0.11.

Recently, Wunno et al. [125] investigated a potentially new sustainable delivery system of Ag NPs for, among other activities, antibiofilm action. In their work, an ex situ thermosensitive hydrogel based on poloxamers loaded with biosynthesized Ag NPs from *Eucalyptus camaldulensis* was created and tested against Gram-positive (*S. aureus* and *S. epidermidis*) and Gram-negative bacterial (*A. baumannii* and *P. aeruginosa*) biofilms. At a

1/2 minimum inhibitory concentration (MIC), the proportion of biofilm inhibition reached 83%. When the mature biofilms were exposed to the Ag NP hydrogel and analyzed by confocal laser scanning, loosening of the biofilm architecture and cell death were revealed after 4 h of co-incubation with the hydrogel formulation at a 2 MIC (μg/mL) concentration. Based on the presented results, it is clear that the tested hydrogel formulation successfully interrupted biofilm formation and eradicated cell viability within the mature biofilms.

3. Gold Nanoparticles (Au NPs)

The wide range of applications of Au NPs is related to their physicochemical properties such as the tunable core size, photothermal [126] and photodynamic properties, high chemical stability, biocompatibility [127], high X-ray absorption coefficient, efficiency in generating ROS, and localized surface plasmon resonance (LSPR) properties [92,128,129]. Furthermore, Au NPs also exhibit antimicrobial properties absent in bulk or ionic gold. For example, Au NPs destroy bacterial membranes and slow down their metabolism [58,93–95]. Due to the NP's small size, gold colloid may be susceptible to NP aggregation. Thus, Au NPs are usually stabilized with additives such as polyelectrolytes or polymers [96,97]. These stabilizers act as capping or protecting agents, and they prevent aggregation due to steric hindrance [97–107].

In the design of advanced hydrogels, the Au NPs are embedded into a hydrogel, or in situ synthesized inside the porous gel structure. Some of the polymers used to incorporate Au NPs, include chitosan [60], alginate [59], gelatin [61], hydroxypropyl methylcellulose [24], silk [26], acrylamide, diethylene glycol, and indole-3-acetic acid, poloxamer 407, Pluronic F-127, carbopol [64], carboxy methyl tamarind, methacrylated gelatin, and metal-organic frameworks (MOFs) [69]. Au NPs encapsulated in hydrogels have shown antimicrobial or bactericidal activity against Gram-positive bacteria, such as *Bacillus cereus*, *S. aureus* [126], and *S. epidermidis* [124]. Additionally, against Gram-negative bacteria such as *P. aeruginosa* [61,126], *E. coli* [59,126], *K. pneumoniae*, and *E. cloacae* [29], and fungus such as *C. albicans* [24].

Some important characteristics of Au NPs, such as size and shape have been tailored and explored to improve the antibacterial activity of the hydrogel. This was evidenced in studies that developed nanospherical of 29.2 nm [130], nanorods of 82.5 nm [24], 54 nm [131], 49.2 nm [130,132], and nanostars with a core diameter of 25 nm and an average size of 50 nm, 70 nm, and 120 nm [126]. The latest advances in gold nanoparticles embedded in hydrogels for the treatment of multidrug-resistant bacterial infections are discussed below. The efficiency of Au NPs against pathogenic bacteria is presented from three relevant aspects: antibacterial activity, biofilm activity, and antibacterial and antibiofilm mechanism of action.

3.1. Antibacterial of Au NPs Loaded into Hydrogels

Au NPs have shown very promising results against multi-resistant bacteria to antibiotics. The incorporation of Au NPs into biocompatible supports, such as liposomes, is one approach used in biomedicine. This structure can interact easily with bacterial membrane and possesses a high-delivery capacity for NPs, antibiotics, enzymes, etc. To treat bacterial infections, Zhang et al. [133] fabricated a hydrogel containing pH-responsive gold nanoparticle-stabilized liposomes as a topical antimicrobial carrier. The authors used carboxyl-modified AuNPs as stabilizers for cationic liposomes and chemically cross-linked polyacrylamide as a hydrogel. The hydrogel viscoelasticity was tailored by the cross-linker concentration, and this resulted in tunable release kinetics of the Au NP liposomes. *S. aureus* was used as a model pathogen, and the hydrogel formulation effectively released nanoparticles into the bacterial culture. No skin reaction was observed when the hydrogel formulation was topically applied to mouse skin over a 7-day treatment period [133].

One of the methodologies used to obtain hydrogels uses natural polymers, such as alginate. Alginate is a hydrophilic linear polysaccharide extracted from the cell wall of some specific species of algae or bacteria. Alginate can form a gel or act as a crosslinker with other polymers due to the exchange of Na^+ ions from the guluronic acids ($C_6H_{10}O_7$) with other cations (Ca^{2+}, Ba^{2+}, and Mg^{2+}) [134]. Alginate-based hydrogels are biocompatible, biodegradable, non-toxic, and exhibit a higher capacity of fluid load, acting as a carrier for NP delivery [135]. Gold nanostars (Au NSts) were loaded into sodium alginate-based hydrogel by Kaul et al. [126]. The sharp spike (size of 120 nm) from the NSts could puncture the bacterial wall and membrane. The antimicrobial activity was 35.4% against *S. aureus* MTCC 1430, as higher as 80% against *P. aeruginosa* MTCC 1934 and *E. coli.* MTCC 443, using 0.3 to 0.6 μg/mL of nanoparticles on the wounds of Sprague Dawley rats. The spike length, as well as the topology, of the Au NSts damaged the surface and the bacterial membrane due to the rupture process. *S. aureus* were more resistance due to the thick peptidoglycan layer outside the bacterial cell wall. In similar study, Zhang et al. [59] prepared an acrylamide (AM) and alginate (SA) hydrogel incorporating Au NPs (8 nm). This Au/AM-SA hydrogel inhibited the growth of *E. coli.* The study suggested that Au NPs interact with the capsule of *E. coli*, cross the cell wall, and attack the proteins of the membrane and cell wall. This process leads to the disruption of the outer membrane, followed by death of the *E. coli.*

Other biocompatible and biodegradable polymers have been used as a base to produce advanced hydrogels with metallic NPs, such as chitosan (CS) and gelatin. Lu et al. [60] synthesized Au NPs in chitosan solution. Then, their surface was functionalized with a shell of 2-mercapto-1-methylimidazole (MMT), resulting in an Au-CS–MMT nanocomposite with size 8 to 10 nm. This system was loaded into gelatin using tannin acid as a crosslinker, the final product was Au-CS–MMT/gelatin. The antimicrobial activity was explored against *S. aureus* ATCC 25923, *E. coli* ATCC 25922, and MRSA using New Zealand rabbits as a model. The minimum inhibitory concentration (MIC) was <20 μM for the three bacteria strains. The antibacterial activity of Au-CS–MMT/gelatin was better than the standard ampicillin treatment used as a control. The results from the surface charge by zeta potential and the content of Au in *E. coli* and *S. aureus*, confirmed the strong electrostatic interactions with the Au-CS–MMT particles. Additionally, scanning electron microscopy (SEM) and transmission electron microscopy (TEM) images showed that Au-CS–MMT damaged the morphology and disrupted the bacterial cell membrane in less than 1 h of contact. Ryan et al. [136] synthesized a chitosan and siloxane hydrogel to incorporate Au NPs. Tetraethyl orthosilicate (TEOS) was used to form an interpenetrating polymer network and improved the hydrogel structural properties, such as flexibility and strength (67–74 mPa). The size distribution of the Au NPs was 19 nm ± 18%. Antimicrobial tests displayed that cross-linking with $SiC_8H_{20}O_4$ (TEOS) reduced the attachment of *E. coli* to the well plate surface by 80%.

Gelatin is a natural, amphoteric, non-inflammatory polymer and it is obtained from the hydrolysis of collagen [60]. It has many functional groups allowing its polymerization with several crosslinking agents [137]. A gelatin-based hydrogel was fabricated by Jiang et al. [61], in which Au NPs (5 nm) were capped by 6-aminopenicillanic acid (APA), and embedded into electrospun fibers of poly(ε-caprolactone)/gelatin. The MIC of Au–APA/gelatin was 2.5 μg/mL against *E. coli* and *K. pneumoniae*, >5 μg/mL against *P. aeruginosa*, 5 μg/mL against MDR *E. coli* and MDR *K. pneumoniae*. It was observed that *E. coli* cell walls were leaky and broken when the concentration of Au–APA/gelatin increased.

Hydroxypropyl methylcellulose (HPMC) is a non-toxic and non-ionic biopolymer which is used as a stabilizer, thickener, and emulsifier in several applications in the food and pharmaceutical industry. HPMC has many polar and non-polar groups which easily interact with nanoparticles by coordination bonds [138]. Recently, Wafaa Soliman et al. [24] obtained embedded Au rod-shaped NPs into a HPMC hydrogel for topical application. The size distribution and surface charge (ζ) of the Au NPs was 82.5 nm and 34.8 mV, respectively. Male Wistar rats were used as a model for in vivo studies. The MICs against *S. aureus* ATCC 10400, *E. coli* ATCC 25922, and *C. albicans* ATCC 90028 were 0.125–0.25 ng/mL. The minimum bactericidal concentrations (MBCs) were 0.1–0.5 ng/mL. This hydrogel was more efficient against *S. aureus* and *E. coli*. The authors suggested that the interactions between Au NPs and bacteria happened by the large cationic surface charge of the Au nanorods leading to membrane disruption, damage to the bacterial cell structures, and consequently death of the pathogenic microorganism.

A treatment for focal infections, based on laser-mediated heating of Au NPs (13 nm) suspended in an injectable and degradable silk hydrogel, has also been suggested [26]. Silk is a natural, biocompatible, cheaper polymer, and the silk hydrogel characteristics can be tailored and controlled by the gelation time [28,139]. The bactericidal procedure consists of injecting the silk hydrogel/Au NPs composite into the subcutaneous infection, and to deliver a laser beam with 150 mW of incident green light (532 nm wavelength) for 10 min [26]. The wavelength light is absorbed by the nanoparticles and converted into heat. This localized heat has a bactericidal effect at the infection site without causing systemic side effects. The in vivo results showed *S. aureus* reduced after one round of laser-exposure, killing 80% of bacteria, demonstrating the potential applicability of this proposal.

In addition to hydrogels obtained from natural polymers, other types of synthetic polymers or monomers have been used for hydrogel preparation for NPs delivery. A pH-sensitive hydrogel with antimicrobial activity and wound-healing properties was produced by Chitra et al. [140]. Au NPs of 17 nm were loaded into a porous hydrogel, obtained by condensation–polymerization reactions with citric acid (CA), diethylene glycol (DEG), and indole-3-acetic acid (IAA, $C_{10}H_9NO_2$). The swelling profile of the hydrogel decreased when the content of Au NPs increased, in basic medium. The antibacterial performance against *S. aureus* showed an inhibition zone of 8.33 to 11.67 mm, using 1000–2000 μg/mL of Au/hydrogel by the diffusion method. In a similar study, Au NPs (8–30 nm) and Ag NPs (4–12 nm) were incorporated into a hydrogel composite based on the condensation–polymerization between citric acid (CA), diethylene glycol (DEG), and indole-3-acetic acid (IAA) [128]. This hydrogel nanocomposite was tested against *S. aureus*, *E. coli*, and *Bacillus cereus* at 2000 μg/well. The inhibition zones of the Au/hydrogel were 14, 16, and 15 mm against *E. coli*, *S. aureus*, and *B. cereus*, respectively. Chitra et al. [128] described that the results were due to the outside structure of Gram-positive bacteria, which allows the entry and absorption of foreign molecules into the bacterial cell membrane.

Poloxamer 407 is a triblock copolymer (poly(ethylene glycol)-block-poly(propylene glycol)-block-poly(ethylene glycol)). It is water-soluble, and has been explored in the pharmaceutical industry as an antibiotic delivery platform [141]. A polymeric hydrogel with Au NPs was reported by Mahmoud et al. [130] to treat wounds. A poloxamer 407 hydrogel was used to support sphere- and rod-shaped Au NPs with different coating agents, such as CTAB ($C_{16}H_{33}N(CH_3)_3Br$), polyacrylic acid (PAA), poly(allylamine hydrochloride) (PAH), and poly(ethylene glycol) (PEG), to endow the surface with negatively, positively and neutrally charged polymers. Rod-shaped Au NPs capped with PEG and positively charged NPs capped with PAH proved to be the most efficient systems for wound healing after 14 days of treatment. Likewise, the two hydrogel systems presented a high reduction in viable bacterial against *S. aureus* and *P. aeruginosa*, the most common skin bacteria.

Pluronic F-127 is a synthetic thermoresponsive polymer which displays sol–gel transition near 37 °C, excellent biocompatibility, good mechanical strength, and the ability to retain water [142]. An injectable hydrogel for muscle regeneration was produced by Ge et al. [68], in which gold–polythyleneimine NPs (10 nm) were embedded into a Pluronic F-127 hydrogel scaffold, named FPAu. This FPAu biomaterial was obtained by the double crosslinking of Pluronic F127, 4-hydroxy benzaldehyde, K_2CO_3, and modified polydopamine NPs. The number of colony-forming unit decreased rapidly after 2 h of contact between bacteria and the hydrogel in in vitro tests. The antibacterial activity against *S. aureus* and *E. coli* was 87.5% and 83%, respectively. This study suggested that the antibacterial property of the FPAu hydrogel is due to branched polyethyleneimine linked to the Au NP surface.

Another approach is to coat the hydrogel with a pretreated macrophage membrane of bacteria. This creates a bacterial receptor able to identify specific sites when it interacts with the target pathogenic bacteria. Li et al. [131] fabricated a photothermal hydrogel of N-acryloyl glycinamide (PNAGA) in which Au nanorods were previously coated in polydopamine (PDA). The hydrogel was also coated with membrane macrophages against *E. coli* and *S. aureus*. The PNAGA enhanced the mechanical properties of the hydrogel, showing a tensile strength of 1.64 MPa and a compressive strength of 12.490 MPa. Au nanorods gave a photothermal ability to the hydrogel under NIR irradiation. When exposed to NIR irradiation for 5 min, the antibacterial activity of this hydrogel, without the macrophage membrane, was 74.2 and 72.5% against *E. coli* and *S. aureus*, respectively. However, the hydrogel coated with the activated membrane of macrophage against *E. coli* and *S. aureus*, led to an antibacterial efficiency of 98.4 and 97.6%, respectively.

Some authors have explored alternative methodologies, for example, by incorporating dual metallic nanoparticles, such as Ag and Au NPs, as core–shell nanoparticles into the porous structure of a carbopol-based hydrogel [64]. The particle size was 5 ± 3 nm and the Ag–Au NPs presented several morphologies into the hydrogel. The inhibition zones against *B. cereus* and *E. coli* were 18.5 and 18.1 mm, respectively. This study showed that Ag–Au NPs inhibited the growth of bacteria by forming pits between the NPs and the cell wall. This interaction resulted in bacterial death due to leakage of molecules and proteins from the wall.

The bimetallic NPs, Au–Ag NPs were incorporated by Kumar et al. [29] in a carboxy methyl tamarind (CMT) hydrogel against MDR *E. coli*, *E. cloacae*, and *S. aureus* MRSA for in vitro tests using mammalian cells. The bimetallic NPs had a hydrodynamic size of 147 nm and a negative surface charge of −31.5 mV. The growth profiles of the cells were studied at different concentrations of Au–Ag NPs. *E. coli* showed an extended lag phase when exposed to Au–Ag NPs at a concentration of 1 to 3 μg/mL, while at a concentration of 3 to 12 μg/mL for MRSA in presence of Au–Ag NPs. The lag phase is the earliest period of the bacterial growth cycle, which the bacteria adjust to their environment, and cells increase in size [143]. The reported MIC values were 3 and 6 μg/mL for *E. coli* and MRSA, respectively. This hydrogel was also tested against clinical isolates of *E. cloacae* EC18, which was efficient at 6 μg/mL. The antimicrobial activity results showed that the Au–Ag/hydrogel was efficient when 20- to 25-fold less concentrated than other drugs, such as gentamicin.

In a similar study, a gelatin sponge hydrogel functionalized with silver/gold clusters (Au/Ag–gelatin and Au–gelatin) was used for antibacterial applications [144]. This system was obtained by a simple one-pot method. Glutathione (GSH) acted as a reducing agent and as a thiol-ligand. The system's biocompatibility, as well as good water absorbency and water retention properties, allows efficient bactericidal effects and presents this hydrogel as a promising material for wound dressing applications. The antibacterial activities of gelatin, Au–gelatin and Au/Ag–gelatin were probed by inhibition zone assays, using *P. aeruginosa* as a model, since it is implicated in wound infection. Gelatin did not present any inhibition zone under any of the conditions, while for Au–gelatin and Au/Ag–gelatin, the inhibition appeared under white light irradiation. According to the authors, the bactericidal activity is due to ROS generated by the excited NCs. The inhibition zone of Au/Ag–gelatin was 31.9 mm higher than the Au–gelatin with 25.1 mm.

Ribeiro et al. synthesized in situ Au and Ag NPs embedded in a silk fibroin-based hydrogel [28]. The size distribution of the Au NPs was 9 to 55 nm and 12 to 69 nm for the Ag NPs. The NP concentration influenced the antimicrobial activity. For example, the hydrogel loaded with a Au NPs concentration > 0.5% was efficient against the *S. aureus* ATCC 33591, MRSA, and *P. aeruginosa* ATCC 27853. When the concentration was >0.1%, it inhibited the cell proliferation of *S. aureus* ATCC 25923 and MSSA, and *E. coli* ATCC 25922. However, the hydrogel showed antimicrobial performance against all previous pathogenic cells, regardless of Ag NPs concentration, even with *S. epidermidis* RP62A ATCC 35984.

Furthermore, Au NPs have been loaded into porous systems such as metal–organic frameworks (MOFs). The zeolitic imidazolate framework-8 (ZIF-8) MOF are crystalline, biocompatible, and biodegradable. ZIF-8 can generate ROS under visible light by means of photocatalysis [145]. The Au NPs/ZIF-8 strategy can improve the antimicrobial activity of

Au NPs. For example, Deng et al. [69] embedded Au NPs into a pristine ZIF-8 network. The nanocomposite was embedded in oxidized sodium alginate (OSA), and carbohydrazide-modified methacrylated gelatin (GelMA-CDH) obtaining an injectable hydrogel, named Au–ZIF–GCOA. The stability of the hydrogel was improved by adding another crosslinking step during the polymerization, leading to a double-network hydrogel. This material displayed high bactericidal activity against *E. coli* and *S. aureus* using mice as an in vivo model. The number of bacteria colonies decreased in more than 99.0% for both strains using 0.2 mg/mL of Au–ZIF–GCOA. The antibacterial activity was due to the photoproduction of hydroxyl radicals by Au–ZIF-8 nanostructures, under visible-light irradiation. Mainly, Au NPs were responsible for the conversion of oxygen to singlet oxygen ($^1O_2$). The interaction of the hydrogel with light prevented the bacteria from acquiring resistance mechanisms to metal nanoparticles.

Other strategies developed against bacteria resistance combine the properties of metallic nanoparticles conjugated with drugs in the same hydrogel structure. Au NPs were incorporated into a cellulose-grafted polyacrylamide hydrogel [146], leading to a PAMC/Au nanocomposite. Afterwards, ciprofloxacin was embedded into this nanocomposite/hydrogel. The antibacterial performance was evaluated against *E. coli*, *S. flexneri*, *B. cereus*, and *Listeria Inuaba*. An increased concentration of Au NPs into the hydrogel was observed to enhance the antibacterial activity from 67 to 95% against the *E. coli*, and from 48 to 79% against the *S. Flexneri*. However, this hydrogel nanocomposite was not as efficient against *B. cereus* and *L. Inuaba* with an antimicrobial activity only between 35–53%. The study suggested that antimicrobial efficacy was influenced by the hydrogen bonding interaction between the Au NPs and the amide acrylamide groups.

Au NPs have been incorporated into the hydrogel network combined with Ag NPs leading into antibacterial hydrogels with dual metallic NPs (Au–Ag). Table 3 summarizes the hydrogels containing Au NPs for antibacterial application. The antibiofilm activity of Au NPs incorporated into hydrogels is shown below.

Table 3. Au NPs loaded into hydrogel for antibacterial application.

System	Materials	Au NP Properties (Size and Surface Charge)	NP Synthesis Method	Bacteria	Target	Antibacterial Properties: Inhibition Zone (mm) and MIC Values	Ref.
AuC/liposome	Cationic phospholipid liposomes, acrylamide, (glycol) dimethacrylate (PEGDMA)	97.1 nm, ζ: −25.3 mV	Chemical reduction with $NaBH_4$	*S. aureus* MRSA252	In vitro; in vivo, mice	No skin reaction after 7-day treatment. Hydrogel activity was influenced by pH	[133]
Au NSt/alginate	Sodium alginate (SA), $CaCl_2$, and polyethylene imine (PEI)	Core diameter: 25 nm; Spikes size: 50 nm, 70 nm, and 120 nm	Chemical reduction with trisodium citrate	*S. aureus* MTCC1430 *P. aeruginosa* MTCC 1934 *E. coli* MTCC 443	In vitro, NIH-3T3; in vivo, rats	The plate count method; the antimicrobial activity: 35.4% (*S. aureus*), and >80% (*P. aeruginosa* and *E. coli*.)	[126]
Au/poly (acrylamide-co-alginate)	Acrylamide (AM), alginate (SA), N,N-methylenebisacrylamide, and $HAuCl_4$	8 nm	In situ, chemical reduction	*E. coli*	in vitro	Optical absorbance around 0.05–0.75. *E. coli* did not growth more after 2 h 30 min	[59]
CS-Au–MMT/gelatin	2-mercapto-1-methylimidazole (MMT), tannin acid, chitosan (CS), and gelatin	10.07 ± 2.34 nm 8.32 ± 1.97 nm	Chemical reduction	*S. aureus* ATCC 25923, *E. coli* ATCC 25922 MRSA	In vitro, L929 and L02; in vivo, rabbits	In situ; the microtiter broth dilution method, and MIC < 20 µM for all bacteria	[60]
Au-Ag/CS/TEOS	$HAuCl_4$, HNO_3, chitosan, and tetraethyl orthosilicate (TEOS)	Ag: 16 ± 25% nm Au: 19 ± 18% nm	Polymerization reaction and drop casting method	*E. coli*	In vitro	Crystal violet attachment; 80% inhibition of *E. coli* on the surface	[136]
Au–APA/gelatin	6-aminopenicillanic acid (APA), gelatin, and $HAuCl_4$,	5 nm	Chemical reduction by $NaBH_4$	*E. coli*, *K. pneumoniae*, *P. aeruginosa*, MDR *E. coli*, and MDR *K. pneumoniae*	In vivo, rats	The microtiter broth dilution method; MIC were 2.5 µg/mL against *E. coli* and *K. pneumoniae*, >5 µg/mL against *P. aeruginosa*, 5 µg/mL against MDR *E. coli* and MDR *K. pneumoniae*	[61]
Au/HPMC	Tetrachloroauric acid, cetyltrimethyl ammonium bromide, ascorbic acid, $NaBH_4$, $AgNO_3$, and hydroxypropyl methylcellulose (HPMC)	82.5 nm; ζ: 34.8 mV	Chemical reduction method using CTAB and $NaBH_4$. Au NPs were embedded into HPMC	*Staph. aureus* ATCC 10400, *E. coli* ATCC 25922, and *C. albicans* ATCC 90028	In vitro; in vivo, rats	Micro broth dilution assay. MIC and MBC: 0.25 and 0.1 nM/mL for *Staph. aureus*, MIC and MBC: 0.125 and 0.125 nM/mL for *E. coli*, and MIC and MBC: 0.25 and 0.5 nM/mL for *C. albicans*,	[24]
Au/Silk	$HAuCl_4$, sodium citrate, bombyx mori cocoons, $NaCO_3$, and LiBr	13 nm	Chemical reduction using sodium citrate	*E. coli* ATCC 25922 *and S. aureus* ATCC 25923	In vitro; in vivo, mice	Killed 80% of bacteria in 10 min; using a laser exposure time of 15 min and 600 mW, the zone inhibition was about 16 mm^2	[26]
Au/CA-DEG-IAA	Citric acid (CA), diethylene glycol (DEG), and indolylacetic acid (IAA)	17 nm	In situ; chemical reduction with $Na_3C_6H_5O_7$	*S. aureus*	In vitro	The diffusion method; Inhibition zone: 8.33–11.6 mm	[140]
Au/CA-DEG-IAA Ag/CA-DEG-IAA	Citric acid (CA), diethyleneglycol (DEG), and indole-3-acetic acid (IAA)	Au NPs: 8–30 nm Ag NPs: 4–12 nm	Condensation polycondensation; chemical reduction with $Na_3C_6H_5O_7$	*S. aureus*, *E. coli*, and *Bacillus cereus*	In vitro	Inhibition zone (mm): 25 and 15 mm, 23 and 14 mm, 25 and 15 mm	[128]
Au/poloxamer 407	CTAB ($C_{16}H_{33}N(CH_3)3Br$), PAA (polyacrylic acid), PAH (poly(allylamine hydrochloride), and PEG (Poly(ethylene glycol)	Rod shape: 49.2 nm Spherical shape: 29.2 nm	Chemical reduction using $Na_3C_6H_5O_7$	*S. aureus ATCC 29213*, and *P. aeruginosa* ATCC 27853	In vitro; in vivo, rats	Reduction in bacterial viable count was >99.5% and 99.0> against *S. aureus* and *P. aeruginosa* using PAH-Au NPs and PEG-Au NPs.	[130]

Table 3. *Cont.*

System	Materials	Au NP Properties (Size and Surface Charge)	NP Synthesis Method	Bacteria	Target	Antibacterial Properties: Inhibition Zone (mm) and MIC Values	Ref.
FPAu	Polyethyleneimine (PEI), Polydopamine (PDA), Pluronic F127, 4-hydroxy benzaldehyde (PHBA), $HAuCl_4$, and K_2CO_3	10 nm	Chemical reduction with $NaBH_4$, and polyvinyl pyrrolidone (PVP)	*E. coli* *S. aureus*	In vitro; in vivo, rats	The plate count method; inhibited bacterial growth in 75% after 2 h	[68]
Au–PDA/PNAGA	$HAuCl_4$, $NaBH_4$, dopamine hydrochloride, and N-acryloyl glycinamide (PNAGA)	Diameter 32 nm and length 54 nm	Seeded growth method Polymerization	*S. aureus* ATCC29213 and *E. coli* ATCC25922	In vitro, L929 cells; in vivo, rats	97.6%; 98.4%	[131]
Ag–Au/carbopol	Carbopol® 980, acrylamide, $AgNO_3$, and $HAuCl_4$	2–8 nm	In situ reduction using mint leaf extract	*Bacillus* *E. coli*	In vitro	The disc method; inhibition zone: 18.5 mm 18.1 mm	[64]
Ag–Au/CMT	$AgNO_3$, $KAuCl_4$, and carboxy methyl tamarind (CMT)	187 nm	Seeded growth method	Clinical *E. cloacae isolate* Ec18, *E. cloacae* BAA-1143, ATCC, and *E. coli* BAA-2469, ATCC	In vitro; in vivo, mice	The disc method. MIC: 6 µg/mL, 6 µg/mL, and 3 µg/mL	[29]
Au/Ag–gelatin	glutathione (GSH), $HAuCl_4$, $AgNO_3$, and N-hydroxysuccinimide (NHS)	Au NCs: 1.5–3.5 nm Au/Ag: 102 nm	Au/Ag NCs was incorporated into gelatin after NPs synthesis	*P. aeruginosa*	In vitro, pigskin	Inhibition zone: 31.9 mm	[144]
Au or Ag/silk fibroin	$AgNO_3$, $HAuCl_4$, and cocoons of *Bombyx mori* silkworm	Au NPs: 9–55 nm Ag NPs: 12–69 nm	In situ chemical reduction	*S. aureus* ATCC 33591, MRSA and *P. aeruginosa* ATCC 27853 *S. aureus* ATCC 25923, MSSA and *E. coli* ATCC 25922 *S. epidermidis* RP62A ATCC 35984.	In vitro, MG63 cells	Using sessile and planktonic bacteria. 0.1% of Au NPs were effective against *S. aureus*, and *E. coli* while 0.5% of Au NPs was antibacterial against *P. aeruginosa*	[28]
Au–ZIF8/OSA-GelMA	HAuCl, $Na_3C_6H_5O_7$, polyvinyl pyrrolidone (PVP), gelatin, $Zn(NO_3)_2.6H_2O$, CH_6N_4O, oxidized sodium alginate (OSA), and carbohydrazide-modified methacrylated gelatin (GelMA-CDH)	15 nm; ζ: −4.8 mV	Chemical reduction with $Na_3C_6H_5O_7$; Schiff-base reaction, and radical polymerization	*E. coli* ATCC 25922 *S. aureus* ATCC 29213	In vitro, NIH-3 T3 cells	The number of bacteria colonies decreased by more than 99%	[69]
Au/C/PAM	Acrylamide monomer, cellulose, $HAuCl_4$, and ciprofloxacin	Length: 5 µm and diameter 70 nm	In situ; chemical reduction with $Na_3C_6H_5O_7$	*E. coli*, *S. flexneri*, *Bacillus cereus*, and *Listeria inuaba*	In vitro, L929 cells	The diffusion method; the antibacterial activity was 95% against the *E. coli*, and 79% against the *S. flexneri*.	[146]

3.2. Antibiofilm Activity of Au NPs Loaded into Hydrogels

Au NPs have previously shown promising results against several microorganisms and biofilms in terms of growth inhibition and cell damage [147,148]. With this background information, Bermúdez-Jiménez et al. [149] embedded gold nanorods (Au NRs) into a non-toxic chitosan hydrogel, exploring its antibiofilm activity against Gram-positive and Gram-negative pathogenic bacteria multi-species biofilms, by photothermal therapy. The authors reported 5 to 8 Log_{10} reductions in bacterial load when the *Streptococcus oralis* and *E. faecalis* biofilm was exposed to the Au NR hydrogel subjected to a 10 °C temperature rise. However, when the hydrogel was subjected to a 5 °C temperature rise, no discernible drop in the bacterial load was seen. These findings suggest that photothermal therapy was essential in antibiofilm activity of this Au NR-loaded hydrogel. Correspondingly, Al-Bakri et al. [132] also investigated the potential of Au NRs incorporated into a hydrogel by photothermal therapy against *P. aeruginosa* biofilms. In this study, the photothermal-based bactericidal activity of the Au NR hydrogel against biofilms showed the same percentage and Log reduction in viable bacterial count under two different modes of laser excitation. The results showed approximately 4 Log cycle, and 1 Log cycle reduction in the viable cells for both, continuous and pulsed laser excitation at 3 W cm^{-2} and 1 W cm^{-2} laser doses, respectively.

Wickramasinghe et al. [150] also investigated the photoactivated Au NRs incorporated into hydrogel composites, to explore their antibiofilm activity against *S. aureus* bacterial biofilms on metal implant materials. A set of 1 W/cm^2 power intensities, with a 1 cm^2 laser spot size, and 15 s of laser pulses was devised to assess the hydrogel's ability to completely eradicate preformed bacterial biofilms on the surface of metal alloy disks. According to crystal violet assays, the hydrogel completely eradicated the biofilms. Additionally, a colony-forming assay was carried out. The results demonstrated that no surviving bacteria established new colonies because of the gel treatment. Finally, SEM analysis showed that this hydrogel did not leave any bacterial cells on the surface of the metal alloy disks, even in the microscale grooves. SEM analysis conducted by Soliman et al. [24] using *S. aureus* and *E.coli* biofilms has also shown the intense reduction in cell number and morphological changes after treatment with Au NPs incorporated into hydrogels.

Recent work also sought to depict this technique as a useful strategy against biofilm formation from different bacterial species. In 2022, Galdámez-Falla et al. [151] developed an *E. faecalis* biofilm on human roots using the static and dynamic method (modified drip flow reactor (MDFR), aiming to use photothermal therapy with a hydrogel solution with Au NRs as an antibiofilm agent. The authors found differences in colony-forming unit (CFU) when comparing the Au NRs-treated biofilm (188.6 $\pm$ 26.7 CFU) with the control group (337.3 $\pm$ 2.82 CFU). The Au NRs successfully eliminated *E. faecalis* biofilms. The laser application time, however, was 20 min, which is longer than would be feasible for an in vivo scenario. The researchers suggested that, in the future, this strategy should be tested with a shorter laser application time with similar positive outcomes.

Another antibiofilm strategy that has been explored used Au NPs hydrogels in association with Ag NPs. Due to the effectiveness in combating bacteria, this strategy has received great attention among many researchers. The synergistic effect is well-described by Kumar et al. [29]. In their study, bimetallic NPs (Au–Ag NPs), capped with complex carbohydrates, and incorporated into a carboxy methyl tamarind (CMT) polysaccharide hydrogel, showed a dose-dependent effect on biofilm formation for both Gram-positive and Gram-negative bacteria. Low concentrations, such as 1 µg/mL for *E. coli* and 1.5 µg/mL for *S. aureus*, were able to eradicate the biofilms.

4. Mechanism of Action of Ag NPs and Au NPs Loaded into Hydrogels

4.1. Mechanism of Antibacterial Action

Ag NPs and Au NPs exhibit antibacterial action against different antimicrobial-resistant bacteria, including Gram-positive and Gram-negative bacteria. In addition, their low reactivity and low toxicity, compared to Au and Ag ions, presents them as a relevant therapeutic strategy for drug-resistant bacterial infections [152,153].

Due to their proven potential, the understanding of the mechanisms of bactericidal action become relevant. In general, Ag NPs and Au NPs act through the following mechanisms: (a) adhesion and alteration to the surface of the microbial membrane; (b) penetration into bacterial cells promoting the breakdown of biomolecules and other intracellular damage; (c) induction of cellular toxicity by the generation of ROS that promote oxidative stress within the cell; and (d) inhibition of intracellular signal transduction pathways [148,154–156], shown in Figure 2.

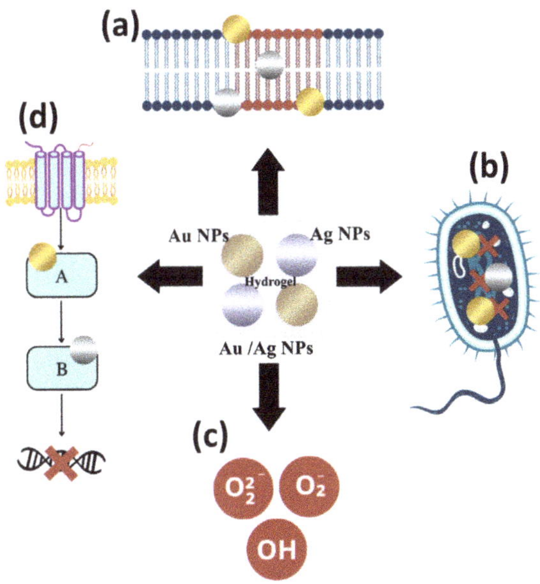

Figure 2. The mechanisms of bactericidal action of Au NPs and Ag NPs loaded into hydrogels.

Ag NPs and Au NPs act at the cell wall or membrane, as they adhere to these structures by electrostatic interactions. NPs release their positively charged ions, generated by metal oxidation, to the negatively charged bacterial cell surface [157,158]. In addition to the possibility of this interaction, Ag NPs also have an affinity for sulfur proteins in the microbial cell wall. The adhesion or accumulation of these nanostructures promotes irreversible morphological changes in the structure of the cell wall and membrane [154,159].

In this sense, it is evident that these nanoparticles interfere with the integrity of the lipid bilayer by denaturation. This can cause cell lysis, which increases the cell membrane permeability, affecting the cell's ability to regulate the transport of substances and causes loss or leakage of cellular contents, such as cytoplasm, proteins, ions and the cellular energy reservoir (i.e., adenosine triphosphate) [160,161]. Thus, Ag NPs and Au NPs increase the permeability of bacterial cell membranes allowing the entry of antibiotics combined with NPs to potentiate the antibacterial effects [162,163].

Ag NPs and Au NPs can penetrate cells through existing porins in the outer or cytoplasmic membrane, promoting changes in cellular activity through the binding of NPs to cellular structures. This includes ribosomes, leading to protein synthesis reduction in the cytoplasm as well as of biomolecules such as proteins, lipids, and DNA. Among the biomolecules that are altered by the binding of Ag and Au ions, proteins and bacterial DNA are the most important [155,156,164].

Ag and Au ions released into the environment will bind to negatively charged protein, altering the protein structure, denaturing it, and interfering with the normal growth and metabolism of bacterial cells [164,165]. In addition, Ag ions bind to DNA via bonds with the sulfur and phosphorus components of the nucleic acid, causing denaturation, problems in DNA replication and stopping cell growth [148]. Au NPs, on the other hand, neutralize the plasmid charge and prevent its movement. These NPs can decrease the stability of the DNA structure by electrostatic repulsion [158,166].

Another mechanism of action of Ag NPs and Au NPs is the production of ROS and free radicals. These radicals promote oxidative stress in bacteria, inducing lipid damage, leakage of cellular biomolecules, protein aggregation, DNA destruction, and eventually, lead to cellular apoptosis. In addition, these NPs are considered the main agents for cell membrane disruption and DNA modification [148,167]. The production of ROS is normally dependent on the concentration of the nanostructures. ROS are generated after the uptake of free Ag and Au ions in the cells, which can alter the respiratory chain in the inner membrane by interacting with thiol groups forming Au-thiol groups. Ag-thiol groups promote the coagulation of respiratory enzymes, interrupting the production of adenosine triphosphate by altering the electron transport systems, and activating the apoptosis pathway [148,156].

Ag NPs and Au NPs can also alter bacterial signaling pathways. The mechanism of the signaling depends on the phosphorylation and dephosphorylation of cascade proteins or enzymes that are essential for cell activity and bacterial growth [164,165,168]. Due to the unique physicochemical properties of NPs, there is a possibility that these nanoparticles act as modulators of signal transduction in microbial cells. They may mediate bacterial cell apoptosis by disrupting the bacterial actin cytoskeletal network causing morphological changes in the bacterial form. Thus, bacterial cell membranes become more fluid, followed by cell rupture [155,169].

Au NPs have other different antipathetic activity mechanisms. Near-infrared radiation can be used to induce Au NPs to convert light into heat, which destroys the cytomembrane structure, and kills bacteria through lysis and disintegration by local heating. This therapy significantly reduces the number of bacteria, even at low concentrations, and includes specific mechanisms such as protein denaturation, cell fluid evaporation, cell structure breakdown, and blister formation. All these mechanisms damage the bacterial cell wall and promote cell wall penetration [157,160].

The dissolution state, the size and shape of the NPs in the exposure medium affect the release of ions and their antibacterial effect and mechanisms. Dissolution efficiency depends on synthesis and processing factors, as well as on the intrinsic characteristics of Ag NPs and the surrounding media [163,170]. Thus, studies claim that smaller NPs with a spherical or near-spherical shape are more prone to release silver. Therefore, reducing particle size and increasing the dispersibility can help to improve the antibacterial properties of the NPs. Additionally, this may facilitate adsorption, and penetration due to the greater surface area, and better dissolution in more acidic environments [152,162]. Thus, it is evident that the different mechanisms of action presented by the NPs increase the effect of their antibacterial action.

4.2. Mechanism of Action for Inhibiting Biofilm Formation

Regarding the literature, it is clear that these NPs are promising agents in inhibiting biofilm formation. However, the mechanism of action for inhibiting biofilm formation and the interrelation between Ag and Au NPs with biofilms is not completely understood. Some therapeutic targets are predicted to inhibit biofilm formation, such as: (a) the EPS network that facilitates the initial attachment of bacteria to the surface and increases bacterial resistance to host immunity and antibiotics; (b) the flagella, crucial structures for the initial communication between cells and the surface; (c) adhesion proteins, which allow for the initial attachment; and (d) quorum sensing (QS), bacterial cell–cell communication, in which bacteria give feedback via extracellular, signaling molecules to manage microbial virulence and to release autoinducers that increase in concentration as a function of bacterial cell density, shown in Figure 3 [168,171,172].

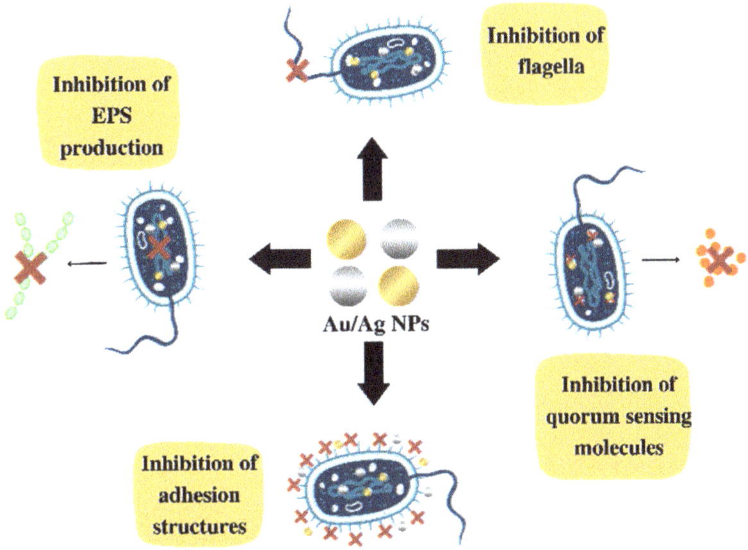

Figure 3. Mechanism of inhibiting biofilm formation by Au NPs and Ag NPs.

Cells treated with NPs may show alterations in their morphology, presenting a crumpled surface morphology, relatively elongated size, and no clear septum. These findings suggest that Ag NPs prevent bacterial cell division, causing membrane destruction, and preventing biofilm formation. With this effect, it can be seen that NPs induce morphological alterations reducing biofilm formation [159,161,173].

The inhibition of biofilm production also occurs by the generation of ROS induced by Ag NPs and by the release of silver. These mechanisms may inhibit the expression of genes related to motility and biofilm formation [156,167]. In a similar way, Au NPs cause mechanical damage to the cell wall through electrostatic interactions. Additionally, Au NPs can stimulate the production of ROS, and damage cellular structures, functions and proteins due to the release of metal ions [164].

Due to the positive correlation between EPS secretion and biofilm formation, EPS inhibition is also considered an alternative target to mitigate the biofilms of pathogenic bacteria [153,165]. For example, Ag NPs and Au NPs inhibit alginate production in a concentration-dependent manner. Alginate is a vital constituent of the EPS matrix. It helps bacteria attach to surfaces, protecting them from the host's immune response, and thus making them resistant to antimicrobials. In this sense, one of the focuses of NP activity is to prevent the production and secretion of EPS matrix components, such as polysaccharides, proteins, and extracellular DNA or eDNA. These components, confer integrity to the biofilm and functional architecture, as well as resistance against antibiotics [171,172].

A fundamental step in biofilm formation is bacterial adhesion to a surface. In this sense, adhesion proteins, fimbriae, and flagella-mediated motility, that regulate the initial attachment of bacteria to a wide range of surfaces are therapeutic targets where NPs enact their antibiofilm activity [154,158]. Thus, some metallic nanoparticles can reduce bacterial adhesion on surfaces through the release of ions, inhibiting the enzymatic activity of proteins involved in peptidoglycan synthesis, delaying biofilm formation [156,165].

In addition to adhesion, another essential process is quorum sensing (QS). Some studies have indicated that metallic NPs can disrupt the production of QS molecules, especially the autoinducer (AI-2). Thus, the production of exopolysaccharides and rhamnolipids, motility, and some virulence factors necessary for QS-regulated biofilm production is substantially altered [171,172,174]. Thus, Ag NPs and Au NPs are considered very promising agents as a therapeutic strategy for coating surfaces and hospital utensils to prevent multidrug-resistant bacterial infections with biofilm production.

4.3. Mechanism for Biofilm Eradication

Most antibacterial agents have difficulty in penetrating the EPS matrix produced in a biofilm, making it difficult to eradicate and treat infections. Thus, it is essential to develop therapeutic strategies to solve this inconvenience. In this sense, the use of Ag NPs and Au NPs with intrinsic antimicrobial potential can act as biofilm-targeting agents, promoting its eradication [170,175]. Some studies have shown the mechanisms of action of these nanoparticles in biofilm eradication (Figure 4).

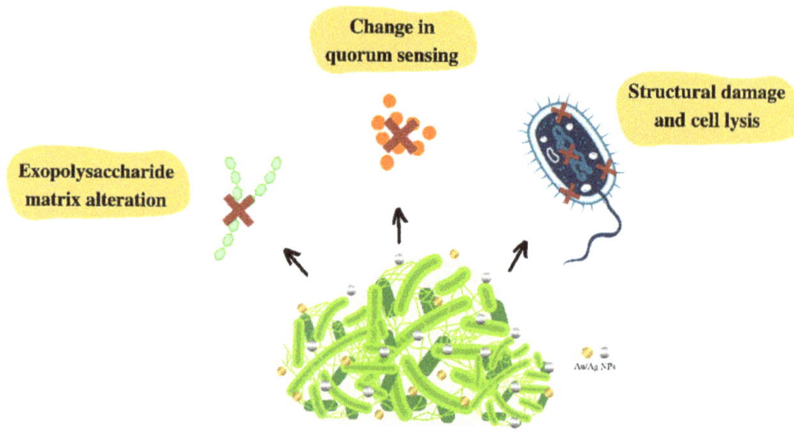

Figure 4. Mechanism of biofilm eradication of Au NPs and Ag NPs.

The level of biofilm–nanoparticle interaction depends on the physicochemical properties of the EPS, nanoparticles, and the environment around the biofilm. Thus, it is necessary to understand the NP transport process that includes movement to the biofilm–fluid interface, attachment to the surface of the biofilm and migration within the biofilm [176,177].

Initially, the penetration of nanoparticles into the EPS matrix is influenced by NP size, and by the interactions with components of the extracellular polymeric matrix with the NP surface properties, such as charge and functional groups [174,176]. Bacterial biofilms, in general, have a polyanionic and negatively charged matrix, which enables them to interact with positively charged Ag and Au metallic ions. Other, physicochemical-modulated characteristics are also important, especially electrostatic ones, such as the zeta potential [176,178,179].

When NPs begin to come into contact with an environment containing organic molecules, a corona-like coating is formed on the surface of the NPs. The nature of this corona influences NP–biofilm interactions [177,180]. After the initial contact, the nanoparticles begin to interact with macromolecules present in the biofilm, changing their known surface properties, related to size, strength, functionalization, and other biological properties [170,177].

Penetration, diffusion and antibiofilm effectiveness depend on physicochemical characteristics, such as adequate size, polydispersity index, purity, and zeta potential. The composition and structure of the biofilm are also important, including pore size, presence of water channels, hydrophobicity of the environment, and the chemical gradient of the matrix, as well as the ionic composition and concentration of the nanoparticle solution [178,180].

In this sense, NPs of 5–500 nm can penetrate into the biofilm water channels, affecting EPS matrix diffusion as a result of surface functional groups or charge interactions. NPs interact with bacterial cells through penetration and intracellular accumulation promoting the inhibition of protein function, DNA damage, translation disorders and/or transcriptional dysregulation. The pathogen viability is reduced by altering bacterial cell wall permeability (Figure 4) [164,166,169,173].

Studies have shown that NPs can bind to the negatively charged bacterial surface and the biofilm interacts electrostatically, promoting changes in quorum sensing and influencing bacterial growth (Figure 4). The results also showed a dose-dependent decrease in exopolysaccharide production, preventing the biofilm from maintaining its standard architecture (Figure 4) [169,171,172,174]. Biofilms treated with the formulation of Ag NPs and Au NPs may exhibit dispersed cell aggregates with acute structural destruction. Disruption of cells through membrane bubble whites is attributed to the close interaction of NPs and Ag and Au ions with the bacterial membrane, in addition to the possible ROS formation [156,167].

Some studies suggest that the main mechanism of biofilm destruction occurs through the binding of Ag NPs and Au NPs to the exopolysaccharide matrix. The biofilm structure is disrupted by recognizing the peptidoglycan structure present in bacterial membranes, causing physical damage, ion release, production of ROS, leading to oxidative stress, and DNA damage [154,164,167,169].

When bacteria are treated with Ag NPs morphological changes are revealed in the biofilm architecture. The irregular cell surface, suggesting cell lysis, relevant morphological damage to the cell wall, damage to membrane corrugation, changes in membrane polarization and/or permeability, and the distinct formation of an EPS matrix around bacteria are observed. In addition, electrostatic interactions between NPs and bacterial membranes cause them to disrupt, so that Ag NPs can penetrate the mature biofilm [156,164,173,174].

In addition, NPs can interfere with the condensation of the condensed cytoplasmic membrane, the pathways of bacterial metabolism, and the production of extracellular polysaccharides, leading to a change in the layout of the biofilm [170,180,181]. Due to these antibiofilm properties, Ag NPs and Au NPs are considered very promising for the treatment of multidrug-resistant bacterial infections and biofilm production [153,179,182].

5. Conclusions

Bacterial infections are a worldwide public health problem, and it has become more urgent due to some bacteria developing mechanisms of resistance to current antibiotics. This is a challenging problem, even more when microorganisms are capable of producing biofilms. However, this global public health challenge has motivated new research to develop therapeutic strategy and antibacterial biomaterials described in this review, such as hydrogels incorporating metallic NPs, in particular Ag and Au.

There is a wide variety of biocompatible polymers used in the production of hydrogels, such as alginate, chitosan, gelatin, konjac glucomannan, carbopol, carboxymethyl cellulose, carboxymethyl chitosan, poly-vinyl alcohol, gelatin methacrylate, polyacrylamide, and polyvinylpyrrolidone. These polymers have been shown to be excellent candidates as carriers of antibacterial NPs for the prevention and treatment of localized bacterial infections. Among those polymers, chitosan, Pluronic F127, gelatin, and poloxamer 407 seem to be the most promising. Some of these systems have achieved an antibacterial efficiency of 99.86%, 99.94%, 99.5%, and 99.0% against *E. coli*, *S. aureus*, *K. pneumoniae*, and *P. aeruginosa*, respectively. These bacteria are the most studied pathogenic microorganisms due to their pathogenic potential, ability to produce biofilms and to acquired antibiotic resistance. In vivo clinical trials, performed on animals, have shown that antibacterial hydrogels also help with healing and re-epithelialization.

To enhance the antibacterial action and inhibit the development of new resistance mechanisms by bacteria, new strategies have been explored, such as: (i) the surface functionalization of nanoparticles with other antimicrobial agents, such as polyethyleneimine, 2-mercapto-1-methylimidazole, 6-aminopenicillins acid, poly(allylamine hydrochloride), poly(ethylene glycol), and quercetin; (ii) drug encapsulation within the nanoparticles; (iii) incorporation of two metallic nanoparticles into the hydrogel; (iv) in situ photosynthesis of NPs into the hydrogel structure; and (v) photo-irradiation of the NP/hydrogel to eradicate bacteria. Therefore, there are many possible ways to fight against multidrug-resistant bacteria with the use of metallic nanoparticles incorporated into hydrogels.

Furthermore, the dispersion of metallic nanoparticles within the matrix of hydrogels significantly improves their ability to prevent and eliminate biofilm formation. This strategy was shown to be highly effective in the eradication of biofilms formed in wounds. Additionally, other techniques can be used alongside hydrogels to amplify their efficiency, such as phototherapy or photothermal therapy. Thus, these hydrogels are promising antibiofilm agents and may be critical in treating bacterial infections associated with biofilms in the future.

Although there are many results in the literature and examples of in vitro and in vivo studies in animals, clinical applications of biopolymers are limited to a few reports. Examples of clinical applications in human patients are even scarcer for Ag NPs and Au NPs incorporated into biopolymers. Real case applications are essential to demonstrate their technical, economical and clinical feasibility. Therefore, great opportunities for the development of such composites are on the horizon.

Author Contributions: Investigation, literature searching, writing—original draft preparation, and conceptualization, Y.P.M.R., L.A.d.A.C. and M.A.A.A.; writing—review and editing, supervision, and resources, A.G. and I.M.F.C. All authors have read and agreed to the published version of the manuscript.

Funding: This research was funded by the Foundation for the Support of Science and Technology of the State of Pernambuco—FACEPE (Brazil), grants number APQ-0229-1.06/19, 316048/2020-8, and Innovative Scientists 2022 (APQ-0287-4.03/22), by Propesqi n° 09/2021-Support for Qualified Production of the Federal University of Pernambuco, and by the National Council for Scientific and Technological Development—CNPq (Brazil).

Institutional Review Board Statement: Not applicable.

Informed Consent Statement: Not applicable.

Data Availability Statement: Not applicable.

Conflicts of Interest: The authors declare that they have no conflict of interest.

References

1. Tommasi, R.; Brown, D.G.; Walkup, G.K.; Manchester, J.I.; Miller, A.A. ESKAPEing the labyrinth of antibacterial discovery. *Nat. Rev. Drug Discov.* **2015**, *14*, 529–542. [CrossRef]
2. Shrivastava, S.; Shrivastava, P.; Ramasamy, J. World health organization releases global priority list of antibiotic-resistant bacteria to guide research, discovery, and development of new antibiotics. *J. Med. Soc.* **2018**, *32*, 76. [CrossRef]
3. Piddock, L.J.V. The crisis of no new antibiotics—What is the way forward? *Lancet Infect. Dis.* **2012**, *12*, 249–253. [CrossRef]
4. Zheng, K.; Setyawati, M.I.; Leong, D.T.; Xie, J. Antimicrobial silver nanomaterials. *Coord. Chem. Rev.* **2018**, *357*, 1–17. [CrossRef]
5. Coates, A.; Hu, Y.; Bax, R.; Page, C. The future challenges facing the development of new antimicrobial drugs. *Nat. Rev. Drug Discov.* **2002**, *1*, 895–910. [CrossRef]
6. Centers for Disease Control and Prevention. *Antibiotic Resistance Threats in the United States, 2013*; Centres for Disease Control and Prevention, US Department of Health and Human Services: Altanta, GA, USA, 2013.
7. Panáček, A.; Kvítek, L.; Smékalová, M.; Večeřová, R.; Kolář, M.; Röderová, M.; Dyčka, F.; Šebela, M.; Prucek, R.; Tomanec, O.; et al. Bacterial resistance to silver nanoparticles and how to overcome it. *Nat. Nanotechnol.* **2018**, *13*, 65–71. [CrossRef]
8. Muhammad, M.H.; Idris, A.L.; Fan, X.; Guo, Y.; Yu, Y.; Jin, X.; Qiu, J.; Guan, X.; Huang, T. Beyond Risk: Bacterial Biofilms and Their Regulating Approaches. *Front. Microbiol.* **2020**, *11*, 928. [CrossRef]
9. Alav, I.; Sutton, J.M.; Rahman, K.M. Role of bacterial efflux pumps in biofilm formation. *J. Antimicrob. Chemother.* **2018**, *73*, 2003–2020. [CrossRef]
10. Sharahi, J.Y.; Azimi, T.; Shariati, A.; Safari, H.; Tehrani, M.K.; Hashemi, A. Advanced strategies for combating bacterial biofilms. *J. Cell Physiol.* **2019**, *234*, 14689–14708. [CrossRef]
11. Khatoon, Z.; McTiernan, C.D.; Suuronen, E.J.; Mah, T.-F.; Alarcon, E.I. Bacterial biofilm formation on implantable devices and approaches to its treatment and prevention. *Heliyon* **2018**, *4*, e01067. [CrossRef]
12. Vestby, L.K.; Grønseth, T.; Simm, R.; Nesse, L.L. Bacterial Biofilm and its Role in the Pathogenesis of Disease. *Antibiotics* **2020**, *9*, 59. [CrossRef] [PubMed]

13. Alshehri, S.M.; Aldalbahi, A.; Al-Hajji, A.B.; Chaudhary, A.A.; Panhuis, M.I.H.; Alhokbany, N.; Ahamad, T. Development of carboxymethyl cellulose-based hydrogel and nanosilver composite as antimicrobial agents for UTI pathogens. *Carbohydr. Polym.* **2016**, *138*, 229–236. [CrossRef] [PubMed]
14. dos Santos, E.M.P.; Martins, C.C.B.; de Oliveira Santos, J.V.; da Silva, W.R.C.; Silva, S.B.C.; Pelagio-Flores, M.A.; Galembeck, A.; Cavalcanti, I.M.F. Silver nanoparticles–chitosan composites activity against resistant bacteria: Tolerance and biofilm inhibition. *J. Nanopart. Res.* **2021**, *23*, 196. [CrossRef]
15. Andrei, S.; Droc, G.; Stefan, G. FDA approved antibacterial drugs: 2018–2019. *Discoveries* **2019**, *7*, e102. [CrossRef]
16. Baym, M.; Lieberman, T.D.; Kelsic, E.D.; Chait, R.; Gross, R.; Yelin, I.; Kishony, R. Spatiotemporal microbial evolution on antibiotic landscapes. *Science* **2016**, *353*, 1147–1151. [CrossRef]
17. Gerência de Vigilância e Monitoramento em Serviços de Saúde; Gerência Geral de Tecnologia em Serviços de Saúde; Agência Nacional de Vigilância Sanitária. *ANVISA: Boletim de segurança do paciente e qualidade em serviços de saúde nº 14: Avaliação dos indicadores nacionais das infecções relacionadas à assistência à saúde (IRAS) e resistência microbiana do ano de 2015*; Agência Nacional de Vigilância Sanitária: Brasília, Brazil, 2016; Volume 14, pp. 1–116.
18. Gerência de Vigilância e Monitoramento em Serviços de Saúde; Gerência Geral de Tecnologia em Serviços de Saúde; Agência Nacional de Vigilância Sanitária. *Plano Nacional para a Prevenção e o Controle da Resistência Microbiana nos Serviços de Saúde*; Agência Nacional de Vigilância Sanitária: Brasília, Brazil, 2017; Volume 84.
19. Perelshtein, I.; Lipovsky, A.; Perkas, N.; Gedanken, A.; Moschini, E.; Mantecca, P. The influence of the crystalline nature of nano-metal oxides on their antibacterial and toxicity properties. *Nano Res.* **2015**, *8*, 695–707. [CrossRef]
20. Jung, W.K.; Koo, H.C.; Kim, K.W.; Shin, S.; Kim, S.H.; Park, Y.H. Antibacterial activity and mechanism of action of the silver ion in *Staphylococcus aureus* and *Escherichia coli*. *Appl. Environ. Microbiol.* **2008**, *74*, 2171–2178. [CrossRef]
21. Lakshminarayanan, R.; Ye, E.; Young, D.J.; Li, Z.; Loh, X.J. Recent Advances in the Development of Antimicrobial Nanoparticles for Combating Resistant Pathogens. *Adv. Healthc. Mater.* **2018**, *7*, 1701400. [CrossRef]
22. Xie, Y.; Liao, X.; Zhang, J.; Yang, F.; Fan, Z. Novel chitosan hydrogels reinforced by silver nanoparticles with ultrahigh mechanical and high antibacterial properties for accelerating wound healing. *Int. J. Biol. Macromol.* **2018**, *119*, 402–412. [CrossRef] [PubMed]
23. Jiang, Y.G.; Huang, J.J.; Wu, X.W.; Ren, Y.H.; Li, Z.A.; Ren, J.A. Controlled release of silver ions from AgNPs using a hydrogel based on konjac glucomannan and chitosan for infected wounds. *Int. J. Biol. Macromol.* **2020**, *149*, 148–157. [CrossRef]
24. Soliman, W.E.; Elsewedy, H.S.; Younis, N.S.; Shinu, P.; Elsawy, L.E.; Ramadan, H.A. Evaluating Antimicrobial Activity and Wound Healing Effect of Rod-Shaped Nanoparticles. *Polymers* **2022**, *14*, 2637. [CrossRef]
25. Zahedi, S.M.; Mansourpanah, Y. Construction of chitosan-carboxymethyl beta-cyclodextrin silver nanocomposite hydrogel to improve antibacterial activity. *Plast. Rubber Compos.* **2018**, *47*, 273–281. [CrossRef]
26. Kojic, N.; Pritchard, E.M.; Tao, H.; Brenckle, M.A.; Mondia, J.P.; Panilaitis, B.; Omenetto, F.; Kaplan, D.L. Focal Infection Treatment using Laser-Mediated Heating of Injectable Silk Hydrogels with Gold Nanoparticles. *Adv. Funct. Mater.* **2012**, *22*, 3793–3798. [CrossRef] [PubMed]
27. Chen, X.S.; Zhang, H.M.; Yang, X.; Zhang, W.H.; Jiang, M.; Wen, T.; Wang, J.; Guo, R.; Liu, H.J. Preparation and Application of Quaternized Chitosan- and AgNPs-Base Synergistic Antibacterial Hydrogel for Burn Wound Healing. *Molecules* **2021**, *26*, 4037. [CrossRef] [PubMed]
28. Ribeiro, M.; Ferraz, M.P.; Monteiro, F.J.; Fernandes, M.H.; Beppu, M.M.; Mantione, D.; Sardon, H. Antibacterial silk fibroin/nanohydroxyapatite hydrogels with silver and gold nanoparticles for bone regeneration. *Nanomed. Nanotechnol. Biol. Med.* **2017**, *13*, 231–239. [CrossRef] [PubMed]
29. Kumar, S.; Majhi, R.K.; Singh, A.; Mishra, M.; Tiwari, A.; Chawla, S.; Guha, P.; Satpati, B.; Mohapatra, H.; Goswami, L.; et al. Carbohydrate-Coated Gold–Silver Nanoparticles for Efficient Elimination of Multidrug Resistant Bacteria and in Vivo Wound Healing. *ACS Appl. Mater. Interfaces* **2019**, *11*, 42998–43017. [CrossRef]
30. Edhari, B.A.; Mashreghi, M.; Makhdoumi, A.; Darroudi, M. Antibacterial and antibiofilm efficacy of Ag NPs, Ni NPs and Al_2O_3 NPs singly and in combination against multidrug-resistant Klebsiella pneumoniae isolates. *J. Trace Elem. Med. Biol.* **2021**, *68*, 126840. [CrossRef]
31. Ali, S.G.; Ansari, M.A.; Alzohairy, M.A.; Alomary, M.N.; AlYahya, S.; Jalal, M.; Khan, H.M.; Asiri, S.M.M.; Ahmad, W.; Mahdi, A.A.; et al. Biogenic Gold Nanoparticles as Potent Antibacterial and Antibiofilm Nano-Antibiotics against Pseudomonas aeruginosa. *Antibiotics* **2020**, *9*, 100. [CrossRef]
32. Hu, J.; Zhang, C.; Zhou, L.; Hu, Q.; Kong, Y.; Song, D.; Cheng, Y.; Zhang, Y. A smart hydrogel for on-demand delivery of antibiotics and efficient eradication of biofilms. *Sci. China Mater.* **2021**, *64*, 1035–1046. [CrossRef]
33. Rezaei, N.; Hamidabadi, H.G.; Khosravimelal, S.; Zahiri, M.; Ahovan, Z.A.; Bojnordi, M.N.; Eftekhari, B.S.; Hashemi, A.; Ganji, F.; Darabi, S.; et al. Antimicrobial peptides-loaded smart chitosan hydrogel: Release behavior and antibacterial potential against antibiotic resistant clinical isolates. *Int. J. Biol. Macromol.* **2020**, *164*, 855–862. [CrossRef]
34. Mishra, R.; Panda, A.K.; De Mandal, S.; Shakeel, M.; Bisht, S.S.; Khan, J. Natural Antibiofilm Agents: Strategies to Control Biofilm-Forming Pathogens. *Front. Microbiol.* **2020**, *11*, 566325. [CrossRef] [PubMed]
35. Lu, L.; Hu, W.; Tian, Z.; Yuan, D.; Yi, G.; Zhou, Y.; Cheng, Q.; Zhu, J.; Li, M. Developing natural products as potential antibiofilm agents. *Chin. Med.* **2019**, *14*, 11. [CrossRef] [PubMed]

36. Ozcelik, B.; Ho, K.K.K.; Glattauer, V.; Willcox, M.; Kumar, N.; Thissen, H. Poly(ethylene glycol)-Based Coatings Combining Low-Biofouling and Quorum-Sensing Inhibiting Properties to Reduce Bacterial Colonization. *ACS Biomater. Sci. Eng.* **2017**, *3*, 78–87. [CrossRef] [PubMed]
37. Pandian, M.; Selvaprithviraj, V.; Pradeep, A.; Rangasamy, J. In-situ silver nanoparticles incorporated N, O-carboxymethyl chitosan based adhesive, self-healing, conductive, antibacterial and antibiofilm hydrogel. *Int. J. Biol. Macromol.* **2021**, *188*, 501–511. [CrossRef]
38. Carpa, R.; Remizovschi, A.; Culda, C.A.; Butiuc-Keul, A.L. Inherent and Composite Hydrogels as Promising Materials to Limit Antimicrobial Resistance. *Gels* **2022**, *8*, 70. [CrossRef] [PubMed]
39. Garg, D.; Matai, I.; Sachdev, A. Toward Designing of Anti-infective Hydrogels for Orthopedic Implants: From Lab to Clinic. *ACS Biomater. Sci. Eng.* **2021**, *7*, 1933–1961. [CrossRef]
40. Ferrag, C.; Li, S.P.; Jeon, K.; Andoy, N.M.; Sullan, R.M.A.; Mikhaylichenko, S.; Kerman, K. Polyacrylamide hydrogels doped with different shapes of silver nanoparticles: Antibacterial and mechanical properties. *Colloids Surf. B-Biointerfaces* **2021**, *197*, 111397. [CrossRef]
41. Fan, Z.; Liu, B.; Wang, J.; Zhang, S.; Lin, Q.; Gong, P.; Ma, L.; Yang, S. A Novel Wound Dressing Based on Ag/Graphene Polymer Hydrogel: Effectively Kill Bacteria and Accelerate Wound Healing. *Adv. Funct. Mater.* **2014**, *24*, 3933–3943. [CrossRef]
42. Kwiatkowska, A.; Drabik, M.; Lipko, A.; Grzeczkowicz, A.; Stachowiak, R.; Marszalik, A.; Granicka, L.H. Composite Membrane Dressings System with Metallic Nanoparticles as an Antibacterial Factor in Wound Healing. *Membranes* **2022**, *12*, 215. [CrossRef]
43. Haidari, S.; FFA, I.J.; Metsemakers, W.J.; Maarse, W.; Vogely, H.C.; Ramsden, A.J.; McNally, M.A.; Govaert, G.A.M. The Role of Negative-Pressure Wound Therapy in Patients with Fracture-Related Infection: A Systematic Review and Critical Appraisal. *BioMed Res. Int.* **2021**, *2021*, 7742227. [CrossRef]
44. Haidari, H.; Bright, R.; Garg, S.; Vasilev, K.; Cowin, A.J.; Kopecki, Z. Eradication of Mature Bacterial Biofilms with Concurrent Improvement in Chronic Wound Healing Using Silver Nanoparticle Hydrogel Treatment. *Biomedicines* **2021**, *9*, 1182. [CrossRef]
45. Banerjee, D.; Shivapriya, P.M.; Gautam, P.K.; Misra, K.; Sahoo, A.K.; Samanta, S.K. A Review on Basic Biology of Bacterial Biofilm Infections and Their Treatments by Nanotechnology-Based Approaches. *Proc. Natl. Acad. Sci. India Sect. B Biol. Sci.* **2020**, *90*, 243–259. [CrossRef]
46. Vazquez-Munoz, R.; Avalos-Borja, M.; Castro-Longoria, E. Ultrastructural analysis of Candida albicans when exposed to silver nanoparticles. *PLoS ONE* **2014**, *9*, e108876. [CrossRef] [PubMed]
47. Bowman, M.C.; Ballard, T.E.; Ackerson, C.J.; Feldheim, D.L.; Margolis, D.M.; Melander, C. Inhibition of HIV fusion with multivalent gold nanoparticles. *J. Am. Chem. Soc.* **2008**, *130*, 6896–6897. [CrossRef]
48. Zhao, Y.; Tian, Y.; Cui, Y.; Liu, W.; Ma, W.; Jiang, X. Small Molecule-Capped Gold Nanoparticles as Potent Antibacterial Agents That Target Gram-Negative Bacteria. *J. Am. Chem. Soc.* **2010**, *132*, 12349–12356. [CrossRef]
49. Bankar, A.; Joshi, B.; Kumar, A.R.; Zinjarde, S. Banana peel extract mediated synthesis of gold nanoparticles. *Colloids Surf. B Biointerfaces* **2010**, *80*, 45–50. [CrossRef]
50. Baram-Pinto, D.; Shukla, S.; Gedanken, A.; Sarid, R. Inhibition of HSV-1 attachment, entry, and cell-to-cell spread by functionalized multivalent gold nanoparticles. *Small (Weinh. Der Bergstr. Ger.)* **2010**, *6*, 1044–1050. [CrossRef]
51. Jones, N.; Ray, B.; Ranjit, K.T.; Manna, A.C. Antibacterial activity of ZnO nanoparticle suspensions on a broad spectrum of microorganisms. *FEMS Microbiol. Lett.* **2008**, *279*, 71–76. [CrossRef]
52. Feng, Y.; Min, L.; Zhang, W.; Liu, J.; Hou, Z.; Chu, M.; Li, L.; Shen, W.; Zhao, Y.; Zhang, H. Zinc Oxide Nanoparticles Influence Microflora in Ileal Digesta and Correlate Well with Blood Metabolites. *Front. Microbiol.* **2017**, *8*, 992. [CrossRef]
53. He, L.; Liu, Y.; Mustapha, A.; Lin, M. Antifungal activity of zinc oxide nanoparticles against Botrytis cinerea and Penicillium expansum. *Microbiol. Res.* **2011**, *166*, 207–215. [CrossRef]
54. Darbari, S.; Abdi, Y.; Haghighi, F.; Mohajerzadeh, S.; Haghighi, N. Investigating the antifungal activity of TiO2nanoparticles deposited on branched carbon nanotube arrays. *J. Phys. D Appl. Phys.* **2011**, *44*, 245401. [CrossRef]
55. Karunakaran, G.; Suriyaprabha, R.; Manivasakan, P.; Yuvakkumar, R.; Rajendran, V.; Kannan, N. Impact of nano and bulk ZrO_2, TiO_2 particles on soil nutrient contents and PGPR. *J. Nanosci. Nanotechnol.* **2013**, *13*, 678–685. [CrossRef] [PubMed]
56. Esteban-Tejeda, L.; Malpartida, F.; Esteban-Cubillo, A.; Pecharromán, C.; Moya, J.S. Antibacterial and antifungal activity of a soda-lime glass containing copper nanoparticles. *Nanotechnology* **2009**, *20*, 505701. [CrossRef]
57. Tang, Z.-X.; Lv, B.-F. MgO nanoparticles as antibacterial agent: Preparation and activity. *Braz. J. Chem. Eng.* **2014**, *31*, 591–601. [CrossRef]
58. Sawai, J.; Yoshikawa, T. Quantitative evaluation of antifungal activity of metallic oxide powders (MgO, CaO and ZnO) by an indirect conductimetric assay. *J. Appl. Microbiol.* **2004**, *96*, 803–809. [CrossRef] [PubMed]
59. Zhang, Y.; Lou, Z.; Zhang, X.; Hu, X.; Zhang, H. A simple strategy to fabricate poly (acrylamide-co-alginate)/gold nanocomposites for inactivation of bacteria. *Appl. Phys. A* **2014**, *117*, 2009–2018. [CrossRef]
60. Lu, B.; Ye, H.; Shang, S.; Xiong, Q.; Yu, K.; Li, Q.; Xiao, Y.; Dai, F.; Lan, G. Novel wound dressing with chitosan gold nanoparticles capped with a small molecule for effective treatment of multiantibiotic-resistant bacterial infections. *Nanotechnology* **2018**, *29*, 425603. [CrossRef] [PubMed]
61. Yang, X.; Yang, J.; Wang, L.; Ran, B.; Jia, Y.; Zhang, L.; Yang, G.; Shao, H.; Jiang, X. Pharmaceutical Intermediate-Modified Gold Nanoparticles: Against Multidrug-Resistant Bacteria and Wound-Healing Application via an Electrospun Scaffold. *ACS Nano* **2017**, *11*, 5737–5745. [CrossRef]

62. Shariatinia, Z. Carboxymethyl chitosan: Properties and biomedical applications. *Int. J. Biol. Macromol.* **2018**, *120*, 1406–1419. [CrossRef]
63. Badhwar, R.; Mangla, B.; Neupane, Y.R.; Khanna, K.; Popli, H. Quercetin loaded silver nanoparticles in hydrogel matrices for diabetic wound healing. *Nanotechnology* **2021**, *32*, 505102. [CrossRef]
64. Varaprasad, K.; Reddy, G.S.M.; Jayaramudu, J.; Sadiku, R.; Ramam, K.; Ray, S.S. Development of microbial resistant Carbopol nanocomposite hydrogels via a green process. *Biomater. Sci.* **2014**, *2*, 257–263. [CrossRef]
65. Ou, Q.; Huang, K.; Fu, C.; Huang, C.; Fang, Y.; Gu, Z.; Wu, J.; Wang, Y. Nanosilver-incorporated halloysite nanotubes/gelatin methacrylate hybrid hydrogel with osteoimmunomodulatory and antibacterial activity for bone regeneration. *Chem. Eng. J.* **2020**, *382*, 123019. [CrossRef]
66. Yang, S.; Zhou, Y.; Zhao, Y.; Wang, D.; Luan, Y. Microwave synthesis of graphene oxide decorated with silver nanoparticles for slow-release antibacterial hydrogel. *Mater. Today Commun.* **2022**, *31*, 103663. [CrossRef]
67. Tejamaya, M.; Römer, I.; Merrifield, R.C.; Lead, J.R. Stability of Citrate, PVP, and PEG Coated Silver Nanoparticles in Ecotoxicology Media. *Environ. Sci. Technol.* **2012**, *46*, 7011–7017. [CrossRef]
68. Ge, J.; Li, Y.; Wang, M.; Gao, C.; Yang, S.; Lei, B. Engineering conductive antioxidative antibacterial nanocomposite hydrogel scaffolds with oriented channels promotes structure-functional skeletal muscle regeneration. *Chem. Eng. J.* **2021**, *425*, 130333. [CrossRef]
69. Deng, Z.; Li, M.; Hu, Y.; He, Y.; Tao, B.; Yuan, Z.; Wang, R.; Chen, M.; Luo, Z.; Cai, K. Injectable biomimetic hydrogels encapsulating Gold/metal–organic frameworks nanocomposites for enhanced antibacterial and wound healing activity under visible light actuation. *Chem. Eng. J.* **2021**, *420*, 129668. [CrossRef]
70. Nešović, K.; Mišković-Stanković, V. Silver/poly(vinyl alcohol)/graphene hydrogels for wound dressing applications: Understanding the mechanism of silver, antibacterial agent release. *J. Vinyl Addit. Technol.* **2022**, *28*, 196–210. [CrossRef]
71. Nešović, K.; Janković, A.; Perić-Grujić, A.; Vukašinović-Sekulić, M.; Radetić, T.; Živković, L.; Park, S.-J.; Yop Rhee, K.; Mišković-Stanković, V. Kinetic models of swelling and thermal stability of silver/poly(vinyl alcohol)/chitosan/graphene hydrogels. *J. Ind. Eng. Chem.* **2019**, *77*, 83–96. [CrossRef]
72. Nesovic, K.; Jankovic, A.; Kojic, V.; Vukasinovic-Sekulic, M.; Peric-Grujic, A.; Rhee, K.Y.; Miskovic-Stankovic, V. Silver/poly(vinyl alcohol)/chitosan/graphene hydrogels—Synthesis, biological and physicochemical properties and silver release kinetics. *Compos. Part B-Eng.* **2018**, *154*, 175–185. [CrossRef]
73. Kujda, M.; Wieja, M.; Adamczyk, Z.; BocheSka, O.; Bra, G.; Kozik, A.; BielaSka, E.; Barbasz, J. Charge Stabilized Silver Nanoparticles Applied as Antibacterial Agents. *J. Nanosci. Nanotechnol.* **2015**, *15*, 3574–3583. [CrossRef]
74. Alavi, M.; Karimi, N. Biosynthesis of Ag and Cu NPs by secondary metabolites of usnic acid and thymol with biological macromolecules aggregation and antibacterial activities against multi drug resistant (MDR) bacteria. *Int. J. Biol. Macromol.* **2019**, *128*, 893–901. [CrossRef]
75. Aurore, V.; Caldana, F.; Blanchard, M.; Kharoubi Hess, S.; Lannes, N.; Mantel, P.-Y.; Filgueira, L.; Walch, M. Silver-nanoparticles increase bactericidal activity and radical oxygen responses against bacterial pathogens in human osteoclasts. *Nanomed. Nanotechnol. Biol. Med.* **2018**, *14*, 601–607. [CrossRef]
76. Beyene, H.D.; Werkneh, A.A.; Bezabh, H.K.; Ambaye, T.G. Synthesis paradigm and applications of silver nanoparticles (AgNPs), a review. *Sustain. Mater. Technol.* **2017**, *13*, 18–23. [CrossRef]
77. Feizi, S.; Taghipour, E.; Ghadam, P.; Mohammadi, P. Antifungal, antibacterial, antibiofilm and colorimetric sensing of toxic metals activities of eco friendly, economical synthesized Ag/AgCl nanoparticles using Malva Sylvestris leaf extracts. *Microb. Pathog.* **2018**, *125*, 33–42. [CrossRef]
78. Lok, C.-N.; Ho, C.-M.; Chen, R.; He, Q.-Y.; Yu, W.-Y.; Sun, H.; Tam, P.K.-H.; Chiu, J.-F.; Che, C.-M. Silver nanoparticles: Partial oxidation and antibacterial activities. *JBIC J. Biol. Inorg. Chem.* **2007**, *12*, 527–534. [CrossRef]
79. Yusuf, M. *Handbook of Ecomaterials*; Martínez, L.M.T., Kharissova, O.V., Kharisov, B.I., Eds.; Springer International Publishing: Cham, Switzerland, 2019; pp. 2343–2356.
80. Sondi, I.; Salopek-Sondi, B. Silver nanoparticles as antimicrobial agent: A case study on E. coli as a model for Gram-negative bacteria. *J. Colloid Interface Sci.* **2004**, *275*, 177–182. [CrossRef]
81. Morones, J.R.; Elechiguerra, J.L.; Camacho, A.; Holt, K.; Kouri, J.B.; Ramírez, J.T.; Yacaman, M.J. The bactericidal effect of silver nanoparticles. *Nanotechnology* **2005**, *16*, 2346–2353. [CrossRef]
82. Ansari, M.A.; Khan, H.M.; Khan, A.A.; Ahmad, M.K.; Mahdi, A.A.; Pal, R.; Cameotra, S.S. Interaction of silver nanoparticles with Escherichia coli and their cell envelope biomolecules. *J. Basic Microbiol.* **2014**, *54*, 905–915. [CrossRef]
83. Mirzajani, F.; Ghassempour, A.; Aliahmadi, A.; Esmaeili, M.A. Antibacterial effect of silver nanoparticles on *Staphylococcus aureus*. *Res. Microbiol.* **2011**, *162*, 542–549. [CrossRef]
84. Kim, J.S.; Kuk, E.; Yu, K.N.; Kim, J.-H.; Park, S.J.; Lee, H.J.; Kim, S.H.; Park, Y.K.; Park, Y.H.; Hwang, C.-Y.; et al. Antimicrobial effects of silver nanoparticles. *Nanomed. Nanotechnol. Biol. Med.* **2007**, *3*, 95–101. [CrossRef]
85. Hwang, E.T.; Lee, J.H.; Chae, Y.J.; Kim, Y.S.; Kim, B.C.; Sang, B.-I.; Gu, M.B. Analysis of the Toxic Mode of Action of Silver Nanoparticles Using Stress-Specific Bioluminescent Bacteria. *Small* **2008**, *4*, 746–750. [CrossRef]
86. Sies, H. Oxidative stress: Oxidants and antioxidants. *Exp. Physiol.* **1997**, *82*, 291–295. [CrossRef]
87. Dibrov, P.; Dzioba, J.; Gosink, K.K.; Häse, C.C. Chemiosmotic Mechanism of Antimicrobial Activity of Ag^+ in *Vibrio cholerae*. *Antimicrob. Agents Chemother.* **2002**, *46*, 2668–2670. [CrossRef]

88. Gnanadhas, D.P.; Ben Thomas, M.; Thomas, R.; Raichur, A.M.; Chakravortty, D. Interaction of Silver Nanoparticles with Serum Proteins Affects Their Antimicrobial Activity In Vivo. *Antimicrob. Agents Chemother.* **2013**, *57*, 4945–4955. [CrossRef]
89. Agnihotri, S.; Mukherji, S.; Mukherji, S. Size-controlled silver nanoparticles synthesized over the range 5–100 nm using the same protocol and their antibacterial efficacy. *RSC Adv.* **2014**, *4*, 3974–3983. [CrossRef]
90. Sadeghi, B.; Garmaroudi, F.S.; Hashemi, M.; Nezhad, H.R.; Nasrollahi, A.; Ardalan, S.; Ardalan, S. Comparison of the anti-bacterial activity on the nanosilver shapes: Nanoparticles, nanorods and nanoplates. *Adv. Powder Technol.* **2012**, *23*, 22–26. [CrossRef]
91. Zhang, Y.; Peng, H.; Huang, W.; Zhou, Y.; Yan, D. Facile preparation and characterization of highly antimicrobial colloid Ag or Au nanoparticles. *J. Colloid Interface Sci.* **2008**, *325*, 371–376. [CrossRef]
92. Ivask, A.; ElBadawy, A.; Kaweeteerawat, C.; Boren, D.; Fischer, H.; Ji, Z.; Chang, C.H.; Liu, R.; Tolaymat, T.; Telesca, D.; et al. Toxicity Mechanisms in Escherichia coli Vary for Silver Nanoparticles and Differ from Ionic Silver. *ACS Nano* **2014**, *8*, 374–386. [CrossRef]
93. Abbaszadegan, A.; Ghahramani, Y.; Gholami, A.; Hemmateenejad, B.; Dorostkar, S.; Nabavizadeh, M.; Sharghi, H. The effect of charge at the surface of silver nanoparticles on antimicrobial activity against gram-positive and gram-negative bacteria: A preliminary study. *J. Nanomater.* **2015**, *16*, 53. [CrossRef]
94. Fayaz, A.M.; Balaji, K.; Girilal, M.; Yadav, R.; Kalaichelvan, P.T.; Venketesan, R. Biogenic synthesis of silver nanoparticles and their synergistic effect with antibiotics: A study against gram-positive and gram-negative bacteria. *Nanomed. Nanotechnol. Biol. Med.* **2010**, *6*, 103–109. [CrossRef]
95. Banerjee, M.; Mallick, S.; Paul, A.; Chattopadhyay, A.; Ghosh, S.S. Heightened Reactive Oxygen Species Generation in the Antimicrobial Activity of a Three Component Iodinated Chitosan−Silver Nanoparticle Composite. *Langmuir* **2010**, *26*, 5901–5908. [CrossRef]
96. Mishra, S.K.; Raveendran, S.; Ferreira, J.M.F.; Kannan, S. In Situ Impregnation of Silver Nanoclusters in Microporous Chitosan-PEG Membranes as an Antibacterial and Drug Delivery Percutaneous Device. *Langmuir* **2016**, *32*, 10305–10316. [CrossRef]
97. Liang, D.; Lu, Z.; Yang, H.; Gao, J.; Chen, R. Novel Asymmetric Wettable AgNPs/Chitosan Wound Dressing: In Vitro and In Vivo Evaluation. *ACS Appl. Mater. Interfaces* **2016**, *8*, 3958–3968. [CrossRef]
98. Jin, R.; Zeng, C.; Zhou, M.; Chen, Y. Atomically Precise Colloidal Metal Nanoclusters and Nanoparticles: Fundamentals and Opportunities. *Chem. Rev.* **2016**, *116*, 10346–10413. [CrossRef]
99. Fang, J.; Zhang, B.; Yao, Q.; Yang, Y.; Xie, J.; Yan, N. Recent advances in the synthesis and catalytic applications of ligand-protected, atomically precise metal nanoclusters. *Coord. Chem. Rev.* **2016**, *322*, 1–29. [CrossRef]
100. Tao, Y.; Li, M.; Ren, J.; Qu, X. Metal nanoclusters: Novel probes for diagnostic and therapeutic applications. *Chem. Soc. Rev.* **2015**, *44*, 8636–8663. [CrossRef]
101. Luo, Z.; Zheng, K.; Xie, J. Engineering ultrasmall water-soluble gold and silver nanoclusters for biomedical applications. *Chem. Commun.* **2014**, *50*, 5143–5155. [CrossRef]
102. Díez, I.; Eronen, P.; Österberg, M.; Linder, M.B.; Ikkala, O.; Ras, R.H.A. Functionalization of Nanofibrillated Cellulose with Silver Nanoclusters: Fluorescence and Antibacterial Activity. *Macromol. Biosci.* **2011**, *11*, 1185–1191. [CrossRef]
103. Wang, X.; Gao, W.; Xu, S.; Xu, W. Luminescent fibers: In situ synthesis of silver nanoclusters on silk via ultraviolet light-induced reduction and their antibacterial activity. *Chem. Eng. J.* **2012**, *210*, 585–589. [CrossRef]
104. Balagna, C.; Irfan, M.; Perero, S.; Miola, M.; Maina, G.; Santella, D.; Simone, A. Characterization of antibacterial silver nanocluster/silica composite coating on high performance Kevlar® textile. *Surf. Coat. Technol.* **2017**, *321*, 438–447. [CrossRef]
105. Willing, B.P.; Pepin, D.M.; Marcolla, C.S.; Forgie, A.J.; Diether, N.E.; Bourrie, B.C.T. Bacterial resistance to antibiotic alternatives: A wolf in sheep's clothing? *Anim. Front.* **2018**, *8*, 39–47. [CrossRef]
106. Fonseca-Santos, B.; Chorilli, M. An overview of carboxymethyl derivatives of chitosan: Their use as biomaterials and drug delivery systems. *Mater. Sci. Eng. C* **2017**, *77*, 1349–1362. [CrossRef]
107. Liu, L.; Wen, H.; Rao, Z.; Zhu, C.; Liu, M.; Min, L.; Fan, L.; Tao, S. Preparation and characterization of chitosan–collagen peptide/oxidized konjac glucomannan hydrogel. *Int. J. Biol. Macromol.* **2018**, *108*, 376–382. [CrossRef]
108. Qin, D.; Zhang, A.; Wang, N.; Yao, Y.; Chen, X.; Liu, Y. Hydroxybutyl chitosan/ oxidized glucomannan self-healing hydrogels as BMSCs-derived exosomes carriers for advanced stretchable wounds. *Appl. Mater. Today* **2022**, *26*, 101342. [CrossRef]
109. Jiang, Y.; Li, G.; Liu, J.; Li, M.; Li, Q.; Tang, K. Gelatin/Oxidized Konjac Glucomannan Composite Hydrogels with High Resistance to Large Deformation for Tissue Engineering Applications. *ACS Appl. Bio Mater.* **2021**, *4*, 1536–1543. [CrossRef]
110. Wu, H.; Bu, N.; Chen, J.; Chen, Y.; Sun, R.; Wu, C.; Pang, J. Construction of Konjac Glucomannan/Oxidized Hyaluronic Acid Hydrogels for Controlled Drug Release. *Polymers* **2022**, *14*, 927. [CrossRef]
111. Chen, H.; Lan, G.; Ran, L.; Xiao, Y.; Yu, K.; Lu, B.; Dai, F.; Wu, D.; Lu, F. A novel wound dressing based on a Konjac glucomannan/silver nanoparticle composite sponge effectively kills bacteria and accelerates wound healing. *Carbohydr. Polym.* **2018**, *183*, 70–80. [CrossRef]
112. Ribeiro, J.S.; Bordini, E.A.F.; Ferreira, J.A.; Mei, L.; Dubey, N.; Fenno, J.C.; Piva, E.; Lund, R.G.; Schwendeman, A.; Bottino, M.C. Injectable MMP-Responsive Nanotube-Modified Gelatin Hydrogel for Dental Infection Ablation. *ACS Appl. Mater. Interfaces* **2020**, *12*, 16006–16017. [CrossRef]
113. Pal, S.; Tak, Y.K.; Song, J.M. Does the Antibacterial Activity of Silver Nanoparticles Depend on the Shape of the Nanoparticle? A Study of the Gram-Negative Bacterium *Escherichia coli*. *Appl. Environ. Microbiol.* **2007**, *73*, 1712–1720. [CrossRef]

114. Huang, T.; Xu, X.-H.N. Synthesis and characterization of tunable rainbow colored colloidal silver nanoparticles using single-nanoparticle plasmonic microscopy and spectroscopy. *J. Mater. Chem.* **2010**, *20*, 9867–9876. [CrossRef]
115. de Lacerda Coriolano, D.; de Souza, J.B.; Bueno, E.V.; Medeiros, S.; Cavalcanti, I.D.L.; Cavalcanti, I.M.F. Antibacterial and antibiofilm potential of silver nanoparticles against antibiotic-sensitive and multidrug-resistant Pseudomonas aeruginosa strains. *Braz. J. Microbiol.* **2021**, *52*, 267–278. [CrossRef]
116. Paladini, F.; Pollini, M. Antimicrobial Silver Nanoparticles for Wound Healing Application: Progress and Future Trends. *Materials* **2019**, *12*, 2540. [CrossRef]
117. Wang, Q.; Qiu, W.; Li, M.; Li, N.; Li, X.; Qin, X.; Wang, X.; Yu, J.; Li, F.; Huang, L.; et al. Multifunctional hydrogel platform for biofilm scavenging and O2 generating with photothermal effect on diabetic chronic wound healing. *J. Colloid Interface Sci.* **2022**, *617*, 542–556. [CrossRef]
118. Imran, M.; Hussain, S.; Mehmood, K.; Saeed, Z.; Parvaiz, M.; Younas, U.; Nadeem, H.A.; Ghalani, S.P.; Saleem, S. Optimization of ecofriendly synthesis of Ag nanoparticles by Linum usitatissimum hydrogel using response surface methodology and its biological applications. *Mater. Today Commun.* **2021**, *29*, 102789. [CrossRef]
119. Alfuraydi, R.T.; Alminderej, F.M.; Mohamed, N.A. Evaluation of Antimicrobial and Antibiofilm Formation Activities of Novel Poly(vinyl alcohol) Hydrogels Reinforced with Crosslinked Chitosan and Silver Nano-Particles. *Polymers* **2022**, *14*, 1619. [CrossRef]
120. Pérez-Díaz, M.; Alvarado-Gomez, E.; Magaña-Aquino, M.; Sánchez-Sánchez, R.; Velasquillo, C.; Gonzalez, C.; Ganem-Rondero, A.; Martínez-Castañon, G.; Zavala-Alonso, N.; Martinez-Gutierrez, F. Antibiofilm activity of chitosan gels formulated with silver nanoparticles and their cytotoxic effect on human fibroblasts. *Mater. Sci. Eng. C Mater. Biol. Appl.* **2016**, *60*, 317–323. [CrossRef]
121. Katas, H.; Mohd Akhmar, M.A.; Suleman Ismail Abdalla, S. Biosynthesized silver nanoparticles loaded in gelatine hydrogel for a natural antibacterial and antibiofilm wound dressing. *J. Bioact. Compat. Polym.* **2021**, *36*, 111–123. [CrossRef]
122. Vazquez-Muñoz, R.; Meza-Villezcas, A.; Fournier, P.G.J.; Soria-Castro, E.; Juarez-Moreno, K.; Gallego-Hernández, A.L.; Bogdanchikova, N.; Vazquez-Duhalt, R.; Huerta-Saquero, A. Enhancement of antibiotics antimicrobial activity due to the silver nanoparticles impact on the cell membrane. *PLoS ONE* **2019**, *14*, e0224904. [CrossRef]
123. Li, X.; Li, B.; Liu, R.; Dong, Y.; Zhao, Y.; Wu, Y. Development of pH-responsive nanocomposites with remarkably synergistic antibiofilm activities based on ultrasmall silver nanoparticles in combination with aminoglycoside antibiotics. *Colloids Surf. B Biointerfaces* **2021**, *208*, 112112. [CrossRef]
124. Lopez-Carrizales, M.; Mendoza-Mendoza, E.; Peralta-Rodriguez, R.D.; Pérez-Díaz, M.A.; Portales-Pérez, D.; Magaña-Aquino, M.; Aragón-Piña, A.; Infante-Martínez, R.; Barriga-Castro, E.D.; Sánchez-Sánchez, R.; et al. Characterization, antibiofilm and biocompatibility properties of chitosan hydrogels loaded with silver nanoparticles and ampicillin: An alternative protection to central venous catheters. *Colloids Surf. B Biointerfaces* **2020**, *196*, 111292. [CrossRef]
125. Wunnoo, S.; Bilhman, S.; Waen-ngoen, T.; Yawaraya, S.; Paosen, S.; Lethongkam, S.; Kaewnopparat, N.; Voravuthikunchai, S.P. Thermosensitive hydrogel loaded with biosynthesized silver nanoparticles using Eucalyptus camaldulensis leaf extract as an alternative treatment for microbial biofilms and persistent cells in tissue infections. *J. Drug Deliv. Sci. Technol.* **2022**, *74*, 103588. [CrossRef]
126. Kaul, S.; Sagar, P.; Gupta, R.; Garg, P.; Priyadarshi, N.; Singhal, N.K. Mechanobactericidal, Gold Nanostar Hydrogel-Based Bandage for Bacteria-Infected Skin Wound Healing. *ACS Appl. Mater. Interfaces* **2022**, *14*, 44084–44097. [CrossRef] [PubMed]
127. Okkeh, M.; Bloise, N.; Restivo, E.; De Vita, L.; Pallavicini, P.; Visai, L. Gold Nanoparticles: Can They Be the Next Magic Bullet for Multidrug-Resistant Bacteria? *Nanomaterials* **2021**, *11*, 312. [CrossRef] [PubMed]
128. Chitra, G.; Franklin, D.S.; Sudarsan, S.; Sakthivel, M.; Guhanathan, S. Noncytotoxic silver and gold nanocomposite hydrogels with enhanced antibacterial and wound healing applications. *Polym. Eng. Sci.* **2018**, *58*, 2133–2142. [CrossRef]
129. Amendola, V.; Pilot, R.; Frasconi, M.; Maragò, O.M.; Iatì, M.A. Surface plasmon resonance in gold nanoparticles: A review. *J. Phys. Condens. Matter* **2017**, *29*, 203002. [CrossRef] [PubMed]
130. Mahmoud, N.N.; Hikmat, S.; Abu Ghith, D.; Hajeer, M.; Hamadneh, L.; Qattan, D.; Khalil, E.A. Gold nanoparticles loaded into polymeric hydrogel for wound healing in rats: Effect of nanoparticles' shape and surface modification. *Int. J. Pharm.* **2019**, *565*, 174–186. [CrossRef]
131. Li, J.; Wang, Y.; Yang, J.; Liu, W. Bacteria activated-macrophage membrane-coated tough nanocomposite hydrogel with targeted photothermal antibacterial ability for infected wound healing. *Chem. Eng. J.* **2021**, *420*, 127638. [CrossRef]
132. Al-Bakri, A.G.; Mahmoud, N.N. Photothermal-Induced Antibacterial Activity of Gold Nanorods Loaded into Polymeric Hydrogel against Pseudomonas aeruginosa Biofilm. *Molecules* **2019**, *24*, 2661. [CrossRef]
133. Gao, W.; Vecchio, D.; Li, J.; Zhu, J.; Zhang, Q.; Fu, V.; Li, J.; Thamphiwatana, S.; Lu, D.; Zhang, L. Hydrogel Containing Nanoparticle-Stabilized Liposomes for Topical Antimicrobial Delivery. *ACS Nano* **2014**, *8*, 2900–2907. [CrossRef]
134. Abasalizadeh, F.; Moghaddam, S.V.; Alizadeh, E.; akbari, E.; Kashani, E.; Fazljou, S.M.B.; Torbati, M.; Akbarzadeh, A. Alginate-based hydrogels as drug delivery vehicles in cancer treatment and their applications in wound dressing and 3D bioprinting. *J. Biol. Eng.* **2020**, *14*, 8. [CrossRef]
135. Li, S.; Dong, S.; Xu, W.; Tu, S.; Yan, L.; Zhao, C.; Ding, J.; Chen, X. Antibacterial Hydrogels. *Adv. Sci.* **2018**, *5*, 1700527. [CrossRef] [PubMed]
136. Ryan, C.; Alcock, E.; Buttimer, F.; Schmidt, M.; Clarke, D.; Pemble, M.; Bardosova, M. Synthesis and characterisation of cross-linked chitosan composites functionalised with silver and gold nanoparticles for antimicrobial applications. *Sci. Technol. Adv. Mater.* **2017**, *18*, 528–540. [CrossRef] [PubMed]

137. Jaipan, P.; Nguyen, A.; Narayan, R.J. Gelatin-based hydrogels for biomedical applications. *MRS Commun.* **2017**, *7*, 416–426. [CrossRef]
138. Jayaramudu, T.; Varaprasad, K.; Pyarasani, R.D.; Reddy, K.K.; Akbari-Fakhrabadi, A.; Carrasco-Sánchez, V.; Amalraj, J. Hydroxypropyl methylcellulose-copper nanoparticle and its nanocomposite hydrogel films for antibacterial application. *Carbohydr. Polym.* **2021**, *254*, 117302. [CrossRef] [PubMed]
139. Makvandi, P.; Ali, G.W.; Della Sala, F.; Abdel-Fattah, W.I.; Borzacchiello, A. Biosynthesis and characterization of antibacterial thermosensitive hydrogels based on corn silk extract, hyaluronic acid and nanosilver for potential wound healing. *Carbohydr. Polym.* **2019**, *223*, 115023. [CrossRef]
140. Chitra, G.; Selvi, M.S.; Franklin, D.S.; Sudarsan, S.; Sakthivel, M.; Guhanathan, S. pH-sensitive biopolymeric hydrogel-based on indole-3-acetic acid for wound healing and anti-cancer applications. *SN Appl. Sci.* **2019**, *1*, 1641. [CrossRef]
141. Giuliano, E.; Paolino, D.; Fresta, M.; Cosco, D. Drug-Loaded Biocompatible Nanocarriers Embedded in Poloxamer 407 Hydrogels as Therapeutic Formulations. *Medicines* **2019**, *6*, 7. [CrossRef]
142. Chatterjee, S.; Hui, P.C.-L.; Kan, C.-W.; Wang, W. Dual-responsive (pH/temperature) Pluronic F-127 hydrogel drug delivery system for textile-based transdermal therapy. *Sci. Rep.* **2019**, *9*, 11658. [CrossRef]
143. Rolfe, M.D.; Rice, C.J.; Lucchini, S.; Pin, C.; Thompson, A.; Cameron, A.D.; Alston, M.; Stringer, M.F.; Betts, R.P.; Baranyi, J.; et al. Lag phase is a distinct growth phase that prepares bacteria for exponential growth and involves transient metal accumulation. *J. Bacteriol.* **2012**, *194*, 686–701. [CrossRef]
144. Wang, X.; Guo, J.; Zhang, Q.; Zhu, S.; Liu, L.; Jiang, X.; Wei, D.-H.; Liu, R.-S.; Li, L. Gelatin sponge functionalized with gold/silver clusters for antibacterial application. *Nanotechnology* **2020**, *31*, 134004. [CrossRef]
145. Zhu, Y.; Yao, Z.; Liu, Y.; Zhang, W.; Geng, L.; Ni, T. Incorporation of ROS-Responsive Substance P-Loaded Zeolite Imidazolate Framework-8 Nanoparticles into a Ca(2+)-Cross-Linked Alginate/Pectin Hydrogel for Wound Dressing Applications. *Int. J. Nanomed.* **2020**, *15*, 333–346. [CrossRef] [PubMed]
146. Prusty, K.; Swain, S.K. Release of ciprofloxacin drugs by nano gold embedded cellulose grafted polyacrylamide hybrid nanocomposite hydrogels. *Int. J. Biol. Macromol.* **2019**, *126*, 765–775. [CrossRef] [PubMed]
147. Tian, E.-K.; Wang, Y.; Ren, R.; Zheng, W.; Liao, W. Gold Nanoparticle: Recent Progress on Its Antibacterial Applications and Mechanisms. *J. Nanomater.* **2021**, *2021*, 2501345. [CrossRef]
148. Abdalla, S.S.I.; Katas, H.; Azmi, F.; Busra, M.F.M. Antibacterial and Antibiofilm Biosynthesised Silver and Gold Nanoparticles for Medical Applications: Mechanism of Action, Toxicity and Current Status. *Curr. Drug Deliv.* **2020**, *17*, 88–100. [CrossRef]
149. Bermúdez-Jiménez, C.; Niño-Martínez, N.; Patiño-Marín, N.; Martínez-Gutiérrez, F.; Ruiz, F.; Bach, H.; Martínez-Castañón, G. Effective control of biofilms by photothermal therapy using a gold nanorod hydrogel. *J. Biomed. Mater. Res. Part B Appl. Biomater.* **2020**, *108*, 333–342. [CrossRef]
150. Wickramasinghe, S.; Ju, M.; Milbrandt, N.B.; Tsai, Y.H.; Navarreto-Lugo, M.; Visperas, A.; Klika, A.; Barsoum, W.; Higuera-Rueda, C.A.; Samia, A.C.S. Photoactivated Gold Nanorod Hydrogel Composite Containing d-Amino Acids for the Complete Eradication of Bacterial Biofilms on Metal Alloy Implant Materials. *ACS Appl. Nano Mater.* **2020**, *3*, 5862–5873. [CrossRef]
151. Galdámez-Falla, V.-M.; Castillo-Martínez, J.-C.; de Alba-Montero, I.; Patiño-Marín, N.; Niño-Martínez, N.; Ruiz, F.; Martínez-Castañón, G.-A. Formation of a Mature Biofilm of Enterococcus Faecalis in Root Canal and Its Treatment Using Gold Nanorods. *J. Mater. Sci. Res. Rev.* **2022**, *9*, 31–43.
152. Kabiri, F.; Aghaei, S.S.; Pourbabaee, A.A.; Soleimani, M.; Komeili Movahhed, T. Antibiofilm and cytotoxic potential of extracellular biosynthesized gold nanoparticles using actinobacteria *Amycolatopsis* sp. KMN. *Prep. Biochem. Biotechnol.* **2022**, 1–14. [CrossRef]
153. Sahoo, B.; Panigrahi, L.L.; Das, R.P.; Pradhan, A.K.; Arakha, M. Biogenic Synthesis of Silver Nanoparticle from *Punica granatum* L. and Evaluation of Its Antioxidant, Antimicrobial and Antibiofilm Activity. *J. Inorg. Organometall. Polym. Mater.* **2022**, *32*, 4250–4259. [CrossRef]
154. Dharul Salam, F.; Nadar Vinita, M.; Puja, P.; Prakash, S.; Yuvakkumar, R.; Kumar, P. Anti-bacterial and antibiofilm efficacies of bioinspired gold nanoparticles. *Mater. Lett.* **2020**, *261*, 126998. [CrossRef]
155. Singh, P.; Pandit, S.; Jers, C.; Joshi, A.S.; Garnæs, J.; Mijakovic, I. Silver nanoparticles produced from Cedecea sp. exhibit antibiofilm activity and remarkable stability. *Sci. Rep.* **2021**, *11*, 12619. [CrossRef] [PubMed]
156. Bhattacharjee, G.; Gohil, J.; Gohil, N.; Chaudhari, H.; Gangapuram, B.; Khambhati, K.; Maurya, R.; Alzahrani, K.J.; Ramakrishna, S.; Singh, V. Biosynthesis and characterization of Serratia marcescens derived silver nanoparticles: Investigating its antibacterial, antibiofilm potency and molecular docking analysis with biofilm-associated proteins. *J. Mol. Liq.* **2022**, *365*, 120094. [CrossRef]
157. Lavaee, F.; Motamedifar, M.; Rafiee, G. The effect of photodynamic therapy by gold nanoparticles on Streptococcus mutans and biofilm formation: An in vitro study. *Lasers Med. Sci.* **2022**, *37*, 1717–1725. [CrossRef] [PubMed]
158. Selem, E.; Mekky, A.F.; Hassanein, W.A.; Reda, F.M.; Selim, Y.A. Antibacterial and antibiofilm effects of silver nanoparticles against the uropathogen Escherichia coli U12. *Saudi J. Biol. Sci.* **2022**, *29*, 103457. [CrossRef]
159. Rajivgandhi, G.N.; Ramachandran, G.; Maruthupandy, M.; Manoharan, N.; Alharbi, N.S.; Kadaikunnan, S.; Khaled, J.M.; Almanaa, T.N.; Li, W.-J. Anti-oxidant, anti-bacterial and antibiofilm activity of biosynthesized silver nanoparticles using Gracilaria corticata against biofilm producing K. pneumoniae. *Colloids Surf. A* **2020**, *600*, 124830. [CrossRef]
160. Aksoy, İ.; Küçükkeçeci, H.; Sevgi, F.; Metin, Ö.; Hatay Patir, I. Photothermal Antibacterial and Antibiofilm Activity of Black Phosphorus/Gold Nanocomposites against Pathogenic Bacteria. *ACS Appl. Mater. Interfaces* **2020**, *12*, 26822–26831. [CrossRef] [PubMed]

161. Govindappa, M.; Tejashree, S.; Thanuja, V.; Hemashekhar, B.; Srinivas, C.; Nasif, O.; Pugazhendhi, A.; Raghavendra, V.B. Pomegranate fruit fleshy pericarp mediated silver nanoparticles possessing antimicrobial, antibiofilm formation, antioxidant, biocompatibility and anticancer activity. *J. Drug Deliv. Sci. Technol.* **2021**, *61*, 102289. [CrossRef]
162. Huang, Y.; Bai, L.; Yang, Y.; Yin, Z.; Guo, B. Biodegradable gelatin/silver nanoparticle composite cryogel with excellent antibacterial and antibiofilm activity and hemostasis for Pseudomonas aeruginosa-infected burn wound healing. *J. Colloid Interface Sci.* **2022**, *608*, 2278–2289. [CrossRef]
163. Mostafa, E.M.; Abdelgawad, M.A.; Musa, A.; Alotaibi, N.H.; Elkomy, M.H.; Ghoneim, M.M.; Badawy, M.S.E.M.; Taha, M.N.; Hassan, H.M.; Hamed, A.A. Chitosan Silver and Gold Nanoparticle Formation Using Endophytic Fungi as Powerful Antimicrobial and Antibiofilm Potentialities. *Antibiotics* **2022**, *11*, 668. [CrossRef]
164. Rini, P.; Sugiharto; Agusniar Furkani, L.; Masfufatun; Lusiani, T.; Noer Kumala, I. Antibacterial and antibiofilm effect of silver and gold nanoparticles in Uropathogenic *Escherichia coli*. *Berk. Penelit. Hayati* **2021**, *27*, 67–72. [CrossRef]
165. Neihaya, H.Z.; Zaman, H.H. Investigating the effect of biosynthesized silver nanoparticles as antibiofilm on bacterial clinical isolates. *Microb. Pathog.* **2018**, *116*, 200–208. [CrossRef]
166. Hetta, H.F.; Al-Kadmy, I.M.S.; Khazaal, S.S.; Abbas, S.; Suhail, A.; El-Mokhtar, M.A.; Ellah, N.H.A.; Ahmed, E.A.; Abd-ellatief, R.B.; El-Masry, E.A.; et al. Antibiofilm and antivirulence potential of silver nanoparticles against multidrug-resistant *Acinetobacter baumannii*. *Sci. Rep.* **2021**, *11*, 10751. [CrossRef]
167. Ghar, S.B.; Das, Y.K. The green & eco-friendly biosynthesized biocompatible metallic silver nanoparticles for anti-bacterial and antibiofilm efficacy. *Eur. J. Biotechnol. Biosci.* **2022**, *10*, 43–49.
168. Majumdar, M.; Biswas, S.C.; Choudhury, R.; Upadhyay, P.; Adhikary, A.; Roy, D.N.; Misra, T.K. Synthesis of Gold Nanoparticles Using Citrus macroptera Fruit Extract: Antibiofilm and Anticancer Activity. *ChemistrySelect* **2019**, *4*, 5714–5723. [CrossRef]
169. Khan, F.; Lee, J.W.; Manivasagan, P.; Pham, D.T.N.; Oh, J.; Kim, Y.M. Synthesis and characterization of chitosan oligosaccharide-capped gold nanoparticles as an effective antibiofilm drug against the *Pseudomonas aeruginosa* PAO1. *Microb. Pathog.* **2019**, *135*, 103623. [CrossRef]
170. Allemailem, K.S.; Khadri, H.; Azam, M.; Khan, M.A.; Rahmani, A.H.; Alrumaihi, F.; Khateef, R.; Ansari, M.A.; Alatawi, E.A.; Alsugoor, M.H.; et al. Ajwa-Dates (Phoenix dactylifera)-Mediated Synthesis of Silver Nanoparticles and Their Anti-Bacterial, Antibiofilm, and Cytotoxic Potential. *Appl. Sci.* **2022**, *12*, 4537. [CrossRef]
171. Hussain, A.; Alajmi, M.F.; Khan, M.A.; Pervez, S.A.; Ahmed, F.; Amir, S.; Husain, F.M.; Khan, M.S.; Shaik, G.M.; Hassan, I.; et al. Biosynthesized Silver Nanoparticle (AgNP) From Pandanus odorifer Leaf Extract Exhibits Anti-metastasis and Antibiofilm Potentials. *Front. Microbiol.* **2019**, *10*, 8. [CrossRef]
172. Qais, F.A.; Ahmad, I.; Altaf, M.; Manoharadas, S.; Al-Rayes, B.F.; Ali Abuhasil, M.S.; Almaroai, Y.A. Biofabricated silver nanoparticles exhibit broad-spectrum antibiofilm and antiquorum sensing activity against Gram-negative bacteria. *RSC Adv.* **2021**, *11*, 13700–13710. [CrossRef]
173. Khan, F.; Park, S.K.; Bamunuarachchi, N.I.; Oh, D.; Kim, Y.M. Caffeine-loaded gold nanoparticles: Antibiofilm and anti-persister activities against pathogenic bacteria. *Appl. Microbiol. Biotechnol.* **2021**, *105*, 3717–3731. [CrossRef]
174. Qais, F.A.; Ahmad, I.; Altaf, M.; Alotaibi, S.H. Biofabrication of Gold Nanoparticles Using Capsicum annuum Extract and Its Antiquorum Sensing and Antibiofilm Activity against Bacterial Pathogens. *ACS Omega* **2021**, *6*, 16670–16682. [CrossRef]
175. El-Telbany, M.; El-Sharaki, A. Antibacterial and antibiofilm activity of silver nanoparticles on multi-drug resistance pseudomonas aeruginosa isolated from dental-implant. *J. Oral Biol. Craniofac. Res.* **2022**, *12*, 199–203. [CrossRef] [PubMed]
176. Fulaz, S.; Vitale, S.; Quinn, L.; Casey, E. Nanoparticle-Biofilm Interactions: The Role of the EPS Matrix. *Trends Microbiol.* **2019**, *27*, 915–926. [CrossRef] [PubMed]
177. Estevez, M.B.; Raffaelli, S.; Mitchell, S.G.; Faccio, R.; Alborés, S. Biofilm Eradication Using Biogenic Silver Nanoparticles. *Molecules* **2020**, *25*, 2023. [CrossRef] [PubMed]
178. Shkodenko, L.; Kassirov, I.; Koshel, E. Metal Oxide Nanoparticles Against Bacterial Biofilms: Perspectives and Limitations. *Microorganisms* **2020**, *8*, 1545. [CrossRef]
179. Sokary, R.; Abu el-naga, M.N.; Bekhit, M.; Atta, S. A potential antibiofilm, antimicrobial and anticancer activities of chitosan capped gold nanoparticles prepared by γ–irradiation. *Mater. Technol.* **2022**, *37*, 493–502. [CrossRef]
180. Wong, C.W.; Chan, Y.S.; Jeevanandam, J.; Pal, K.; Bechelany, M.; Abd Elkodous, M.; El-Sayyad, G.S. Response Surface Methodology Optimization of Mono-dispersed MgO Nanoparticles Fabricated by Ultrasonic-Assisted Sol–Gel Method for Outstanding Antimicrobial and Antibiofilm Activities. *J. Clust. Sci.* **2020**, *31*, 367–389. [CrossRef]
181. Hussein, N.; Khadum, M.M. Evaluation of the Biosynthesized Silver Nanoparticles" Effects on Biofilm Formation. *J. Appl. Sci. Nanotechnol.* **2021**, *1*, 23–31. [CrossRef]
182. Dridi, R.; Essghaier, B.; Hannachi, H.; Khedher, G.B.; Chaffei, C.; Zid, M.F. Biosynthesized silver nanoparticles using Anagallis monelli: Evaluation of antioxidant activity, antibacterial and antifungal effects. *J. Mol. Struct.* **2022**, *1251*, 132076. [CrossRef]

Disclaimer/Publisher's Note: The statements, opinions and data contained in all publications are solely those of the individual author(s) and contributor(s) and not of MDPI and/or the editor(s). MDPI and/or the editor(s) disclaim responsibility for any injury to people or property resulting from any ideas, methods, instructions or products referred to in the content.

antibiotics

MDPI

Article

Antimicrobial Properties of Newly Developed Silver-Enriched Red Onion–Polymer Composites

Judita Pušo [1,*], Jonas Žvirgždas [2], Algimantas Paškevičius [2], Shirin Arslonova [3] and Diana Adlienė [1,*]

1 Department of Physics, Kaunas University of Technology, Studentų Str. 50, LT-51368 Kaunas, Lithuania
2 Laboratory of Biodeterioration Research, Institute of Botany, Nature Research Centre, Akademijos Str. 2, LT-08412 Vilnius, Lithuania; jonas.zvirgzdas@gamtc.lt (J.Ž.); algimantas.paskevicius@gamtc.lt (A.P.)
3 Tashkent City Branch of Republican Specialized Scientific—Practical Medical Centre of Oncology and Radiology, Boguston Str. 1, Tashkent P.O. Box 100070, Uzbekistan; shirinhon@gmail.com
* Correspondence: judita.puiso@ktu.lt (J.P.); diana.adliene@ktu.lt (D.A.); Tel.: +370-610-04-812 (J.P.); +370-612-08-716 (D.A.)

Abstract: Simple low-cost, nontoxic, environmentally friendly plant-extract-based polymer films play an important role in their application in medicine, the food industry, and agriculture. The addition of silver nanoparticles to the composition of these films enhances their antimicrobial capabilities and makes them suitable for the treatment and prevention of infections. In this study, polymer-based gels and films (AgRonPVA) containing silver nanoparticles (AgNPs) were produced at room temperature from fresh red onion peel extract ("Ron"), silver nitrate, and polyvinyl alcohol (PVA). Silver nanoparticles were synthesized directly in a polymer matrix, which was irradiated by UV light. The presence of nanoparticles was approved by analyzing characteristic local surface plasmon resonance peaks occurring in UV-Vis absorbance spectra of irradiated experimental samples. The proof of evidence was supported by the results of XRD and EDX measurements. The diffusion-based method was applied to investigate the antimicrobial activity of several types of microbes located in the environment of the produced samples. Bacteria *Staphylococcus aureus* ATCC 29213, *Acinetobacter baumannii* ATCC BAA 747, and *Pseudomonas aeruginosa* ATCC 15442; yeasts *Candida parapsilosis* CBS 8836 and *Candida albicans* ATCC 90028; and microscopic fungi assays *Aspergillus flavus* BTL G-33 and *Aspergillus fumigatus* BTL G-38 were used in this investigation. The greatest effect was observed on Staphylococcus aureus, Acinetobacter baumannii, and Pseudomonas aeruginosa bacteria, defining these films as potential candidates for antimicrobial applications. The antimicrobial features of the films were less effective against fungi and the weakest against yeasts.

Keywords: red onion; peel; polyvinyl alcohol; gels; films; silver; nanoparticles

Citation: Puišo, J.; Žvirgždas, J.; Paškevičius, A.; Arslonova, S.; Adlienė, D. Antimicrobial Properties of Newly Developed Silver-Enriched Red Onion–Polymer Composites. *Antibiotics* **2024**, *13*, 441. https://doi.org/10.3390/antibiotics13050441

Academic Editors: Marc Maresca, Sotiris K Hadjikakou and Christina N. Banti

Received: 12 March 2024
Revised: 6 May 2024
Accepted: 10 May 2024
Published: 14 May 2024

Copyright: © 2024 by the authors. Licensee MDPI, Basel, Switzerland. This article is an open access article distributed under the terms and conditions of the Creative Commons Attribution (CC BY) license (https://creativecommons.org/licenses/by/4.0/).

1. Introduction

Scientific research on the antibacterial properties of plant materials was started in the second half of the 19th century when Louis Pasteur noted the first antibacterial properties of garlic in 1858 [1]. Since that time, a vast number of different studies have been performed investigating the antimicrobial properties of plants. It was shown [2,3], that one of the oldest cultivated plants Allium cepa L. (Liliaceae) or onion indicates antimicrobial properties and can be used for the treatment of many diseases. Antibacterial and antifungal properties of onion and onion extracts were discussed in several papers [4–8]. It was found that onion peel has 3–5 times higher amounts of isolated phenolic compounds and quercetin and provides approximately 3–5 times higher activity against bacteria than the edible part of the onion [9]. Investigation of antibacterial assays for the isolated compounds containing water extract of onion peel (Allium cepa) has shown that 2-(3,4-dihydroxy phenyl)-4,6-dihydroxy-2-methoxybenzofuran-3-one indicated selective activity against Helicobacter pylori strains, and 3-(quercetin-8-yl)-2,3-epoxy flavanone was effective against MRSA (multidrug-resistant *Staphylococcus aureus*) and *H. pylori* strains [8]. High antimicrobial activity of red onion peel extracts against bacteria *Escherichia coli*, *Pseudomonas fluorescens*, and *Bacillus cereus* and fungi *Aspergillus niger*, *Trichoderma viride*, and *Penicillium cyclopium* was found as well [9]. This indicated the potential of onion peels to be applied in antimicrobial compositions. On the other hand, the utilization of red onion peel as an antimicrobial substance may contribute to the reduction in the problem related to the accumulation of a high amount of waste produced due to the increased market of processed onion [10–13].

It is known [14] that some metal-based nanoparticles possess antimicrobial activity as well. Among different metallic NPs, silver nanoparticles (AgNPs) play an enormous role as antimicrobial agents [15]. However, there are some issues with the fabrication of AgNPs. The synthesis of AgNPs can be achieved both chemically and physically. Both approaches include drawbacks: physical operations usually fail to regulate particle sizes in the nanoscale range and irregularly sized particles are created. Chemical synthesis of NPs considers the application of toxic materials and requires high energy resources. Despite the indicated drawbacks, the antimicrobial activity of silver nanoparticles is amazing and attracts the huge attention of researchers. Due to this, antimicrobial properties of composites containing silver nanoparticles have been thoroughly investigated during the last decades [16–18]. Analyzed research information revealed that there is a possibility to enhance the antimicrobial activity of compounds containing AgNPs, exploring environmentally friendly green synthesis methods for gaining silver nanoparticles [19–21]. The green synthesis method has gained a lot of attraction since metallic NPs can be synthesized biologically using various plants and their extracts which are easily available in huge quantities. The plants and their extracts are safe to handle, less toxic, and eco-friendly. Created nanoparticles are biocompatible and efficient against microbial intrusion. It should also be noted that plant extracts act as NP reductants and stabilizers. Several studies discussed the application of onion peel for green synthesis of silver nanoparticles [22,23].

Taking into account that polymer films originally containing AgNPs may be of advantage for their antimicrobial applications, the use of red onion peel extract for green synthesis of silver nanoparticles in polymer films directly was introduced in our previous work [24]. Gelatine (natural polymer) as a matrix for AgNPs has been considered due to its biodegradability, biocompatibility, and unique biological properties. However, gelatine samples exhibited poor mechanical and adhesive properties and were temperature-sensitive, thus limiting their end-use applications [25–34]. To overcome this problem, PVA, a non-toxic, biodegradable, flexible synthetic polymer possessing characteristics of hydrophilic nature, was suggested as a substitute for gelatine [29,35]. PVA has a good film formation capability, solid conglutination, and excellent thermal stability [36–38]. It indicates good forming and manufacturing features and is already adopted for soft tissue replacement (in its hydrogel form) [38]. It is noticed that PVA itself and also gelatin do not possess any antioxidant and antimicrobial properties and do not impact the antimicrobial properties of bioactive materials [39–45].

Taking into account the results of the conducted literature analysis, our study is aimed at the fabrication and characterization of red onion peel–PVA compositions containing silver nanoparticles and assessment of their antimicrobial efficacy when interacting with different types of microorganisms (bacteria, yeast, fungi).

2. Results and Discussion

2.1. Optical Properties of Red Onion Peel–PVA Gels Containing Green Synthesized Silver Nanoparticles

Red onion extract was prepared following the procedure described in our previous paper [24]. The composition of the prepared samples is indicated in the Materials and Methods chapter. UV-Vis absorbance spectra were used for the identification of the optical properties of different material compositions (Figure 1). Dark-red-colored aqueous red onion peel extract "Ron" indicated a broad low-intensity peak with a maximum of 538 nm. The addition of a small amount of Ron extract to a colorless PVA–water solution resulted in a color change to light pink confirming the formation of a RonPVA hydrogel with an almost negligible UV-Vis peak at 562 nm due to the high dilution of the Ron extract. The concentration of the red onion extract was chosen so that the UV-Vis absorbance peaks corresponding to red onion anthocyanins did not obscure the local surface resonance peaks that indicate the presence of silver nanoparticles in the gels [24]. The addition and dissolving of silver nitrate in the RonPVA hydrogel solution under continuous stirring has not introduced significant color changes, and the formed AgRonPVA gel demonstrated almost the same transparent color, but the observed UV-Vis absorbance peak at 447 nm was shifted towards a lower wavelength (Figure 1).

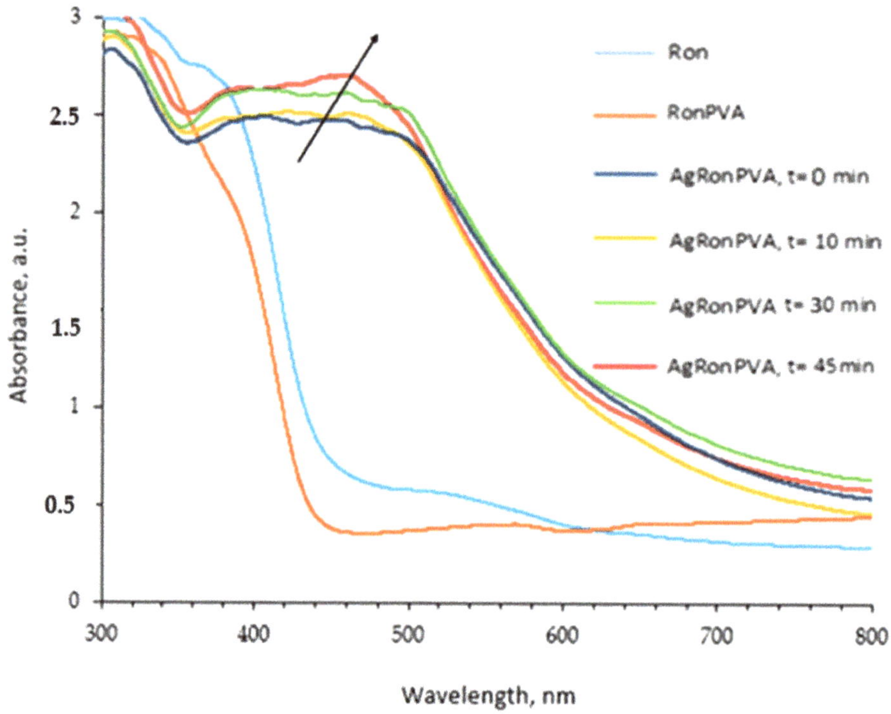

Figure 1. UV-Vis spectra of experimental samples: red onion extract (Ron); RonPVA hydrogel; and AgRonPVA gel after UV exposure for 0 min, 10 min, 30 min, and 45 min.

A possible green-synthesis mechanism was provided by H. Yang et al. [46]. Based on the findings provided in the mentioned article, it was suggested that red onion peel extract can contribute to the green synthesis of nano seeds/nanoparticles and can be used to reduce Ag^+ and form $Ag/Ag_2O/AgO$ nanocomposites. It should be noted that the strong oxidation property of nanocomposites can be better addressed by providing an ion-associated form of the composite: $Ag/Ag^+/Ag^{3+}$. Our suggestion was based on the fact that red onion peel comprises different molecular substances. Proteins and polyphenols and others containing strong reducible hydroxyl groups are among them. The hydroxyl group can reduce silver ions into silver. On the other hand, biological macromolecules containing amino and carbonyl groups exhibit a strong complexation for silver and silver ions, which can be wrapped on the surface of AgNPs providing the effect of dispersion and protection.

The free energy (surface tension) on the surface of nanoparticles, especially when newly formed, is very high, so the nanoparticles are unstable and can easily agglomerate into larger particles. In our case, silver nanoparticles are wrapped by biological macromolecules of red onion peel extract, resulting in reduced surface free energy of the particles and avoiding to some extent particle aggregation and their growth. It should be noticed that in the case of the same volume, the lowest energy of the system is provided when spherical particles are present because spherical-shaped particles occupy the smallest surface area [46].

Based on the information provided above, we assumed that red onion peel extract contributed to the formation of spheric AgNPs as silver composites containing silver (Ag) and silver oxides, $Ag_2O(Ag^+)$ and $AgO(Ag^{3+})$, after the addition of $AgNO_3$ to the peel extract, as it can be proven by the appearance of the LSPR peak at 447 nm of the UV-Vis spectrum (Figure 1).

However, it should be noted that the influence of the PVA matrix for accommodation of AgNPs is not fully disclosed. Also, the mechanism of dispersion and stabilization of AgNPs coated by biological macromolecules needs further development [47–49], especially when speaking about two opposite processes: formation and preservation of AgNP growth by plant extracts and initiation of AgNP growth by photoreduction (UV exposure).

$$Ag^+ + \text{Phytochemicals of red onion} \overset{h\gamma}{\rightarrow} Ag^0(\text{Ag NPs with } Ag_xO_y \text{ coating}) \quad (1)$$

Polymer hydrogels containing $AgNO_3$ were exposed to UV light for 10 min, 30 min, and 45 min. UV-Vis absorbance spectra of differently exposed samples were used to follow up the formation of AgNPs (Figure 1).

It was found that UV-exposed gel samples obtained a reddish-brown color. The appearance of a broad UV-Vis absorbance peak with the maximum intensity at 448 nm was observed after gel exposure for 10 min. The intensity of the peak was slightly increasing with the prolonged duration of UV exposure, and the peak maximum position itself was slightly shifted to a longer wavelength (462 nm) after gel exposure for 45 min. No significant color changes of exposed gels were observed. The observed peaks were identified as local surface plasmon resonance (LSPR) peaks that were characteristic for metal particles introduced in a polymer matrix. It was suggested that the formation of silver nanoparticles (AgNPs) starts from silver seeds which can be created when silver ions bind hydrated electrons and reactive water radicals produced as a consequence of UV exposure or interact with fragments of the red onion since plant extracts are known as reductants and stabilizers for nanoparticles [19]. Created silver seeds grew to silver nanoparticles by binding silver ions on single seeds. The growth of silver nanoparticles is reflected by the broadening of the LSPR peak with a shift towards longer wavelengths. Based on the analysis of the obtained UV-Vis spectra it was assumed that in this experiment, silver nanoparticles of approximately the same size, but in different amounts, were formed depending on the duration of the exposure, rather than growing in size.

To identify the influence of formed AgNPs on the optical properties of newly composed hydrogels, different amounts of AgRonPVA gels exposed to UV for 45 min were diluted in distilled water, and the UV-Vis absorbance spectra of these solutions were investigated (Figure 2). It was found that the intensity of the absorption peak was decreasing depending on the amount of AgRonPVA in distilled water. Dilution of the AgRonPVA gel led to the appearance of well-expressed LSPR peaks. The maximum of the LSPR peak of diluted gels remained at the same initial position of (462) nm.

It is known [50,51] that the increasing number of formed silver nanoparticles may change the dielectric properties of gels, causing overlapping of SPR peaks from individual nanoparticles and initiating the shift of LSPR peak to longer wavelengths. Results obtained in this study supported the hypothesis that synthesized spherical AgNPs are accommodated in the cells of the RonPVA polymer networks. Since AgNPs do not aggregate into local metallic silver derivatives in distilled water (dielectric), they can be distributed in the AgRonPVA water solution without the original size change.

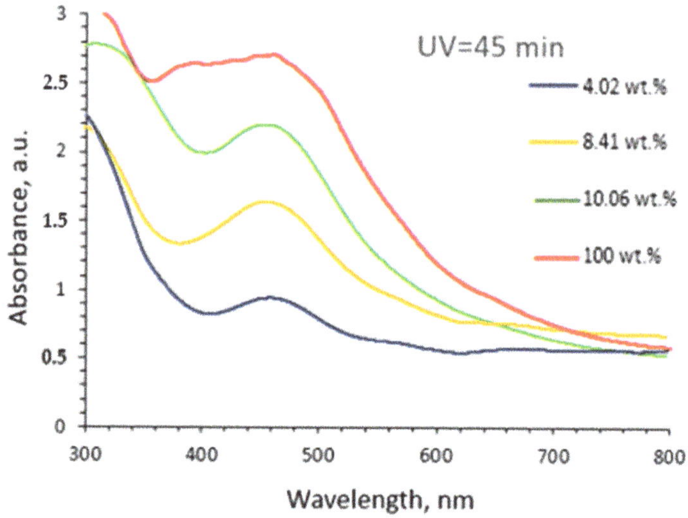

Figure 2. UV-Vis spectra of the diluted in distilled water AgRonPVA gels: 4.02 wt.%; 8.41 wt.%; 10.06 wt.%; 100 wt.% (not diluted AgRonPVA gel).

2.2. Optical Properties of Red Onion Peel–PVA Films Containing Silver Nanoparticles

Antibacterial films containing AgNPs can be fabricated following preset requirements: (1) the avoidance of AgNP agglomeration and formation of metallic silver; (2) preservation of antimicrobial properties of the AgRonPVA gels during the film formation procedure. Also, ~10% loss in mass during the drying process of gels (formation of films) was taken into account. Photographs of the UV-exposed AgRonPVA films are provided in Figure 3 alongside a photograph of the colorless RonPVA film (without AgNPs) which is shown for comparison.

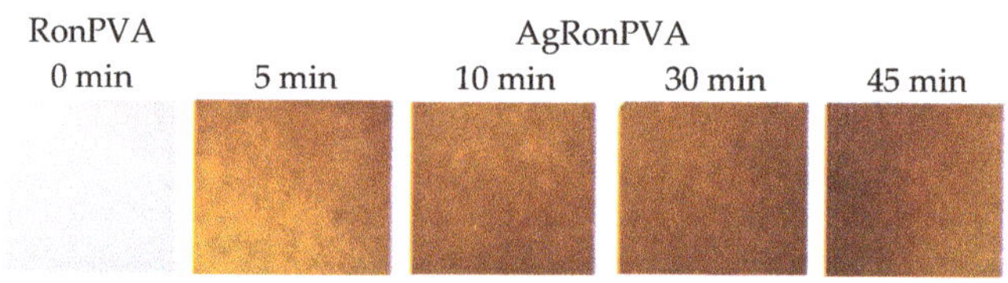

Figure 3. Photographs of fabricated experimental films.

Due to the loss of some water amount by drying of exposed gels, the UV-Vis spectra of films (Figure 4) indicated broadening of the LSPR peaks that were better pronounced and slightly shifted towards longer wavelength as compared to those of the gels shown in (Figure 1). LSPR peak locations of the differently exposed AgRonPVA films were found by eliminating baseline and fitting: locations of LSPR peak maximum at 486 nm and 492 nm were found for 5 min UV exposed film and 45 min exposed film, respectively. A weak LSPR peak with a maximum of 507 nm was found for the RonPVA films. It was slightly shifted towards a lower wavelength as compared to the RonPVA gels indicating the contribution of anthocyanins from red onion peel extract that appeared due to the loss of water during the drying process and were responsible for color changes of the films [24,52]. Taking into account the appearance of only one single LSPR peak and its location in the UV-Vis spectra, it is assumed that only spherical or near-spherical particles were synthesized. In the case of anisotropic particles, two or three LSPR peaks would be seen by their shapes [53]. As the AgRonPVA gel dries, water evaporates from the samples, changing the environment of the silver nanoparticles to an optically denser one (polymer network), limiting particles' mobility in the volume and reducing agglomeration possibility. This results in an increase in the SPR peak intensity as compared to the LSPR peak measured in the gels (Figure 2).

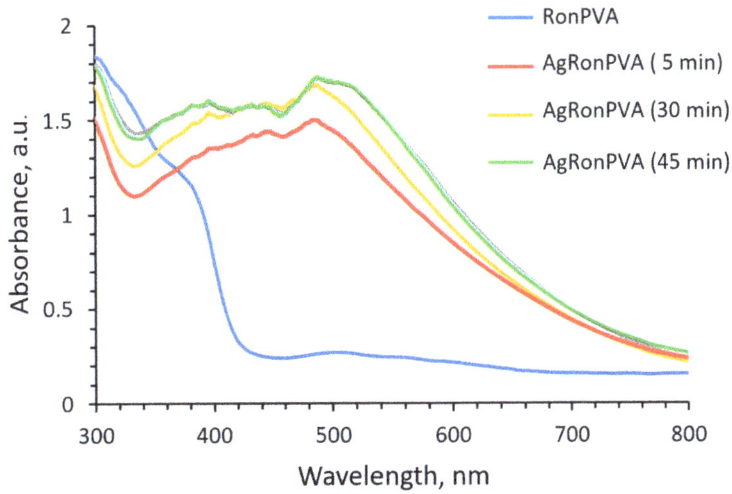

Figure 4. UV-Vis spectra AgRonPVA films obtained from 100 wt.% AgRonPVA gels exposed to UV.

It is assumed that during drying of the film, the water amount in gels will be reduced, which leads to the changes in the polymer matrix. Since the matrix network becomes denser, the growth in the AgNPs from the seeds is restricted. Possible growth and aggregation of silver nanoparticles were simulated using open access "MiePlot" software, version 4.6 [54]. The algorithm of this software is based on calculations of Mie scattering from a sphere (scattering of light by Ag NPs in our case). "MiePlot" software, version 4.6 allows us to calculate scattering efficiency (Q_{ext}—extinction; Q_{abs}—absorption; Q_s—scattering) as a function of wavelength. Light absorption dominates in the extinction (extinction = absorption + scattering) spectrum for particles with a relatively small (radius <20 nm), and light scattering becomes the dominant process for larger particles. In our calculations, we used PVA as a medium for the dispersion of nanoparticles having a diameter from 10 to 90 nm with a 5% standard deviation in a log-normal distribution. Q_{ext} as a function of wavelength is presented in Figure 5.

The performed simulation revealed that only small particles (<40 nm) are indicated by strong well-defined peaks. Increasing the particle size leads to a decrease in the SPR peak intensity, peak broadening, and shift towards a longer wavelength. A comparison of simulation and experimental results leads to the conclusion that >40 nm sized AgNPs were mainly formed in our experiment. However, performed modeling has not accounted for the water content in the films and also the presence of the red onion peel extract, which may affect the SPR peak position.

It is known that the crystalline structure of polymer composites can be estimated from the XRD patterns. The XRD pattern of the 45 min UV-exposed AgRonPVA film is presented in Figure 6.

The XRD peaks corresponding to silver nanoparticles located in the PVA matrix are of a lower intensity in comparison with AgNPs produced in gels, because Ag nanoparticles are integrated into the PVA network [55]. Silver nanocrystals are usually characterized by a face-centered cubic (fcc) structure with a dominating orientation of crystallites along the (111) direction (38.45°). The XRD peaks of lower intensity are seen at 42.30° and 60.70° corresponding to (200) and (220) crystalline planes of the face-centered cubic crystalline structure of silver [JCPDS No. 04-0783, JCPDS file no. 84-0713]. An additional broad peak at 30.6° and distinct peaks at 22° and 41.8° were assigned to corresponding PVA planes

(110) and (200), as was indicated in [55–59]. A Ag crystallite size of 53.96 nm in AgRonPVA films was determined using the Scherrer method [60–62].

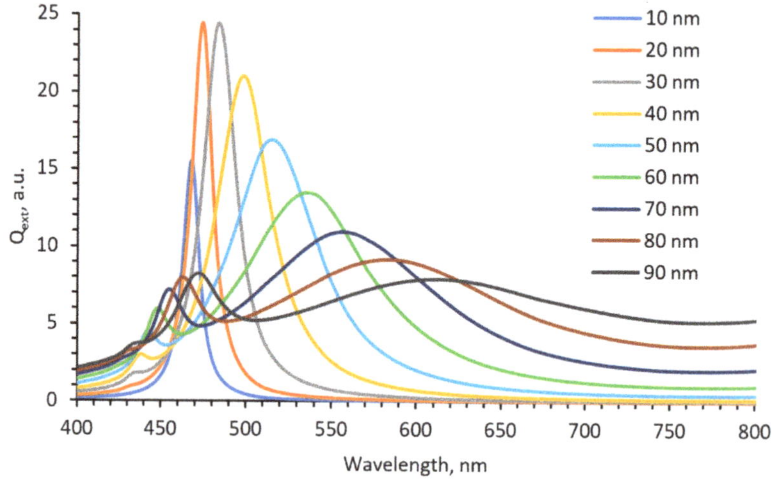

Figure 5. Simulated UV-Vis spectra for differently sized spherical silver nanoparticles in PVA.

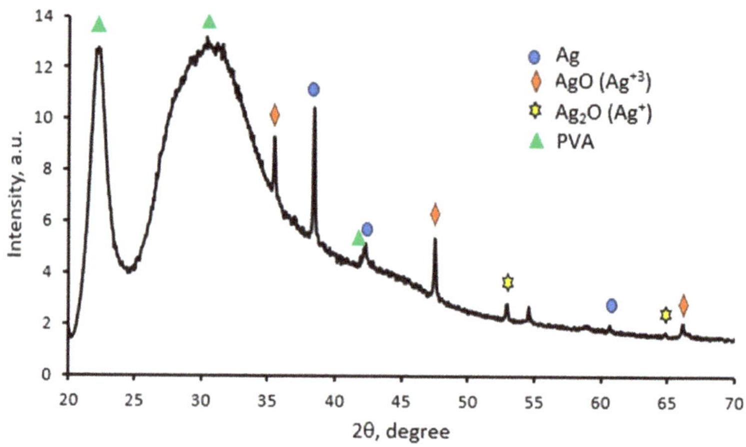

Figure 6. XRD pattern red onion PVA films containing silver nanoparticles (AgRonPVA45).

H.Yang et al. [46] indicated that silver derivatives, Ag, Ag_2O, and AgO, may be registered, suggesting their green synthesis in the bio-organic phase and crystallization on the surface of Ag nanoparticles. This explains the presence of the additional diffraction peaks observed in our XRD pattern (Figure 6): AgO(Ag^{3+}) (JCPDS, file No. 84-1108) and Ag_2O(Ag^+) (JCPDS, file No. 42-0874).

Analysis of SEM images and results of EDX evaluation of unexposed (as prepared) and UV-exposed AgRonPVA films revealed that the observation of AgNPs was very problematic, due to the low amount of seeds/particles in the film and their embedding in the network of the polymer matrix. Due to the very small amount of silver, it was not detected in as-prepared AgRonPVA films. An example of the obtained results of differently UV-exposed AgRonPVA films is provided in Figure 7.

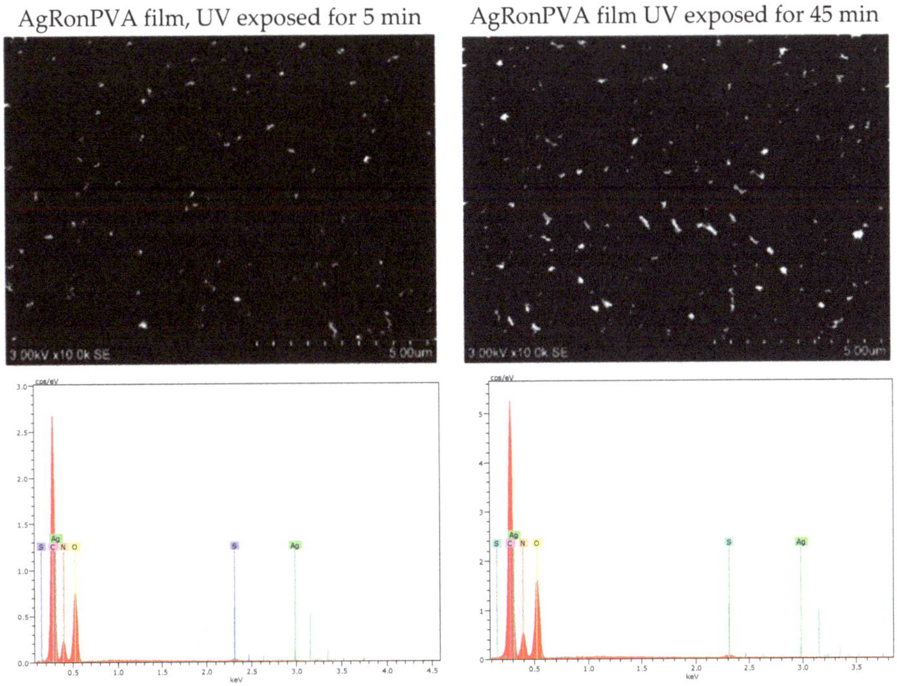

Spectrum: (AgRonPVA film, UV exposed for 5 min)

El	AN	Series	Net unn. [wt.%]	C norm. [wt.%]	C Atom. [at.%]	C Error (1 Sigma) [wt.%]
C	6	K-series 23447	48.26	48.26	54.41	6.11
O	8	K-series 8388	32.38	32.38	27.40	4.68
N	7	K-series 2513	18.52	18.52	17.91	3.36
S	16	K-series 380	0.59	0.59	0.25	0.07
Ag	47	L-series 64	0.24	0.24	0.03	0.06
		Total:	100.00	100.00	100.00	

Spectrum: (AgRonPVA film UV exposed for 45 min)

El	AN	Series	Net unn. [wt.%]	C norm. [wt.%]	C Atom. [at.%]	C Error (1 Sigma) [wt.%]
C	6	K-series 33012	47.77	47.77	54.02	5.85
O	8	K-series 11853	32.18	32.18	27.32	4.43
N	7	K-series 3679	18.89	18.88	18.31	3.16
S	16	K-series 608	0.71	0.71	0.30	0.07
Ag	47	L-series 176	0.46	0.46	0.06	0.07
		Total:	100.00	100.00	100.00	

Figure 7. SEM images and EDX analysis results.

It was found that the amount of silver in as-prepared and 5 min UV-exposed films was low—0.24 wt.%—indicating the possible contribution of red onion peel extract to green synthesis of nano seeds. The registered amount was slightly increasing with the increased duration of UV exposure and was 0.29 wt.% after film UV exposure for 10 min, 0.39 wt.% after 30 min, and 0.46 wt.% after 45 min.

2.3. Antimicrobial Activity of the AgRonPVA Films Containing Silver Nanoparticles

Bacteria *Staphylococcus aureus* ATCC 29213, *Acinetobacter baumannii* ATCC BAA 747, and *Pseudomonas aeruginosa* ATCC 15442; yeasts *Candida parapsilosis* CBS 8836 and *Candida albicans* ATCC 90028; and microscopic fungi *Aspergillus flavus* BTL G-33 and *Aspergillus fumigatus* BTL G-38 were used for the assessment of the antimicrobial activity of experimental AgRonPVA films. The agar diffusion method was used for antimicrobial activity testing of red onion skin extract films containing silver nanoparticles. This was realized by the uniform spreading of the prepared microorganism suspensions (see Section 3) with a swab on Mueller–Hinton agar (bacteria) or on Sabouraud dextrose agar (yeasts and microscopic fungi) in Petri dishes. The pieces of AgRonPVA films (5 mm × 5 mm) containing silver nanoparticles were placed on the surface of each media as the microbial suspension was absorbed into it.

It is to point out that red onion peel extract itself (the main bioactive agents are phenolic compounds) or in combination with silver nanoparticles provides inhibition properties against different bacteria as was shown by several authors: *Streptococcus agalactiae* (Gram-positive), *Pseudomonas aeruginosa* (Gram-negative), *Salmonella typhimurium* (Gram-negative), and *Staphylococcus aureus* (Gram-positive) [63]. The antimicrobial compounds made of red onion peel have various mechanisms to inhibit bacteria: inhibition of phagocytosis in macrophages; increasing the production of Interleukin-12 (IL-12) by quercetin and thus increasing the phagocytic ability of macrophages; altering the membrane potential of bacterial cells and disrupting their performance; and binding to bacteria pili and inhibiting bacterial adhesion [64]. The red onion skin extract also indicates antifungal properties against some fungi, however to a lesser extent and with lower effectiveness as compared with bacteria [9,65,66].

Three antibacterial mechanisms are known when discussing the antimicrobial properties of silver nanoparticles [18,67,68]; however, it should be noted that the chemical size, charge, and surface structure of silver nanoparticles influence their antibacterial capacity as well [69].

The first mechanism proposes that silver nanoparticles might penetrate the outer membrane accumulating in the inner membrane, where the adhesion of the nanoparticles to the cell generates their destabilization and damage, increasing membrane permeability and inducing leakage of cellular content and subsequently its death [70].

The second mechanism proposes that nanoparticles not only can break and cross the cell membrane altering its structure and permeability but can also enter the cell where silver nanoparticles may interact with sulfur or phosphorus groups, present in intracellular content such as DNA and proteins, altering their structure and functions [70]. Silver nanoparticles also may alter the respiratory chain in the inner membrane by interacting with thiol groups in the enzymes inducing reactive oxygen species and free radicals, generating damage to intracellular machinery and activating the apoptosis pathway.

The third mechanism may be present in parallel with two others and accounts for the release of silver ions from the nanoparticles, which due to their size and charge can interact with cellular components altering metabolic pathways, membranes, and even genetic material [71–76].

Based on the results obtained investigating experimental AgRonPVA films containing composite-type silver NPS, it was assumed that the third mechanism of the antimicrobial activity was responsible for the inhibition of different microorganisms.

The results of the performed study have shown (Table 1 and Figure 8) that all differently exposed (duration of exposure was responsible for the formation of a certain number of AgNPs) AgRonPVA films indicated antimicrobial activity against applied bacteria, yeasts, and fungi. The greatest effect was observed on *Staphylococcus aureus*, *Acinetobacter baumannii*, and *Pseudomonas aeruginosa* bacteria. The diameter of the inhibition zones of the tested bacteria ranged from 20.0 mm (UV 5 min) to 18.7 mm (UV 45 min) for *Staphylococcus aureus*, from 18.3 mm (UV 5 min) to 18.0 mm (UV 45 min) for *Acinetobacter baumannii*, and from 21.0 mm (UV 5 min) to 19.7 mm (UV 45 min) for *Pseudomonas aeruginosa* and indicated decreasing tendency with the increasing duration of UV exposure. The antimicrobial features of the AgRonPVA films were less effective against fungi, where the diameter of inhibited zones varied between 13.0 mm (UV 5 min) and 12.6 mm (UV 45 min) for *Aspergillus flavus*, and between 16.0 mm (UV 5 min) and 11.7 mm (UV 45 min) for *Aspergillus fumigatus* which indicated the highest sensitivity to the film exposure duration. The antimicrobial effect of the films on yeasts (*Candida parapsilosis* and *Candida albicans*) was the weakest.

Obtained results are in line with the findings of Pereira et al. [77], who showed that a combination of plant extracts, possessing antimicrobial properties and polymers containing AgNPs, leads to an enhanced antimicrobial effect compared with these compounds alone, thus indicating conjugates as promising candidates for biocidal treatments.

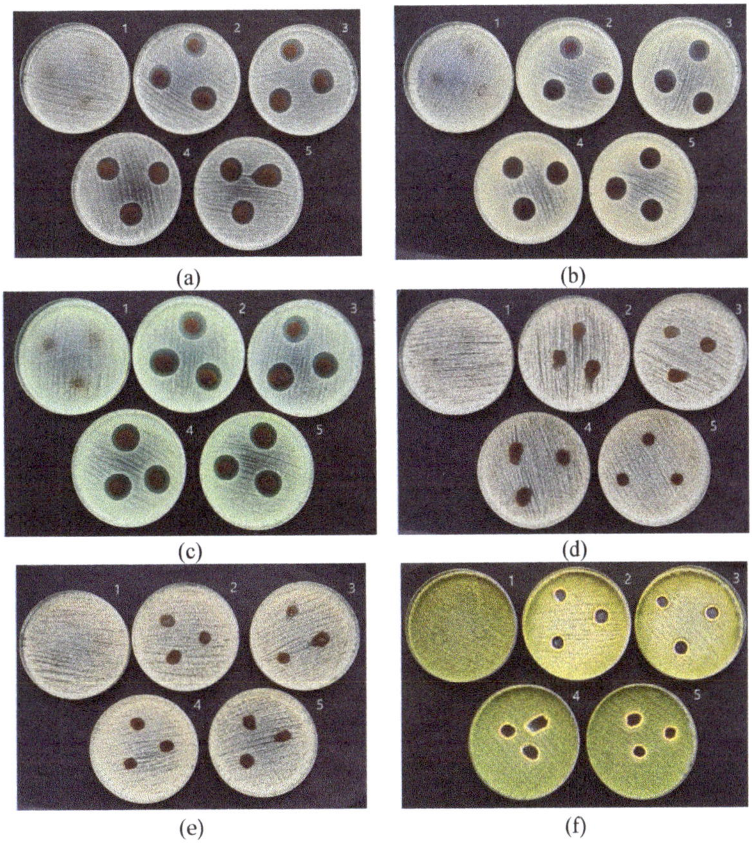

Figure 8. Antimicrobial activity of AgRonPVA films containing silver nanoparticles against (**a**) *Staphylococcus aureus* ATCC 29213, (**b**) *Acinetobacter baumannii* ATCC BAA 747, (**c**) *Pseudomonas aeruginosa* ATCC 15442, (**d**) *Candida parapsilosis* CBS 8836, (**e**) *Candida albicans* ATCC 90028, and (**f**) *Aspergillus flavus* BTL G-33 fungus. Numbering of Petri dishes with immersed films: (1) control; (2) AgRonPVA (UV 5 min); (3) AgRonPVA (UV 10 min); (4) AgRonPVA (UV 30 min); (5) AgRonPVA (UV 45 min).

Table 1. Antimicrobial activity of the AgRonPVA films containing silver nanoparticles.

Microorganism	**Films**				
	1	**2**	**3**	**4**	**5**
	RonPVA (Control)	**AgRonPVA (UV 5 min)**	**AgRonPVA (UV 10 min)**	**AgRonPVA (UV 30 min)**	**AgRonPVA (UV 45 min)**
	Zone Diameter, mm				
Staphylococcus aureus ATCC 29213	0	20.0 ± 1.0	19.7 ± 0.6	18.7 ± 0.6	18.7 ± 0.6
Acinetobacter baumannii BAA 747	0	18.3 ± 0.6	18.0 ± 0.0	18.0 ± 0.0	18.0 ± 0.0
Pseudomonas aeruginosa ATCC 15442	0	21.0 ± 0.0	20.7 ± 0.6	20.7 ± 0.6	19.7 ± 0.6

Table 1. *Cont.*

Microorganism	Films				
	1	2	3	4	5
	RonPVA (Control)	AgRonPVA (UV 5 min)	AgRonPVA (UV 10 min)	AgRonPVA (UV 30 min)	AgRonPVA (UV 45 min)
	Zone Diameter, mm				
Candida parapsilosis CBS 8836	0	10.7 ± 0.4	11.3 ± 0.5	9.3 ± 0.4	9.5 ± 0.3
Candida albicans ATCC 90028	0	9.7 ± 0.4	9.3 ± 0.3	9.7 ± 0.4	9.3 ± 0.4
Aspergillus flavus BTL G-33	0	13.0 ± 0.5	12.7 ± 0.6	12.0 ± 0.0	12.6 ± 0.4
Aspergillus fumigatus BTL G-38	0	16.0 ± 0.0	13.3 ± 0.6	14.3 ± 0.6	11.7 ± 0.5

3. Materials and Methods

3.1. Preparation of AgRonPVA Gel

Fresh bulbs of red onions (Allium cepa 'Red Karmen') were purchased from a local vegetable market. Onion peel extract was prepared following instructions provided in our previous work [24]. Particularly, peels of fresh red onion were removed from the bulbs and dried. A total of 2.65 g of dried red onion peel was immersed in 53 g of distilled water and heated for 20 min at 62.5 °C. The prepared extract was filtered twice using a white-band filter (Filtrak, Germany, size of pores 8–12 µm) and was ready to use.

$AgNO_3$ (CAS No 7761-88-8, purity ≥99.9%, Sigma–Aldrich, Poznan, Poland) was in distilled water to achieve a 1 molar (M) concentration of silver nitrate solution.

PVA $(C_2H_4O)_n$ powder (CAS 25213-24-5, Chempur®, Merck KGaA, Darmstadt, Germany). All was dissolved in distilled water to achieve 10% concentration under continuous stirring keeping a constant temperature of 60 °C. Later, the PVA solution was cooled down to room temperature.

3.2. Synthesis of AgNPs in RonPVA Gels

In total, 200 µL of 1 M $AgNO_3$ water-based solution was dripped into 4 g of freshly prepared "Ron" extract and filled with 76 g of 10 wt.% of PVA water-based solution under continuous stirring. In total, 20 g of the AgRonPVA gel was poured into glass bottle (22 ± 0.5 m, LBG SVSN-C26-121), and the samples were ready for investigation. The composition of experimental samples is provided in Table 2.

Table 2. Composition of the experimental samples.

Solutions	AgRonPVA (Gel)
Red onion extract, wt.%	4.99
PVA gel wt.%	94.73
1 M $AgNO_3$, wt.%	0.28
Total, wt.%	100.00

Photoreduction of silver ions in the solution was achieved by exposing experimental samples to a UV lamp (36 W black light source with a UV emission peak at 365 nm and weak blue light emission peak at 404 nm). The exposure time varied from 5 to 45 min. Exposed AgRonPVA gels were stored in the dark for at least 24 h.

3.3. Preparation of AgRonPVA Films

AgRonPVA films were formed via drying of a small amount (5 g) of the UV-exposed pure (100 wt.%) AgRonPVA gel deposited as a thin layer in a Petri dish. To avoid thermal and photo effects that may contribute to the loss of antimicrobial properties and initiate the growth and aggregation of silver nanoparticles, Petri dishes (55 mm) with the deposited gels were dried in the dark box at room temperature (20 ± 2 °C) for 24 h.

The thickness of the dried films prepared from 5 g of gels was (156 ± 15) μm. The average loss in mass of AgRonPVA films related to the drying process was (10.1 ± 0.3) wt.%.

3.4. Characterization of the AgRonPVA Gels and Films

Optical properties (absorbance) of the UV exposed samples were evaluated by analyzing UV-ViS absorption spectra of experimental gels and films obtained using Photospectrometer Ocean Optics (measurement range from 200 to 1100 UV/Vis, bandwidth 1.5 nm Ocean Optics, Ocean Optics, Inc., Orlando, FL, USA).

The thickness of the films was obtained by measuring with a digital micrometer 0-25MM IP65 (measurement range 0–25 mm, resolution ±0.002 mm, accuracy: ± 4μm (HEBDA Werksvertretungen E.K., Freiburg im Breisgau, Germany).

The structure of synthesized Ag nanoparticles was investigated using a D8 Discover X-ray diffractometer (Bruker AXS GmbH, Karlsruhe, Germany) operating at 40 kV and 40 mA with a Cu Kα radiation source (λ = 1.5418 Å) and parallel beam geometry with 60 mm Göbel mirror. Diffraction patterns were recorded using a fast-counting LynxEye detector with an opening angle of 2.475° and a slit opening of 6 mm. The peak intensities were scanned over the range of 20–7° (coupled 2θ–θ scans) using a step size of 0.05° and a collection time of 60 s per step.

The surface morphology of red-onion-PVA films without and with silver nanoparticles were investigated by a scanning electron microscope (SEM, Hitachi S-3400 N, Tokyo, Japan) using a secondary electron detector.

Elemental mapping of red onion PVA films without and with silver nanoparticles was performed using energy-dispersive X-ray spectroscopy (EDS, Bruker Quad 5040, Hamburg, Germany).

3.5. Microorganisms and Inoculum Preparation

Bacteria *Staphylococcus aureus* ATCC 29213, *Acinetobacter baumannii* ATCC BAA 747, and *Pseudomonas aeruginosa* ATCC 15442, yeasts *Candida parapsilosis* CBS 8836 and *Candida albicans* ATCC 90028, and microscopic fungi *Aspergillus flavus* BTL G-33 and Aspergillus fumigatus BTL G-38 were used in assays. Microorganisms were stored at −70 °C in a freezer in the Laboratory of Biodeterioration Research of the Nature Research Centre (Vilnius, Lithuania). Bacteria for antimicrobial tests were grown on Tryptone Soy Agar (Merck, KGaA, Darmstadt, Germany), and yeasts and fungi were grown on Sabouraud dextrose agar (Merck, KGaA, Darmstadt, Germany). Inoculums were obtained from overnight bacterial cultures grown at 37 ± 1 °C. Yeasts were cultured for 2 days, and fungi were cultured for 4 days at 28 ± 1 °C. The optical density of the microorganism cell suspensions was measured with a spectrophotometer (Thermo Scientific, Waltham, MA, USA) at 530 nm for yeasts and microscopic fungi and at 625 nm for bacteria. The resulting microorganism suspensions were vortexed for 15 s.

3.6. Evaluation of the Antimicrobial Activity of the AgRonPVA Films Containing Silver Nanoparticles

The agar diffusion method was used for antimicrobial activity testing of AgRonPVA films containing silver nanoparticles. For the agar diffusion assay, each microorganism suspension was uniformly spread with a swab on Mueller–Hinton agar (bacteria) or on Sabouraud dextrose agar (yeasts and microscopic fungi) in 90 mm Petri dishes. The pieces of AgRonPVA films (5 mm × 5 mm) containing silver nanoparticles were placed on the surface of each media as the microbial suspension was absorbed into it. The plates were incubated at 28 ± 1 °C for 2 days (yeasts) or 4 days (microscopic fungi), while the plates with bacteria were incubated at 37 ± 1 °C for 1 day only. All tests were triplicated for all strains. The pure RonPVA film was used as a negative control. The inhibition zone's diameters were measured in millimeters after the incubation period. Data were collected and processed using open access program packages. Mean values, standard errors, and confidence intervals were estimated.

4. Conclusions

AgRonPVA nanocomposites containing silver nanoparticles have been produced at room temperature in the form of gels and thin films. The formation of silver nanoparticles in these composites was achieved by the photoreduction method via exposing experimental samples to UV light and was verified by detecting characteristic SPR peaks in the UV-Vis absorbance spectra. It was shown that green synthesis of silver seeds/nano particles was possible, and red onion peel extract contributed to the final reduction in AgNPs. Mainly spheric AgNPs as silver composites containing silver (Ag) and silver oxides, $Ag_2O(Ag^+)$ and $AgO(Ag^{3+})$, were formed.

The antimicrobial activity of the UV exposed AgRonPVA films was investigated using the agar diffusion method. The greatest effect was observed on *Staphylococcus aureus*, *Acinetobacter baumannii*, and *Pseudomonas aeruginosa* bacteria. The antimicrobial features of the films were less effective against fungi and the weakest against yeasts. In general, AgRonPVA films containing silver nanoparticles indicated higher antifungal activity as compared to the previously indicated antimicrobial activity of silver NP-enriched red onion extract–gelatine films [24], thus defining these films as potential candidates for antimicrobial applications.

Author Contributions: Conceptualization, J.P.; methodology, J.P. and A.P.; investigation, J.P., J.Ž. and S.A.; resources, writing—original draft preparation, J.P., J.Ž. and A.P.; writing—review and editing, J.P., A.P. and D.A.; supervision, D.A. All authors have read and agreed to the published version of the manuscript.

Funding: This research received no external funding.

Institutional Review Board Statement: Not applicable.

Informed Consent Statement: Not applicable.

Data Availability Statement: Data are contained with the article.

Conflicts of Interest: The authors declare no conflicts of interest.

References

1. Sivam, G.P. Protection against Helicobacter pylori and other bacterial infections by garlic. *J. Nutr.* **2001**, *131*, 1106S–1108S. [CrossRef] [PubMed]
2. Cowan, M.M. Plant products as antimicrobial agents. *Clin. Microbiol. Rev.* **1999**, *12*, 564–582. [CrossRef] [PubMed]
3. Chassagne, F.; Samarakoon, T.; Porras, G.; Lyles, J.T.; Dettweiler, M.; Marquez, L.; Salam, A.M.; Shabih, S.; Farrokhi, D.R.; Quave, C.L. A Systematic Review of Plants with Antibacterial Activities: A Taxonomic and Phylogenetic Perspective. *Front. Pharmacol.* **2021**, *11*, 586548. [CrossRef] [PubMed]
4. Lanzotti, V. The analysis of onion and garlic. *J. Chromatogr. A* **2006**, *1112*, 3. [CrossRef] [PubMed]
5. Ye, C.L.; Dai, D.H.; Hu, W.L. Antimicrobial and antioxidant activities of the essential oil from onion (*Allium cepa* L.). *Food Control* **2013**, *30*, 48–53. [CrossRef]
6. Shon, M.Y.; Choi, S.D.; Kahng, G.G.; Nam, S.H.; Sung, N.J. Antimutagenic, antioxidant and free radical scavenging activity of ethyl acetate extracts from white, yellow and red onions. *Food Chem. Toxicol.* **2004**, *42*, 659–666. [CrossRef]
7. Sagar, N.A.; Pareek, S.; Benkeblia, N.; Xiao, J. Onion (*Allium cepa* L.) bioactives: Chemistry, pharmacotherapeutic functions, and industrial applications. *Food Front.* **2022**, *3*, 380–412. [CrossRef]
8. Ramos, F.A.; Takaishi, Y.; Shirotori, M.; Kawaguchi, Y.; Tsuchiya, K.; Shibata, H.; Higuti, T.; Tadokoro, T.; Takeuchi, M. Antibacterial and antioxidant activities of quercetin oxidation products from yellow onion (*Allium cepa*) skin. *J. Agric. Food Chem.* **2006**, *54*, 3551–3557. [CrossRef] [PubMed]
9. Škerget, M.; Majhenič, L.; Bezjak, M.; Knez, Ž. Antioxidant, Radical Scavenging and Antimicrobial Activities of Red Onion (*Allium cepa* L) Skin and Edible Part Extracts. *Chem. Biochem. Eng. Q.* **2009**, *23*, 435–444.
10. Roldán, E.; Sánchez-Moreno, C.; de Ancos, B.; Cano, M.P. Characterisation of onion (*Allium cepa* L.) by-products as food ingredients with antioxidant and antibrowning properties. *Food Chem.* **2008**, *108*, 907–916. [CrossRef]

11. Sharma, K.; Mahato, N.; Nile, S.H.; Lee, E.T.; Lee, Y.R. Economical and environmentally-friendly approaches for usage of onion (*Allium cepa* L.) waste. *Food Funct.* **2016**, *7*, 3354–3369. [CrossRef]
12. Santiago, B.; Arias Calvo, A.; Gullón, B.; Feijoo, G.; Moreira, M.T.; González-García, S. Production of flavonol quercetin and fructooligosaccharides from onion (*Allium cepa* L.) waste: An environmental life cycle approach. *Chem. Eng. J.* **2020**, *392*, 123772. [CrossRef]
13. Celano, R.; Docimo, T.; Piccinelli, A.L.; Gazzerro, P.; Tucci, M.; Di Sanzo, R.; Carabetta, S.; Campone, L.; Russo, M.; Rastrelli, L. Onion Peel: Turning a Food Waste into a Resource. *Antioxidants* **2021**, *10*, 304. [CrossRef] [PubMed]
14. Sánchez-López, E.; Gomes, D.; Esteruelas, G.; Bonilla, L.; Lopez-Machado, A.L.; Galindo, R.; Cano, A.; Espina, M.; Ettcheto, M.; Camins, A.; et al. Metal-Based Nanoparticles as Antimicrobial Agents: An Overview. *Nanomaterials* **2020**, *10*, 292. [CrossRef] [PubMed]
15. Yin, I.X.; Zhang, J.; Zhao, I.S.; Mei, M.L.; Li, Q.; Chu, C.H. The Antibacterial Mechanism of Silver Nanoparticles and Its Application in Dentistry. *Int. J. Nanomed.* **2020**, *15*, 2555–2562. [CrossRef] [PubMed] [PubMed Central]
16. Shanmuganathan, R.; Karuppusamy, I.; Saravanan, M.; Muthukumar, H.; Ponnuchamy, K.; Ramkumar, V.S.; Pugazhendhi, A. Synthesis of Silver nanoparticles and their biomedical applications—A comprehensive review. *Curr. Pharm. Des.* **2019**, *25*, 2650–2660. [CrossRef] [PubMed]
17. Crisan, C.M.; Mocan, T.; Manolea, M.; Lasca, L.I.; Tăbăran, F.-A.; Mocan, L. Review on Silver Nanoparticles as a Novel Class of Antibacterial Solutions. *Appl. Sci.* **2021**, *11*, 1120. [CrossRef]
18. Marambio-Jones, C.; Hoek, E.M.V. A Review of the Antibacterial Effects of Silver Nanomaterials and Potential Implications for Human Health and the Environment. *J. Nanoparticle Res.* **2010**, *12*, 1531–1551. [CrossRef]
19. Asif, M.; Yasmin, R.; Asif, R.; Ambreen, A.; Mustafa, M.; Umbreen, S. Green Synthesis of Silver Nanoparticles (AgNPs), Structural Characterization, and their Antibacterial Potential. *Dose Response* **2022**, *20*, 15593258221088709. [CrossRef]
20. Siddiqi, K.S.; Husen, A.; Rao, R.A.K. A review on biosynthesis of silver nanoparticles and their biocidal properties. *J. Nanobiotechnol.* **2018**, *16*, 14. [CrossRef]
21. Giri, A.K.; Jena, B.; Biswal, B.; Pradhan, A.K.; Arakha, M.; Acharya, S.; Acharya, L. Green synthesis and characterization of silver nanoparticles using Eugenia roxburghii DC. extract and activity against biofilm-producing bacteria. *Sci. Rep.* **2022**, *12*, 8383. [CrossRef] [PubMed]
22. Abdullah, H.S.T.S.H.; Asseri, S.N.A.R.M.; Mohamad, W.N.K.W.; Kan, S.-Y.; Azmi, A.A.; Julius, F.S.Y.; Chia, P.W. Green synthesis, characterization and applications of silver nanoparticle mediated by the aqueous extract of red onion peel. *Environ. Pollut.* **2021**, *271*, 116295. [CrossRef] [PubMed]
23. Baran, M.F.; Keskin, C.; Baran, A.; Hatipoğlu, A.; Yildiztekin, M.; Küçükaydin, S.; Kurt, K.; Hoşgören, H.; Sarker, M.M.R.; Sufianov, A.; et al. Green Synthesis of Silver Nanoparticles from *Allium cepa* L. Peel Extract, Their Antioxidant, Antipathogenic, and Anticholinesterase Activity. *Molecules* **2023**, *28*, 2310. [CrossRef] [PubMed]
24. Puišo, J.; Paškevičius, A.; Žvirgždas, J.; Dimitrova, T.L.; Litvakas, A.; Adliene, D. Application of Red Onion Peel Extract for Green Synthesis of Silver Nanoparticles in Hydrogels Exhibiting Antimicrobial Properties. *Gels* **2023**, *9*, 498. [CrossRef] [PubMed]
25. Chen, F.-M.; Liu, X. Advancing biomaterials of human origin for tissue engineering. *Prog. Polym. Sci.* **2016**, *53*, 86–168. [CrossRef] [PubMed]
26. Joyce, K.; Fabra, G.T.; Bozkurt, Y.; Pandit, A. Bioactive potential of natural biomaterials: Identification, retention and assessment of biological properties. *Signal Transduct. Target. Ther.* **2021**, *6*, 122. [CrossRef] [PubMed]
27. Bello, A.B.; Kim, D.; Kim, D.; Park, H.; Lee, S.-H. Engineering and functionalization of gelatin biomaterials: From cell culture to medical applications. *Tissue Eng. Part B Rev.* **2020**, *26*, 164–180. [CrossRef] [PubMed]
28. Fauzi, M.; Lokanathan, Y.; Aminuddin, B.; Ruszymah, B.; Chowdhury, S. Ovine tendon collagen: Extraction, characterisation and fabrication of thin films for tissue engineering applications. *Mater. Sci. Eng. C* **2016**, *68*, 163–171. [CrossRef] [PubMed]
29. Zulkiflee, I.; Fauzi, M.B. Gelatin-Polyvinyl Alcohol Film for Tissue Engineering: A Concise Review. *Biomedicines* **2021**, *9*, 979. [CrossRef]
30. Djagny, K.B.; Wang, Z.; Xu, S. Gelatin: A valuable protein for food and pharmaceutical industries. *Crit. Rev. Food Sci. Nutr.* **2001**, *41*, 481–492. [CrossRef]
31. Davidenko, N.; Schuster, C.F.; Bax, D.V.; Farndale, R.W.; Hamaia, S.; Best, S.M.; Cameron, R.E. Evaluation of cell binding to collagen and gelatin: A study of the effect of 2D and 3D architecture and surface chemistry. *J. Mater. Sci. Mater. Med.* **2016**, *27*, 148. [CrossRef] [PubMed]
32. Reddy, P.R.; Varaprasad, K.; Sadiku, R.; Ramam, K.; Reddy, G.V.S.; Raju, K.M.; Reddy, N.S. Development of gelatin based inorganic nanocomposite hydrogels for inactivation of bacteria. *J. Inorg. Organomet. Polym. Mater.* **2013**, *23*, 1054–1060. [CrossRef]
33. Mandri, A.; Nordin, A.; Hwei, N.M.; Chin, K.-Y.; Abd Aziz, I.; Fauzi, M.B. Natural 3D-Printed Bioinks for Skin Regeneration and Wound Healing: A Systematic Review. *Polymers* **2020**, *12*, 1782. [CrossRef] [PubMed]
34. Zhang, W.J.; Liu, W.; Cui, L.; Cao, Y. Tissue engineering of blood vessel. *J. Cell Mol. Med.* **2007**, *11*, 945–957. [CrossRef] [PubMed]
35. Lutolf, M.; Hubbell, J. Synthetic biomaterials as instructive extracellular microenvironments for morphogenesis in tissue engineering. *Nat. Biotechnol.* **2005**, *23*, 47–55. [CrossRef]
36. Qiu, K.; Netravali, A.N. A composting study of membrane-like polyvinyl alcohol based resins and nanocomposites. *J. Polym. Environ.* **2013**, *21*, 658–674. [CrossRef]
37. Barui, A. *Polymeric Gels*; Elsevier: Amsterdam, The Netherlands, 2018; pp. 55–90.

38. Sarwar, M.S.; Niazi, M.B.K.; Jahan, Z.; Ahmad, T.; Hussain, A. Preparation and characterization of PVA/nanocellulose/Ag nanocomposite films for antimicrobial food packaging. *Carbohydr. Polym.* **2018**, *184*, 453–464. [CrossRef]
39. Singh, S.; Gaikwad, K.K.; Lee, Y.S. Antimicrobial and antioxidant properties of polyvinyl alcohol bio composite films containing seaweed extracted cellulose nano-crystal and basil leaves extract. *Int. J. Biol. Macromol.* **2018**, *107*, 1879–1887. [CrossRef] [PubMed]
40. Kannat, S.; Jethwa, T.; Sawant, K.; Chawla, S. PVA-Gelatin films incorporated with tomato pulp: A potential primary food packaging film. *Int. J. Curr. Microbiol. Appl. Sci.* **2017**, *6*, 1428–1441.
41. Meshram, J.; Koli, V.; Phadatare, M.R.; Pawar, S. Anti-microbial surfaces: An approach for deposition of ZnO nanoparticles on PVA-Gelatin composite film by screen printing technique. *Mater. Sci. Eng. C* **2017**, *73*, 257–266. [CrossRef]
42. Lian, R.; Cao, J.; Jiang, X.; Rogachev, A.V. Prolonged release of ciprofloxacin hydrochloride from chitosan/gelatin/poly (vinyl alcohol) composite films. *Mater. Today Commun.* **2021**, *27*, 102219. [CrossRef]
43. Chen, J.; Wei, D.; Gong, W.; Zheng, A.; Guan, Y. Hydrogen-Bond Assembly of Poly(vinyl alcohol) and Polyhexamethylene Guanidine for Nonleaching and Transparent Antimicrobial Films. *ACS Appl. Mater. Interfaces* **2018**, *10*, 37535–37543. [CrossRef] [PubMed]
44. Gutha, Y.; Pathak, J.L.; Zhang, W.; Zhang, Y.; Jiao, X. Antibacterial and wound healing properties of chitosan/poly(vinyl alcohol)/zinc oxide beads (CS/PVA/ZnO). *Int. J. Biol. Macromol.* **2017**, *103*, 234–241. [CrossRef]
45. Suganthi, S.; Vignesh, S.; Sundar, J.K.; Raj, V. Fabrication of PVA polymer films with improved antibacterial activity by fine-tuning via organic acids for food packaging applications. *Appl. Water Sci.* **2020**, *10*, 100. [CrossRef]
46. Yang, H.; Ren, Y.-Y.; Wang, T.; Wang, C. Preparation and antibacterial activities of $Ag/Ag^{+}/Ag^{3+}$ nanoparticle composites made by pomegranate (*Punica granatum*) rind extract. *Results Phys.* **2016**, *6*, 299–304. [CrossRef]
47. Semaltianos, N.G.; Perrie, W.; Romani, S.; Potter, R.J.; Dearden, G.; Watkins, K.G. Polymer-nanoparticle composites composed of PEDOT: PSS and nanoparticles of Ag synthesised by laser ablation. *Colloid Polym. Sci.* **2012**, *290*, 213–220. [CrossRef]
48. Suzuki, R.O.; Ogawa, T.; Ono, K. Use of ozone to prepare silver oxides. *J. Am. Ceram. Soc.* **1999**, *82*, 2033–2038. [CrossRef]
49. Sajti, C.L.; Sattari, R.; Chichkov, B.N.; Barcikowski, S. Gram scale synthesis of pure ceramic nanoparticles by laser ablation in liquid. *J. Phys. Chem.* **2010**, *114*, 2421–2427. [CrossRef]
50. Costa, J.S.; Zaman, Q.; da Costa, K.Q.; Dmitriev, V.; Pandoli, O.; Fontes, G.; Del Rosso, T. Limits of the Effective Medium Theory in Particle Amplified Surface Plasmon Resonance Spectroscopy Biosensors. *Sensors* **2019**, *19*, 584. [CrossRef]
51. Šileikaitė, A.; Prosyčevas, I.; Puišo, J.; Juraitis, A.; Guobienė, A. Analysis of Silver Nanoparticles Produced by Chemical Reduction of Silver Salt Solution. *Mater. Sci.* **2006**, *12*, 287–291.
52. Tziolas, N.; Ordoudi, S.A.; Tavlaridis, A.; Karyotis, K.; Zalidis, G.; Mourtzinos, I. Rapid Assessment of Anthocyanins Content of Onion Waste through Visible-Near-Short-Wave and Mid-Infrared Spectroscopy Combined with Machine Learning Techniques. *Sustainability* **2021**, *13*, 6588. [CrossRef]
53. Mahdieh, M.; Zo-lanvari, A.; Azimee, A.S.; Mahdieh, M. Green biosynthesis of silver nanoparticles by Spirulina platen-sis. *Sci. Iran.* **2012**, *19*, 926–929. [CrossRef]
54. *The "MiePlot" Software*, Version 4.6; Philip Laven: Bradenton, FL, USA, 2024. Available online: http://www.philiplaven.com/mieplot.htm (accessed on 12 March 2024).
55. Mostafa, A.M.; Menazea, A.A. Polyvinyl Alcohol/Silver nanoparticles film prepared via pulsed laser ablation: An eco-friendly nano-catalyst for 4-nitrophenol degradation. *J. Mol. Struct.* **2020**, *1212*, 128125. [CrossRef]
56. Liu, T.; Jiao, C.; Peng, X.; Chen, Y.-N.; Chen, Y.; He, C.; Liu, R.; Wang, H. Super-strong and tough poly(vinyl alcohol)/poly(acrylic acid) hydrogels reinforced by hydrogen bonding. *J. Mater. Chem. B* **2018**, *6*, 8105–8114. [CrossRef] [PubMed]
57. Islam, A.; Bari, L.; Salem, K.S. Morphology, Thermal Stability, Electrical, and Mechanical Properties of Graphene Incorporated Poly(vinyl alcohol)-Gelatin Nanocomposites. *Int. J. Compos. Mater.* **2016**, *6*, 172–182. [CrossRef]
58. Atta, M.M.; Taha, E.O.; AbdelReheem, A.M. Nitrogen plasma effect on the structural, thermal, and dynamic mechanical properties of PVA/starch/graphene oxide nanocomposite. *Appl. Phys. A* **2021**, *127*, 532. [CrossRef]
59. Aji, M.M.; Bijaksana, S.; Khairurrijal, K.; Abdullah, M. A General Formula for Ion Concentration-Dependent Electrical Conductivities in Polymer Electrolytes. *Am. J. Appl. Sci.* **2012**, *9*, 946–954. [CrossRef]
60. Ali, I.A.M.; Ben Ahmed, A.; Al-Ahmed, H.I. Green synthesis and characterization of silver nanoparticles for reducing the damage to sperm parameters in diabetic compared to metformin. *Sci. Rep.* **2023**, *13*, 2256, Erratum in *Sci. Rep.* **2023**, *13*, 6659. [CrossRef] [PubMed] [PubMed Central]
61. Prathna, T.C.; Chandrasekaran, N.; Raichur, A.M.; Mukherjee, A. Biomimetic synthesis of silver nanoparticles by *Citrus limon* (lemon) aqueous extract and theoretical prediction of particle size. *Colloids Surf. B Biointerfaces* **2011**, *82*, 152–159. [CrossRef]
62. XRD Crystallite (Grain) Size Calculator (Scherrer Equation)—InstaNANO. Available online: https://instanano.com/all/characterization/xrd/crystallite-size/ (accessed on 27 March 2024).
63. Santhosh, A.; Theertha, V.; Prakash, P.; Chandran, S. From waste to a value added product: Green synthesis of silver nanoparticles from onion peels together with its diverse applications. *Mater. Today Proc.* **2020**, *46*, 4460–4463. [CrossRef]
64. Nugraheni, I.P.A.; Widyastika, D.; Maulida, S.; Susilowati, H.; Jonarta, A.L. Effect of Red Onion (*Allium cepa var ascalonicum*) Skin Ethanolic Extract on the Motility and the Adhesion Index of Pseudomonas aeruginosa and Macrophage Phagocytosis Index. *Maj. Obat Tradis.* **2019**, *24*, 40–46. [CrossRef]
65. Genatrika, E.; Sundhani, E.; Oktaviana, M.I. Gel Potential of Red Onion (*Allium cepa* L.) Ethanol Extract as Antifungal Cause Tinea Pedis. *J. Pharm. Bioallied Sci.* **2020**, *12*, S733–S736. [CrossRef] [PubMed] [PubMed Central]

66. Albandary, A. Phenolic compounds content, antioxdidant, antibacterial and antifungal activities of red onions skin. *Iraq J. Agric. Sci.* **2023**, *54*, 1050–1057. [CrossRef]
67. Qing, Y.; Cheng, L.; Li, R.; Liu, G.; Zhang, Y.; Tang, X.; Wang, J.; Liu, H.; Qin, Y. Potential antibacterial mechanism of silver nanoparticles and the optimization of orthopedic implants by advanced modification technologies. *Int. J. Nanomed.* **2018**, *13*, 3311–3327. [CrossRef] [PubMed]
68. Dakal, T.C.; Kumar, A.; Majumdar, R.S.; Yadav, V. Mechanistic basis of antimicrobial actions of silver nanoparticles. *Front. Microbiol.* **2016**, *7*, 1831. [CrossRef] [PubMed]
69. Lu, Z.; Rong, K.; Li, J.; Yang, H.; Chen, R. Size-dependent antibacterial activities of silver nanoparticles against oral anaerobic pathogenic bacteria. *J. Mater. Sci. Mater. Med.* **2013**, *24*, 1465–1471. [CrossRef]
70. Ivask, A.; Elbadawy, A.; Kaweeteerawat, C.; Boren, D.; Fischer, H.; Ji, Z.; Chang, C.H.; Liu, R.; Tolaymat, T.; Telesca, D.; et al. Toxicity mechanisms in *Escherichia coli* vary for silver nanoparticles and differ from ionic silver. *ACS Nano* **2014**, *8*, 374–386. [CrossRef] [PubMed]
71. Seil, J.T.; Webster, T.J. Antimicrobial applications of nanotechnology: Methods and literature. *Int. J. Nanomed.* **2012**, *7*, 2767–2781.
72. Li, W.R.; Xie, X.B.; Shi, Q.S.; Zeng, H.Y.; Ou-Yang, Y.S.; Chen, Y. Ben Antibacterial activity and mechanism of silver nanoparticles on *Escherichia coli*. *Appl. Microbiol. Biotechnol.* **2010**, *85*, 1115–1122. [CrossRef]
73. Gomaa, E.Z. Silver nanoparticles as an antimicrobial agent: A case study on Staphylococcus aureus and *Escherichia coli* as models for Gram-positive and Gram-negative bacteria. *J. Gen. Appl. Microbiol.* **2017**, *63*, 36–43. [CrossRef] [PubMed]
74. Quinteros, M.A.; Cano Aristizábal, V.; Dalmasso, P.R.; Paraje, M.G.; Páez, P.L. Oxidative stress generation of silver nanoparticles in three bacterial genera and its relationship with the antimicrobial activity. *Toxicol. Vitr.* **2016**, *36*, 216–223. [CrossRef] [PubMed]
75. Agnihotri, S.; Mukherji, S.; Mukherji, S. Immobilized silver nanoparticles enhance contact killing and show highest efficacy: Elucidation of the mechanism of bactericidal action of silver. *Nanoscale* **2013**, *5*, 7328–7340. [CrossRef] [PubMed]
76. Bruna, T.; Maldonado-Bravo, F.; Jara, P.; Caro, N. Silver Nanoparticles and Their Antibacterial Applications. *Int. J. Mol. Sci.* **2021**, *22*, 7202. [CrossRef]
77. Pereira, D.; Ferreira, S.; Ramírez-Rodríguez, G.B.; Alves, N.; Sousa, Â.; Valente, J.F.A. Silver and Antimicrobial Polymer Nanocomplexes to Enhance Biocidal Effects. *Int. J. Mol. Sci.* **2024**, *25*, 1256. [CrossRef]

Disclaimer/Publisher's Note: The statements, opinions and data contained in all publications are solely those of the individual author(s) and contributor(s) and not of MDPI and/or the editor(s). MDPI and/or the editor(s) disclaim responsibility for any injury to people or property resulting from any ideas, methods, instructions or products referred to in the content.

antibiotics

MDPI

Article

Differential Antimicrobial Effect of Three-Sized Biogenic Silver Nanoparticles as Broad-Spectrum Antibacterial Agents against Plant Pathogens

Munirah F. Aldayel [1,*,†], Nermin El Semary [1,2,†] and David G. Adams [3]

[1] Biological Sciences Department, College of Science, King Faisal University, Al-Ahsa 31982, Saudi Arabia; nelsemary@kfu.edu.sa
[2] Botany and Microbiology Department, Faculty of Science, Helwan University, Cairo 11795, Egypt
[3] Faculty of Biological Sciences, Leeds University, Leeds LS2 9JT, UK; d.g.adams@leeds.ac.uk
* Correspondence: maldayel@kfu.edu.sa
† These authors contributed equally to this work.

Abstract: Background: Massive fruit losses are caused by microbial pathogens of unknown identities. Therefore, ecofriendly biocontrol measures are well sought after, and biogenic silver nanoparticles are plausible candidates. Here we investigate the antimicrobial effect of three different sized AgNPs samples on those pathogens. Methodology: Identities of three local pathogenic bacteria were investigated using molecular methods. Three different-sized samples of silver nanoparticles were bio-synthesized in the external solution of a cyanobacterial culture, characterized, and used in antimicrobial bioassay. Results: The pathogens were identified as *Erwinia pyrifoliae*, *Staphylococcus warneri*, and *Xanthomonas citri*. UV-vis. and FTIR spectroscopy confirmed the biosynthesis of AgNPs. and their three different sizes were confirmed using Scanning electron microscopy. Growth of bacterial pathogens was inhibited by all three samples of AgNPs, but the largest inhibition zone was for the smallest sized AgNPs against *Staphylococcus warneri* (1.7 cm). Discussion: The identity of the pathogens infecting different local fruits is reported for the first time. They belong to different bacterial lineages. The fact that biogenic AAgNPs were effective against all of them shows their broad-spectrum of antibacterial effect. Customized biosynthesis was successful in yielding different-sized AgNPs. The smaller the AgNPs, the stronger the antimicrobial impact. Conclusion: Local bacterial species infecting fruits are diverse. Customized biogenic AgNPs are effective broad-spectrum biocontrol agents against bacterial pathogens of local fruits and thereby help maintain food security and environmental sustainability.

Keywords: antimicrobial; fruits; environmental sustainability; plant pathogenic bacteria; cyanobacteria; silver nanoparticles

Citation: Aldayel, M.F.; El Semary, N.; Adams, D.G. Differential Antimicrobial Effect of Three-Sized Biogenic Silver Nanoparticles as Broad-Spectrum Antibacterial Agents against Plant Pathogens. *Antibiotics* **2023**, *12*, 1114. https://doi.org/10.3390/antibiotics12071114

Academic Editors: Sotiris K Hadjikakou, Christina N. Banti and Serena Riela

Received: 25 March 2023
Revised: 22 June 2023
Accepted: 25 June 2023
Published: 28 June 2023

Copyright: © 2023 by the authors. Licensee MDPI, Basel, Switzerland. This article is an open access article distributed under the terms and conditions of the Creative Commons Attribution (CC BY) license (https://creativecommons.org/licenses/by/4.0/).

1. Introduction

Nanoparticles are minute particles of size 10^{-9} m whose chemical and physical characteristics are different from bulk particles [1]. Cyanobacteria are wide-spread photosynthetic prokaryotes that represent bio-factories for the production of nanoparticles [2,3]. Cyanobacteria grow rapidly in inexpensive growth media, require only minerals and light energy for growth and can produce nanoparticles without toxic byproducts or energy consumption [4,5]. Therefore, their synthesis of nanoparticles is both cost-effective and eco-friendly [2]. Currently, there is a surge to find new biological sources for antimicrobial agents. In this regard, cyanobacteria are considered promising sources of silver nanoparticles that are effective antimicrobial agents [4,5]. They are reported to reduce silver ions via chemical entities or reductase enzymes such as NADH-dependent reductases and electron carriers such as quinones [6]. The shape and size of the nanoparticles vary according to the synthetic methods, reductant, and biological source [3,7]. Hong et al. [8] used the microwave-assisted method for the preparation of three different shapes of silver nanoparticles: nanocubes, nanospheres, and nanowires. They found that silver nanowires exhibited the weakest antibacterial activity due to poor contact with the bacterial cells as compared with silver nanocubes and nanospheres. Silver nanocubes showed stronger antibacterial activity than nanospheres. They concluded that the shape effect of AgNPs is attributed to the specific reactivity of the surface areas and facets. Indeed, [9] reported that nanoscale size and the presence of a {111} plane increase the biocidal action of silver nanoparticles. With regard to the effect of size, [10] showed that the smaller the size of nanosilver particles, the more effective they are as antibacterial agents. However, little is known about the biogenically-synthesized nanosilver particles and the relation between their size, shape, and their antimicrobial activity, as well as the extent of their antimicrobial bioactivity whether wide-spectrum or narrow. Recently, Oves et al., 2023 [11] showed that by using the flower extract of *Bougainvillea glabra* as a biological reducing agent, silver nanoparticles were generated. The biological synthesis of biogenic silver nanoparticles may proceed internally and externally and enzymes [12], such as NADH-dependent nitrate reductase, act as an electron shuttle by taking electrons from nitrate molecules and transforming them into silver metal ions leading silver nanoparticles formation. Also [13] suggested that the negatively charged carboxylic group in bacterial cell walls may provide an electrostatic interaction and reduction. Moreover, some amino acids such as arginine, aspartic acid, cystine, glutamic acid, lysine, and methionine may be involved. They act as catalysts and produce hydroxyl ions that react with the reducing agent's aldehyde. Interestingly, [14] showed that biogenic silver nanoparticles generated through fungi-mediated synthesis were small (15.56 ± 9.22 nm), spherical, and stable (zeta potential of—38.43 mV) AgNPs with good crystallinity with the FTIR spectroscopy indicating the presence of hydroxyl, amino, and carboxyl groups, on the surface of silver nanoparticles which in turn showed antimicrobial and antibiofilm activities against Gram-positive and Gram-negative bacteria. In addition, they found the MIC and MBC values to range between 16–64 and 32–512 $\mu g\,mL^{-1}$. Using cyanobacterial extract from novel cyanobacterium called *Desterifilum* sp., Hamida et al., 2020 [15] performed green synthesis of silver nanoparticles by adding $AgNO_3$ to cyanobacterial filtrate at room temperature under direct illumination (2000 ± 200 Lux) for 24 h. Indeed, silver nanoparticles were formed which exhibited antimicrobial activity against human pathogenic bacteria. However, they used a single concentration of the precursor silver nitrate and they prepared the cyanobacterial filtrate by filtering through a Whatman filter No. 42. The shape of the biosynthesized silver nanoparticles was spherical but with some aggregation and they suggested that polysaccharides and proteins may have contributed to the bio-fabrication of the synthesis of silver nanoparticles. In the present study, we aim to investigate the effect of biosynthesized silver nanoparticles that were biosynthesized by three different concentrations of silver nitrate. Additionally, the cell-free filtrates were prepared by filtration using micromillipore filtration units. The resulting different sized silver nanoparticles were used as antimicrobial agents against different plant pathogenic bacteria. Those different plant pathogens (both gram-positive and gram-negative) were isolated and identified using molecular methods as part of the study. The factors of the size and shape of silver nanoparticles are also addressed. Antibiosis is addressed using several tests to select for the most efficient silver nanoparticle size as antimicrobial agents. Moreover, the study also documents some of the different plant pathogenic bacteria found in Al Ahsa, Kingdom of Saudi Arabia.

2. Materials and Methods

Materials used in this investigation including silver nitrate and other reagents mostly purchased from Sigma. The bacterial strains were isolated from different infected fruits bought from the local market in Al-Ahsa, KSA. Symptoms of infection included the following: (1) Apple: The fruit was shriveled and brown; (2) Dates: fruit had spots and (3) Mandarin had large corky lesions with centered cracks on a fruit.

2.1. Isolation of the Cyanobacterial Strain and Pathogenic Bacteria

Water samples were collected from rice field Al Ahsa, centrifuged at 3000 rpm to get rid of some of the contaminating bacteria, the cyanobacterial biomass was spread on solid BG 11 medium. The agar streaking was repeated several times and the isolated colonies were used to establish a liquid culture. The culture was subjected to antibiotic treatment (ampicillin 200 uL/L) in the dark overnight. The culture was washed with BPS, recentrifuged and the biomass used to establish a liquid culture. The last two steps were repeated until an axenic culture was obtained and identified by light microscopy [16]. On the other hand, the process of serial dilution was used to isolate pathogenic bacteria from fruits. The fruits were smashed into a suspension with BPS in a pre-sterile mortar and pestle, and the mixture was serially diluted from 10^{-1} to 10^{-6} CFU/mL. A spreader was used to spread 100 µL of each diluted fruit suspension onto nutrient agar medium that was clearly labeled. Pure colonies of bacteria were inoculated into 70% broth media (HIMEDIA) containing 30% Glycerol (*v/v*), incubated at 30 °C for 24 h, and then kept at 4 °C. To isolate bacterial DNA, isolates were grown in 5 mL tubes containing 2 mL of Luria broth. To precipitate bacterial pellets, 2 mL of bacterial suspension was centrifuged at 5000 rpm for 5 min.

2.2. DNA Extraction

DNA was extracted according to [17] as follows: 20 mg of freshly harvested bacterial pellet was ground with 500 µL of Dellaporta buffer (100 mM Tris pH 8, 50 mM ethylenediaminetetraacetate EDTA, 500 mM NaCl, 10 mM beta-mercaptoethanol, BME). A total of 33 µL of 20% sodium dodecyl sulfate (SDS, *w*/*v*) was added, vortexed and incubated for 10 min at 65 °C. Moreover, 160sixty µL of 5 M potassium acetate was added, vortexed and centrifuged for 10 min at 10,000 rpm. Following which, 450 µL of supernatant was transferred to a new tube and 450 µL phenol, chloroform and isoamyl-alcohol (PCI) 25:24:1 was added. The mixture was vortexed then centrifuged for 5 min at 10,000 rpm. After, 400 µL of the upper phase was removed and added to 0.5 volume of isopropanol and centrifuged for 10 min at 14,000 rpm. The nucleic acid pellet, after discarding the supernatant, was washed with 70% ethanol and centrifuged for 5 min at 10,000 rpm. The pellet was resuspended in 100 µL ddH_2O.

2.3. Polymerase Chain Reaction (PCR)

Amplification of the 16S rRNA gene from bacterial isolates was carried out using the universal primers; 27F 5′-AGA GTT TGA TCM TGG CTC AG-3′ and 1492R 5′-TACGGYTACCTTGTTACGACTT using the following program; initial denaturation at 94 °C for 5 min, 30 cycles of the amplification using the following stages; denaturation at 94 °C for 45 s, annealing at 55 °C for 60 s, and extension at 72 °C for 60 s in 30 amplification cycles, followed by a final extension step at 72 °C for 10 min. PCR was done in a 25 µL reaction containing 1 µL of the DNA extract (40 ng of total DNA), 2 mM $MgCl_2$, 2.5 µL of 10× PCR buffer,1.5 µL of 10 µM of each primer, 2.5 µL of 10 mM dNTPs, 0.3 µL of 5U Taq DNA Polymerase and the reaction was made to 25 µL with nuclease-free water. PCR products were purified using the QIAquick® PCR purification kit according to the manufacturer's instructions. The purified PCR products were sequenced by Macrogen Inc. (Seoul, Republic of Korea), and sequencing of the purified isolates was performed in both directions using a primer pair. The sequences were uploaded on BLASTn of Genbank and a similarity check was run. For analysis of phylogenetic inference, the closely related sequences were used alongside the sequences retrieved for all pathogenic bacteria to establish phylogenetic trees using the tree construction within BLASTn.

2.4. Biosynthesis of Silver Nanoparticles by Cyanobacterial Cultures

To synthesize silver nanoparticles of different sizes, the method of [17] was followed (Scheme 1). Briefly, 5 ml of mid-logarithmic *Cyanothece*-like sp. culture diluted with BPS to 18 mL was incubated with 0.2, 1, or 2 mL of 10 mM silver nitrate solution and the mixture was completed to 20 mL. The samples were allowed to stand in the dark at 25 °C until the appearance of brown colour, but with varying intensity for the three treatments. This colour is indicative of the formation of silver nanoparticles. The external solution of the cyanobacterial cultures was filtered through millipore filters of a pore diameter size of 0.2 μm.

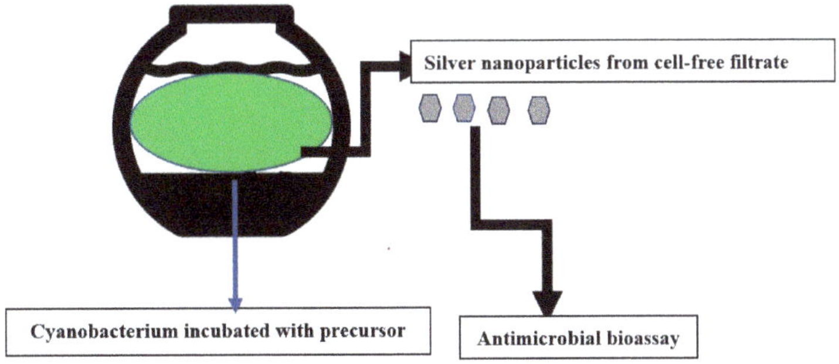

Scheme 1. Method of preparation.

2.5. *Characterization of Silver Nanoparticles*

The steps detailed in [17] were followed. The plasmon resonance absorption spectrum was determined using a UV-visible spectrophotometer in the wavelength range 200–600 nm. FTIR chemical functional groups were identified by the banding pattern in the FTIR spectrum.

2.6. *Scanning Electron Microscopy*

Synthesis of different-sized silver nanoparticles was confirmed by Scanning electron microscopy. Both control and aqueous external solutions that were filtered from the three treatments of biogenic nanosilver were examined by SEM. A drop of each sample was mounted on metallic stub, air dried, and fixed using double face adhesive carbon. Sputter-coating with gold was carried out three times to increase conductivity, reduce thermal damage, and improve secondary electron signals. Samples were visualized using a Joel JSM-5510LV scanning electron microscope.

2.7. *Antimicrobial Bioassay*

Two pathogenic bacteria were isolated from infected apple and mandarin fruits. They were purified triple times and kept as pure cultures for identification by molecular methods using PCR, sequencing, and phylogenetic inference. The third bacterium was previously characterized as the gram-positive *Staphylococcus warneri* that was isolated from date fruit. The antibacterial potency of silver nanoparticles biosynthesized by *Cyanothece*-like sp. was evaluated against pathogenic bacteria. The sensitivity of pathogenic strains to the extract was assayed through a modified Kirby Bauer Disk Diffusion Susceptibility method. Sterilized paper discs were saturated with 30 μL of a silver nanoparticle suspension. Discs were dried and placed on the surface of a nutrient agar medium which was inoculated with bacterial suspensions prepared in physiological saline. Plates were kept for 2 h at 4 °C to ensure diffusion of bioactive material, after which, the plates were incubated at 37 °C. Discs containing 30 μL of sterilized distilled water were left to dry and used as negative controls, whereas positive control sets included discs of the antibiotic chloramphenicol at a concentration of 50 μg/L. Plates were incubated for 24 h and the diameter of inhibition zones (mm) were measured in triplicate and the average standard deviation was recorded. Statistical analysis (one-way ANOVA) was performed using Minitab package, version 13, to evaluate the effect of size of biogenic silver nanoparticles on their antimicrobial activity. To ascertain the antibacterial activity of silver nanoparticles, several methods were used [18,19].

2.7.1. Antibacterial Action and Measurement of Minimum Inhibitory Concentration (MIC)

In dimethylsulfoxide (DMSO, 0.1%), silver nanoparticles 9.2×10^{-7} M were placed. Disks made of filter paper containing silver nanoparticles each were utilized in the experiment. After diluting the tested bacterial species' culture to 1×10^6 CFU, the diameter of the inhibition zones was measured to assess the silver nanoparticles' antibacterial activity. Using the serial dilution approach, the minimum inhibitory concentration (MIC) was determined according to [20].

2.7.2. Time-Kill Test of AgNPs

For *Erwinia amylovora*, *Staphylococcus warneri*, and *Xanthomonas citri*, AgNPs were diluted with the nutritional broth medium (containing 1.5×10^8 CFU/mL bacterial inoculum) to concentrations of 0 × MIC, 0.5 × MIC, 1 × MIC, 2 × MIC, and 4 × MIC. A volume of 0.1 mL of the medium was grown on Müller-Hinton agar and incubated at various times (0, 0.5, 1, 2, 4, and 6 h) for 24 h at 37 °C. All experiments were performed in triplicates.

3. Results

3.1. Molecular Characterization and Phylogenetic Inference of Pathogenic Bacteria

The bacterium isolated from apples was identified as the gram-negative rod-shaped bacterium *Erwinia pyrifoliae* whereas that from mandarins was identified as *Xanthomonas citri,* which is a gram-negative, rod-shaped proteobacterium. The Neighbor-joining phylogenetic tree showed the co-clustering of the first isolate with strains of *Erwinia pyrifoliae* thereby confirming its identity (Figure 1). Similarly, a Neighbor-joining phylogenetic tree (Figure 2) confirmed the identity of *Xanthomonas citri.*

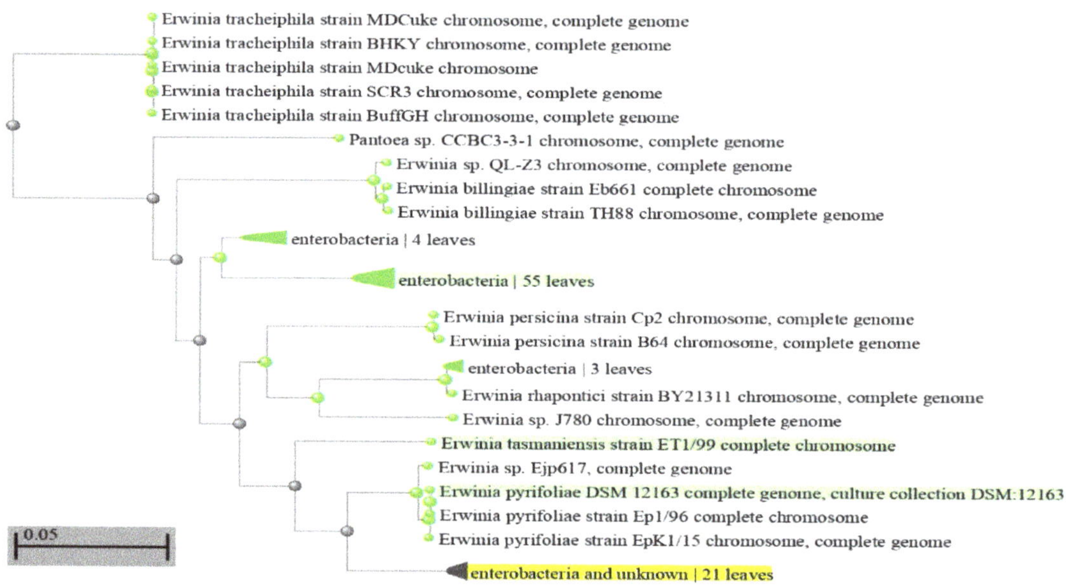

Figure 1. A Neighbor-joining phylogenetic tree for *Erwinia pyrifoliae (pyrifoliae).* The sequence retrieved is highlighted in yellow. Bootstrap value is indicated by the scale bar. The tree was constructed using the Phylogenetic tree reconstruction tool in BLASTn.

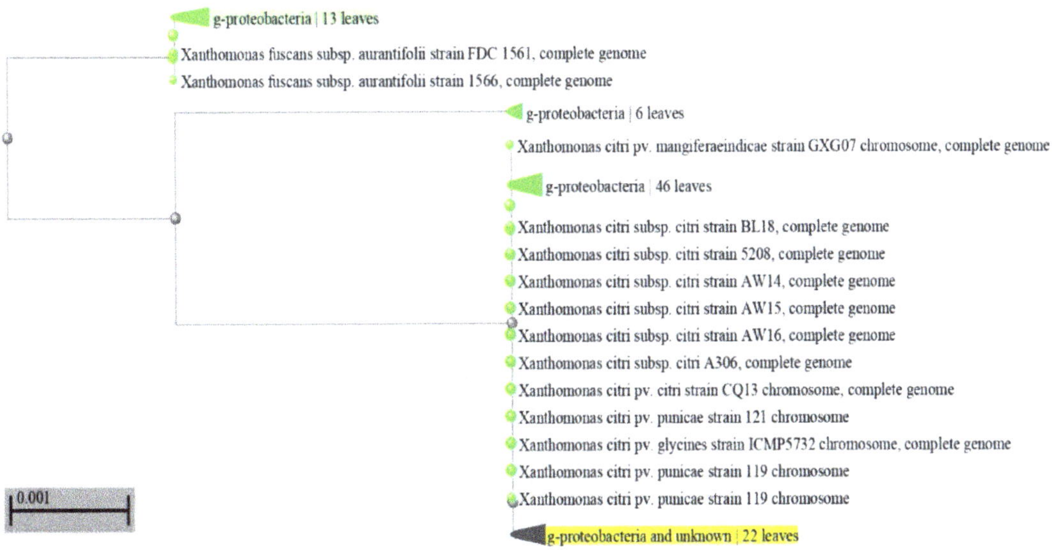

Figure 2. A Neighbor-joining phylogenetic tree for *Xanthomonas citri*. The sequence retrieved is highlighted in yellow. Bootstrap value is indicated by the scale bar. The tree was constructed using the Phylogenetic tree reconstruction tool in BLASTn.

3.2. UV-Visible Spectroscopy

The formation of silver nanoparticles in the cell-free filtered external solution was confirmed by the UV- visible absorbance spectra of the three samples. The three samples of biogenic nanosilver showed a strong absorbance peak at 450 nm (Figure 3 and Supplementary Materials, Figures S1 and S2). This is within the range of plasmon resonance that is characteristic of silver nanoparticles [3]. The characteristic plasmon resonance of silver nanoparticles is due to the vibrational modes of electrons based on the size and shape of nanoparticles [17].

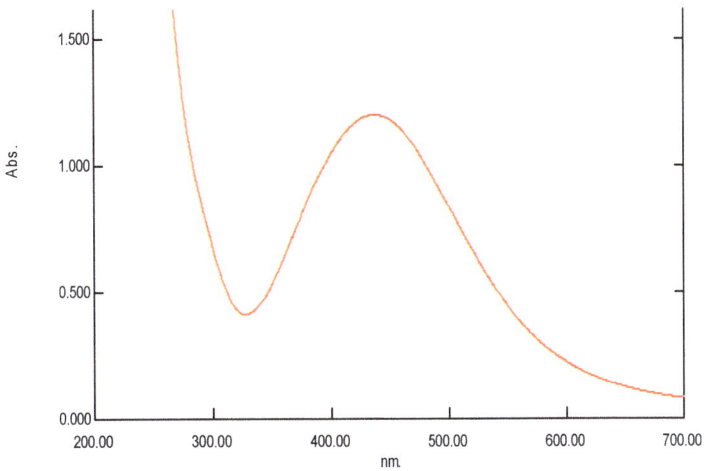

Figure 3. UV-visible spectrum of the largest sized biogenic silver nanoparticles (average size 92 nm). The X axis denotes wavelength, whereas the Y axis denotes absorbance.

3.3. FTIR Spectroscopy

The Fourier transmission infrared spectrum of silver nanoparticles showed the characteristic signal of silver nanoparticles present at 3356–3350 cm^{-1} overlapping with the OH signal, possibly resulting from phenolic compounds leaking from cyanobacteria. Moreover, at 1636–1637 cm^{-1} there was a clear C-H stretching (Supplementary Materials, Figures S3–S5). No other functional groups were detected. The pattern of spectrum of silver nanoparticles is similar to that reported by Zhang et al. (2016) [21], thus confirming the formation of silver nanoparticles. With regard to the confirmation of corona absence, the lack of bands at 1209 cm^{-1} and at 1533 cm^{-1} indicated its absence. The presence of signals at wave numbers of 1427, 1271, and 842 cm^{-1} corresponds to the C–C in the ring, C–N stretching and C–C in ring vibrations of covalent bonds, respectively, most likely related to the extracellular polymeric substances such as polysaccharides and peptides (Mota et al., 2021. [22]. The peak at 513 cm^{-1} indicates the vibration frequency of the Ag-O ionic bond group according to Gharibshahi et al., 2017) [23]. We indeed detected a signal within that range as in Supplementary Materials Figure S3.

3.4. Scanning Electron Microscopy

SEM was used to detect particle shape and size. The electro-micrograph confirmed the angular nature of the AgNPs (Supplementary Materials; Figure S6A–C). The Size detected by SEM was congruent with the size previously reported by Dynamic Light Scattering [13] when using the same synthesis method detailed in El Semary and Bakir, 2022 [17] where the average of the three size ranges was 34 nm for the smallest, 67 nm for the medium, and 93 nm for the largest nanoparticles.

3.5. The Antimicrobial Bioassay of Biogenic Nanosilver Samples against Plant Pathogenic Bacteria

The three samples of biogenic silver nanoparticles were analysed for their ability to adversely inhibit these diverse pathogenic bacteria in order to investigate their broad-spectrum antimicrobial activity. The antimicrobial bioassay showed broad spectrum action of silver nanoparticles of different sizes. They were effective against gram-positive *Staphylococcus warneri* and gram negative *Xanthomonas citri* and *Erwinia pyrifoliae* (Table 1). The statistical analysis showed significant differences among the different treatments as compared to negative control. All three samples of silver nanoparticles were effective as antimicrobial agents against the three tested local plant pathogens but with varying degrees of efficacy. The smallest sized silver nanoparticle was the most active. This is in total agreement with what Dong F. et al. [20] reported that AgNPs were effective at different sizes and concentrations and that the smaller the particle size, the greater the antibacterial activity.

Table 1. Antimicrobial bioassay of three different-sized silver nanoparticles against three plant pathogenic bacteria. Smallest silver nanoparticle average size is 33.9 nm, Medium average size is 67 nm and largest sized is 92.5 nm. Negative control (water) had no inhibition zone whereas positive control inhibition zone diameter was 2.5 cm.

Bacteria	Inhibition Zone of Smallest AgNPs (cm)	Inhibition Zone of Medium AgNPs (cm)	Inhibition Zone of Largest AgNPs (cm)
Erwinia pyrifoliae	1.0 ± 0.2	0.8 ± 0.1	0.6 ± 0.1
Staphylococcus warneri	1.7 ± 0.1	1.5 ± 0.3	1.0 ± 0.1
Xanthomonas citri	1.0 ± 0.1	0.9 ± 0.3	0.9 ± 0.1

The results of the disc diffusion method-based antibacterial test of the PT-AgNPs against *Erwinia amylovora, Staphylococcus warneri, and Xanthomonas citri* are shown in Figure 4 and Supplementary Materials Figure S7. Except for *Staphylococcus warneri*, for which the MIC was 50 g/mL, the data showed that PT-AgNPs had outstanding antibacterial efficacy against all tested bacterium strains. Additionally, the outcomes of the time-kill studies demonstrated that all the examined bacteria were killed by the AgNPs at 4 MIC after 2 h (Figure 4, Tables 2–4).

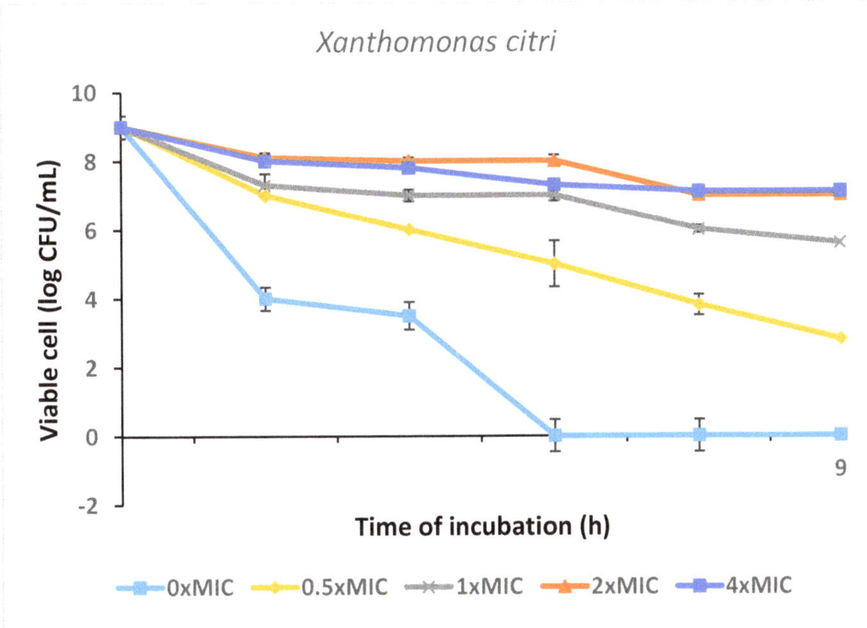

(a)

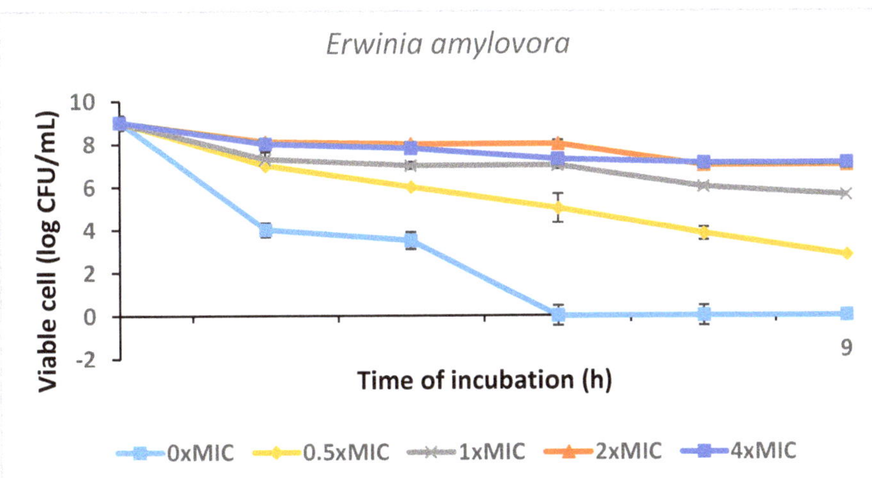

(b)

Figure 4. *Cont.*

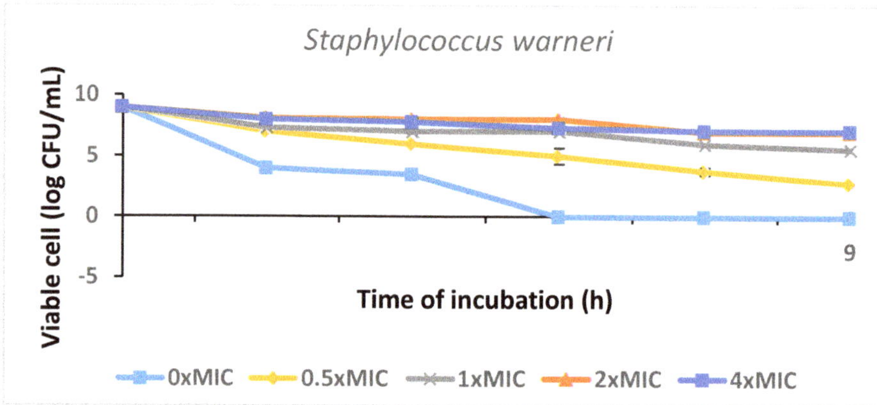

(c)

Figure 4. **(a–c)**. The time-kill curve plots of *S. enteritidis*, after the exposure to the PT-AgNPs at 0 × MIC, 0.5 × MIC, 1 × MIC, 2 × MIC, and 4 × MIC.

Table 2. *Erwinia amylovora*.

Incubation Time (Hours)	MIC (% *w*/*v*)				
	0	**0.5**	**1**	**2**	**4**
0	$_a$ 8 ^{a} ± 0.9	$_a$ 8 ^{a} ± 0.87	$_a$ 8 ^{a} ± 0.77	$_a$ 8.1 ^{a} ± 0.91	$_a$ 8 ^{a} ± 0.78
0.5	$_b$ 7.5 ^{a} ± 0.6	$_b$ 7.1 ^{b} ± 0.68	$_b$ 6.3 ^{c} ± 0.35	$_b$ 5.2 ^{d} ± 0.35	$_b$ 3.1 ^{e} ± 0.32
1	$_b$ 7.5 ^{a} ± 0.4	$_b$ 7 ^{b} ± 0.56	$_c$ 5.7 ^{c} ± 0.39	$_c$ 4.6 ^{d} ± 0.69	$_c$ 2.5 ^{e} ± 0.21
2	$_b$ 7.2 ^{a} ± 0.4	$_b$ 7 ^{b} ± 0.43	$_c$ 5.7 ^{c} ± 0.21	$_d$ 3.3 ^{d} ± 0.28	$_d$ 0 ^{e} ± 0.0
4	$_d$ 7 ^{a} ± 0.22	$_c$ 6 ^{b} ± 0.2	$_d$ 4.7 ^{c} ± 0.11	$_e$ 2.8 ^{d} ± 0.12	$_d$ 0 ^{e} ± 0.0
6	$_d$ 7 ^{a} ± 0.35	$_b$ 6 ^{c} ± 0.4	$_e$ 4 ^{c} ± 0.22	$_f$ 1.7 ^{d} ± 0.13	$_d$ 0 ^{e} ± 0.0

Table 3. *Staphylococcus warneri*.

Incubation Time (Hours)	MIC (% *w*/*v*)				
	0	**0.5**	**1**	**2**	**4**
0	$_a$ 8 ^{a} ± 0.8	$_a$ 8 ^{a} ± 0.91	$_a$ 8 ^{a} ± 0.7	$_a$ 8.1 ^{a} ± 0.81	$_a$ 8 ^{a} ± 0.8
0.5	$_b$ 7.4 ^{a} ± 0.5	$_b$ 7 ^{b} ± 0.68	$_b$ 6 ^{c} ± 0.5	$_b$ 5 ^{d} ± 0.3	$_b$ 3 ^{e} ± 0.1
1	$_b$ 7.4 ^{a} ± 0.3	$_b$ 6.7 ^{b} ± 0.56	$_c$ 5.5 ^{c} ± 0.29	$_c$ 4.4 ^{d} ± 0.9	$_c$ 2.1 ^{e} ± 0.14
2	$_c$ 7 ^{a} ± 0.2	$_b$ 6.7 ^{b} ± 0.43	$_c$ 5.5 ^{c} ± 0.21	$_d$ 3.1 ^{d} ± 0.2	$_d$ 0 ^{e} ± 0.0
4	$_d$ 6 ^{a} ± 0.12	$_c$ 6 ^{b} ± 0.2	$_d$ 4.2 ^{c} ± 0.1	$_e$ 2.2 ^{d} ± 0.1	$_d$ 0 ^{e} ± 0.0
6	$_d$ 6 ^{a} ± 0.44	$_b$ 6 ^{c} ± 0.4	$_e$ 4 ^{c} ± 0.12	$_f$ 1.5 ^{d} ± 0.1	$_d$ 0 ^{e} ± 0.0

Table 4. *Xanthomonas citri.*

Incubation Time (Hours)	MIC (% *w/v*)				
	0	0.5	1	2	4
0	$_{a}8^{a} \pm 0.9$	$_{a}8^{a} \pm 0.8$	$_{a}8^{a} \pm 0.1$	$_{a}8.1^{a} \pm 0.1$	$_{a}8^{a} \pm 0.71$
0.5	$_{b}7.1^{a} \pm 0.4$	$_{b}7^{b} \pm 0.2$	$_{b}6.1^{c} \pm 0.4$	$_{b}5^{d} \pm 0.5$	$_{b}4.1^{e} \pm 0.4$
1	$_{b}7.1^{a} \pm 0.3$	$_{b}6.6^{b} \pm 0.4$	$_{c}5.1^{c} \pm 0.29$	$_{c}4.5^{d} \pm 0.9$	$_{c}3.5^{e} \pm 0.3$
2	$_{c}7^{a} \pm 0.3$	$_{b}6.6^{b} \pm 0.3$	$_{c}5.1^{c} \pm 0.22$	$_{d}3.2^{d} \pm 0.8$	$_{d}0^{e} \pm 0.0$
4	$_{d}6.5^{a} \pm 0.1$	$_{c}5.5^{b} \pm 0.1$	$_{d}4.3^{c} \pm 0.1$	$_{e}2.4^{d} \pm 0.3$	$_{d}0^{e} \pm 0.0$
6	$_{d}6.5^{a} \pm 0.5$	$_{b}5.5^{c} \pm 0.3$	$_{e}4^{c} \pm 0.3$	$_{f}1.1^{d} \pm 0.2$	$_{d}0^{e} \pm 0.0$

Data are expressed as the mean zone of inhibition in mm followed by SD. The values with different subscript letters in the same column and those with different superscript letters in the same row are significantly different according to an ANOVA and Duncan's multiple range tests.

4. Discussion

Molecular and phylogenetic analyses confirmed the identity of *Erwinia pyrifoliae,* which is a gram-negative bacterium that causes shoot blight, a disease that affects trees of apples and pears. The disease can kill fruit, shoots, and entire trees. The molecular and phylogenetic analyses also confirmed the identity of *Xanthomonas citri* which is a gram-negative bacterium that causes citrus canker. *Staphylococcus warneri* on the other hand is a gram- positive bacterium. These pathogenic bacteria, alongside the previously identified *Staphylococcus warneri,* were test organisms in the antimicrobial bioassay using biogenic silver nanoparticles. The biological synthesis of nanoparticles helps avoid the chemical and physical synthetic methods with their lengthy procedures and hazardous by-products. Biological synthesis also allows the cost-effective synthesis of nanoparticles at physiological pH which renders them biocompatible with cellular systems [17,20]. Cyanobacteria have been found to be an eco-friendly, cost-effective bio-factory of biogenic nanoparticles [4]. Indeed, the *Cyanothece*-like isolate used here in the current study proved to be an efficient nanoparticle producer and was employed to biosynthesize three different sized silver nanoparticles efficiently. The filtration of the external solution using Millipore filters of 0.2 μm pore diameter resulted in the removal of macromolecules mostly leaving ions in the cell free filtrate. According to [7], those ions are involved in the reduction of silver ions and formation of silver nanoparticles. The success in the synthesis of three-sized silver nanoparticles was confirmed by chemical analyses including UV, FTIR and DLS as well as SEM. The UV-Visible spectrum showed the characteristic plasmon resonance of silver nanoparticles is due to the vibrational modes of electrons based on the size and shape of nanoparticles [17].

The lack of functional groups associated with silver nanoparticles as FTIR results show indicates the effectiveness of filtration in preventing macromolecules from associating with silver nanoparticles, thereby preventing the formation of corona. This corona usually develops from interfacial interactions between silver nanoparticles and biological fluids including proteins. Corona formation can lead to limited cytotoxicity of silver nanoparticles where it interferes with the dissolution of AgNPs to toxic silver ions [24,25]. In our protocol, the microfiltration was apparently effective in preventing corona formation and hence the cytotoxicity of AgNPs was preserved. Nevertheless, the FTIR consistently showed the characteristic band indicative of AgNPs which overlaps with the OH band [17].

A scanning electromicroscopic analysis was used (SEM) for the determination of shape and size. SEM showed that silver nanoparticles were angular in shape. and this may contribute to the high broad spectrum bactericidal effect. Indeed, it was reported that silver nanocubes showed greater antibacterial effect against *Escherichia coli* (Gram negative) as compared to silver nanospheres [8]. Previously, DLS analysis revealed the size range for the smallest sized sample which was derived from 0.2 mL $AgNO_3$ was of average 33.9 nm, the middle with an average of 67 nm, and the largest average was 92.5 nm [17]. The synthesis of three sizes of silver nanoparticles was further confirmed in this study by scanning electron microscopy which also confirmed the shape of the produced particles. The biogenic AgNPs had angular form and their size coincided with the DLS results previously reported [17].

The proposed mechanisms by which cyanobacteria produce silver nanoparticles was the focus of several studies. For example, Ali et al. [26] reported that filamentous cyanobacteria reduce silver ions using ß-NADPH-dependent nitrate reductase. Lengke et al. [4] reported that the biosynthesis of silver nanoparticles in cyanobacteria occurs both inside the cells (with size < 10 nm) and in solution (with size 1–200 nm), forming spherical and octahedral platelets over the time course. This is in agreement with our results where the different sized nanoparticles in the external solution fell within the range reported by [4]. The shape of our particles also corresponds to the octahedral platelets [4].

Ferdous and Nemmar [24] reviewed the proposed mechanisms of antibacterial action of the silver nanoparticles. One scenario proposes the interaction of silver nanoparticles with bacteria, either through interacting with bacterial membrane, damaging it and killing the bacteria [27] or by inducing reactive oxygen species (ROS) via interaction with respiratory chain proteins on the membrane and interrupting oxygen reduction. Eventually this ROS causes cellular oxidative stress and bacterial death [28]. Another scenario is based on the adhesion of AgNPs to the bacterial wall, following [27]. It is also reported that small-sized AgNPs cause cytotoxicity in both gram-positive and gram-negative bacteria [29]. Smaller silver nanoparticles have a larger surface area to volume ratio, and they show a faster rate of silver ion dissolution in their microenvironment, thereby inducing greater cytotoxicity than larger NPs [24,30]. The released Ag^+ ions interact with respiratory chain proteins and induce the production of reactive oxygen species (ROS), thereby causing cellular oxidative stress in microbes and death [24,31] Inside the bacterial cell, silver nanoparticles interact with sulphur and phosphorus, causing DNA replication inhibition. They also inhibit protein synthesis by denaturing ribosomes [7]. Future application of the current results will include the use of nanosilver in coating films used for packaging, in order to keep them sterile and hygienic [32].

5. Conclusions

The broad-spectrum antimicrobial effect is due to ease of penetration of AgNPs. The smaller the AgNPs, the more easily they penetrate cells. This clearly shows the potentials of using these biogenic nanoparticles against plant pathogens as an alternative to harmful pesticides.

Supplementary Materials: The following supporting information can be downloaded at: https://www.mdpi.com/article/10.3390/antibiotics12071114/s1, Figure S1: UV spectrum of the small sized nanosilver; Figure S2: Uv spectrum of the medium-sized nanosilver; Figure S3: FTIR spectrum of smallest sized (34 nm) biogenic silver nanoparticles; Figure S4: FTIR spectrum of medium-sized (67 nm) biogenic silver nanoparticles; Figure S5: FTIR spectrum of Largest sized (92 nm) biogenic silver nanoparticles; Figure S6: Scanning electron micrograph showing (A) small (average 34 nm), (B) medium (average 67 nm) and (C) large sized (average 92 nm) nanoparticle; Figure S7: The time-kill curve plots of Staphylococcus warneri, Erwinia amylovora and Xanthomonas citri after the exposure to the PT-AgNPs at 0 × MIC, 0.5 × MIC, 1 × MIC, 2 × MIC, and 4 × MIC.

Author Contributions: Conceptualization, N.E.S.; Methodology, N.E.S., M.F.A. software, N.E.S.; validation, N.E.S., M.F.A.; formal analysis, N.E.S., M.F.A.; investigation, N.E.S., M.F.A.; resources, N.E.S., M.F.A.; data curation N.E.S., M.F.A.; writing—original draft preparation N.E.S.; writing—review and editing, N.E.S., D.G.A.; visualization, N.E.S., M.F.A.; supervision, N.E.S.; project administration, N.E.S.; funding acquisition, M.F.A. All authors have read and agreed to the published version of the manuscript.

Funding: The research was funded by Deanship of Scientific Research, Vice presidency for Graduate studies and Scientific Research, King Faisal University, Al-Ahsa 31982, Kingdom of Saudi Arabia, Grant number Annual grant 3141.

Institutional Review Board Statement: Not applicable.

Informed Consent Statement: Not applicable.

Data Availability Statement: Not applicable.

Acknowledgments: Nermin El Semary, would like to thank Aziza AlIbrahim for help with the UV-Visible spectroscopy and Al Anoud AlHomaidy for help with the FTIR.

Conflicts of Interest: The authors declare no conflict of interest.

Abbreviations

AgNPs (Silver nanoparticles); MIC (minimum inhibitory concentration), DMSO (dimethylsulfoxide).

References

1. Abou El-Nour, K.M.; Eftaiha, A.A.; Al-Warthan, A.; Ammar, R.A. Synthesis and applications of silver nanoparticles. *Arab. J. Chem.* **2010**, *3*, 135–140. [CrossRef]
2. Younis, N.S.; El Semary, N.A.; Mohamed, M.E. Silver nanoparticles green synthesis via cyanobacterium *Phormidium* sp.: Characterization, wound healing, antioxidant, antibacterial, and anti-Inflammatory activities. *Eur. Rev. Med. Pharam. Sci.* **2021**, *25*, 3083–3096.
3. Mohamed, M.E.; El Semary, N.A.; Younis, N.S. Silver nanoparticle production by the cyanobacterium *Cyanothece* sp.: De Novo manipulation of nano-biosynthesis by phytohormones. *Life* **2022**, *12*, 139. [CrossRef] [PubMed]
4. Lengke, M.F.; Fleet, M.E.; Southam, G. Biosynthesis of silver nanoparticles by filamentous cyanobacteria from a silver(I) nitrate complex. *Langmuir* **2007**, *23*, 2694–2699. [CrossRef]
5. Patel, V.; Berthold, D.; Puranik, P.; Gantar, M. Screening of cyanobacteria and microalgae for their ability to synthesize silver nanoparticles with antibacterial activity. *Biotechnol. Rep.* **2015**, *5*, 112–119. [CrossRef]
6. Parial, D.; Patra, H.K.; Roychoudhury, P.; Dasgupta, A.K.; Pal, R. Gold nanorod production by cyanobacteria—A green chemistry approach. *J. Appl. Phycol.* **2012**, *24*, 55–60. [CrossRef]
7. Hamida, R.S.; Ali, M.A.; Redhwan, A.; Bin-Meferij, M.M. Cyanobacteria—A promising platform in green nanotechnology: A Review on nanoparticles fabrication and their prospective applications. *Int. J. Nanomed.* **2020**, *15*, 6033–6066. [CrossRef] [PubMed]
8. Hong, X.; Wen, W.J.; Xiong, X.; Hu, Y. Silver nanowire-carbon fiber cloth nanocomposites synthesized by UV curing adhesive for electrochemical point-of-use water disinfection. *Chemosphere* **2016**, *154*, 537–545. [CrossRef]
9. Pal, S.; Yu, K.T.; Song, J.M. Does the antibacterial activity of silver nanoparticles depend on the shape of the nanoparticle? A study of the gram-negative bacterium Escherichia coli. *Appl. Environ. Microbiol.* **2007**, *73*, 1712–1720. [CrossRef] [PubMed]
10. Jeong, Y.; Lim, D.; Choi, J. Assessment of Size-Dependent Antimicrobial and Cytotoxic Properties of Silver Nanoparticles. *Adv. Mater. Sci. Eng.* **2014**, *2014*, 763807. [CrossRef]
11. Oves, M.; Rauf, M.A.; Qari, H.A. Therapeutic Applications of Biogenic Silver Nanomaterial Synthesized from the Paper Flower of *Bougainvillea glabra* (Miami, Pink). *Nanomaterials* **2023**, *13*, 615. [CrossRef]
12. More, P.R.; Pandit, S.; Filippis, A.; Franci, G.; Mijakovic, I.; Galdiero, M. Silver Nanoparticles: Bactericidal and Mechanistic Approach against Drug Resistant Pathogens. *Microorganisms* **2023**, *11*, 369. [CrossRef] [PubMed]
13. Sampath, G.; Chen, Y.-Y.; Rameshkumar, N.; Krishnan, M.; Nagarajan, K.; Shyu, D.J.H. Biologically Synthesized Silver Nanoparticles and Their Diverse Applications. *Nanomaterials* **2022**, *12*, 3126. [CrossRef]
14. Trzcińska-Wencel, J.; Wypij, M.; Rai, M.; Golińska, P. Biogenic nanosilver bearing antimicrobial and antibiofilm activities and its potential for application in agriculture and industry. *Front. Microbiol.* **2023**, *14*, 1125685. [CrossRef]
15. Hamida, R.S.; Abdelmeguid, N.E.; Ali, M.A.; Bin-Meferij, M.M.; Khalil, M.I. Synthesis of Silver Nanoparticles Using a Novel Cyanobacteria *Desertifilum* sp. extract: Their Antibacterial and Cytotoxicity Effects. *Int. J. Nanomed.* **2020**, *15*, 49–63. [CrossRef] [PubMed]
16. Rippka, R.; Deruelles, J.; Waterbury, J.B.; Herdman, M.; Stanier, R.Y. Generic assignments, strain histories and properties of pure cultures of cyanobacteria. *J. Gen. Microbiol.* **1979**, *111*, 1–61. [CrossRef]
17. El Semary, N.; Bakir, E. Multidrug-resistant bacterial pathogens and public health: The antimicrobial effect of cyanobacterial-biosynthesized silver nanoparticles. *Antibiotics* **2022**, *11*, 1003. [CrossRef] [PubMed]
18. Patra, J.K.; Baek, K.-H. Antibacterial Activity and Synergistic Antibacterial Potential of Biosynthesized Silver Nanoparticles against Foodborne Pathogenic Bacteria along with its Anticandidal and Antioxidant Effects. *Front. Microbiol.* **2017**, *8*, 167. [CrossRef] [PubMed]
19. Kubo, I.; Fujita, K.; Kubo, A.; Nihei, K.; Ogura, T. Antibacterial activity of coriander volatile compounds against Salmonella choleraesuis. *J. Agric. Food Chem.* **2004**, *52*, 3329–3332. [CrossRef]

20. Dong, Y.; Zhu, H.; Shen, Y.; Zhang, W.; Zhang, L. Antibacterial activity of silver nanoparticles of different particle size against Vibrio natriegens. *PLoS ONE* **2019**, *14*, e0222322. [CrossRef]
21. Zhang, X.-F.; Liu, Z.-G.; Shen, W.; Gurunathan, S. Silver Nanoparticles: Synthesis, Characterization, Properties, Applications, and Therapeutic Approaches. *Int. J. Mol. Sci.* **2016**, *17*, 1534. [CrossRef]
22. Mota, R.; Flores, C.; Tamagnini, P. Cyanobacterial Extracellular Polymeric Substances (EPS). In *Polysaccharides of Microbial Origin*; Oliveira, J.M., Radhouani, H., Reis, R.L., Eds.; Springer: Cham, Switzerland, 2021. [CrossRef]
23. Gharibshahi, L.; Saion, E.; Gharibshahi, E.; Shaari, A.H.; Matori, K.A. Structural and Optical Properties of Ag Nanoparticles Synthesized by Thermal Treatment Method. *Materials* **2017**, *10*, 402. [CrossRef] [PubMed]
24. Ferdous, Z.; Nemmar, A. Health impact of silver nanoparticles: A Review of the biodistribution and toxicity following various routes of exposure. *Int. J. Mol. Sci.* **2020**, *21*, 2375. [CrossRef] [PubMed]
25. Durán, N.; Silveira, C.P.; Durán, M.; Martinez, D.S.T. Silver nanoparticle protein corona and toxicity: A mini-review. J. *Nanobiotechnol.* **2015**, *13*, 55. [CrossRef] [PubMed]
26. Ali, D.M.; Sasikala, M.; Gunasekaran, M.; Thajuddin, N. Biosynthesis and characterization of silver nanoparticles using marine cyanobacterium, Oscillatoria willei NTDM01. *Dig. J. Nanomater. Biostruct.* **2011**, *6*, 385–390.
27. Qing, Y.; Cheng, L.; Li, R.; Liu, G.; Zhang, Y.; Tang, X.; Wang, J.; Liu, H.; Qin, Y. Potential antibacterial mechanism of silver nanoparticles and the optimization of orthopedic implants by advanced modification technologies. *Int. J. Nanomed.* **2018**, *13*, 3311. [CrossRef]
28. Lee, S.H.; Jun, B.-H. Silver nanoparticles: Synthesis and application for nanomedicine. *Int. J. Mol. Sci.* **2019**, *20*, 865. [CrossRef]
29. Das, R.; Gang, S.; Nath, S.S. Preparation and antibacterial activity of silver nanoparticles. *J. Biomater. Nanobiotechnol.* **2011**, *2*, 472. [CrossRef]
30. Shang, L.; Nienhaus, K.; Nienhaus, G.U. Engineered nanoparticles interacting with cells: Size matters. *J. Nanobiotechnol.* **2014**, *12*, 5. [CrossRef]
31. Long, Y.-M.; Hu, L.-G.; Yan, X.-T.; Zhao, X.-C.; Zhou, Q.-F.; Cai, Y.; Jiang, G.-B. Surface ligand controls silver ion release of nanosilver and its antibacterial activity against Escherichia coli. *Int. J. Nanomed.* **2017**, *12*, 3193. [CrossRef]
32. Baysal, G.; Demirci, C.; Özpinar, H. Proporties and synthesis of biosilver nanofilms for antimicrobial food packaging. *Polymers* **2023**, *15*, 689. [CrossRef] [PubMed]

Disclaimer/Publisher's Note: The statements, opinions and data contained in all publications are solely those of the individual author(s) and contributor(s) and not of MDPI and/or the editor(s). MDPI and/or the editor(s) disclaim responsibility for any injury to people or property resulting from any ideas, methods, instructions or products referred to in the content.

antibiotics

Article

Antifungal Activity of Mycogenic Silver Nanoparticles on Clinical Yeasts and Phytopathogens

Luiz Gustavo Ribeiro [1,2], Gabriella Sales Calaço Roque [1,2], Rafael Conrado [1] and Ana Olívia De Souza [1,2,*]

1 Development and Innovation Laboratory, Instituto Butantan, Avenida Vital Brasil, 1500, São Paulo 05503-900, SP, Brazil
2 Department of Surgery, Faculty of Veterinary Medicine and Animal Science, University of São Paulo, São Paulo 05508-270, SP, Brazil
* Correspondence: ana.souza@butantan.gov.br

Abstract: In this study, seven different silver nanoparticles (AgNPs) were obtained using the fungi species from the phylum Ascomycota, *Aspergillus tubingensis*, *Aspergillus* spp., *Cladosporium pini-ponderosae*, *Fusarium proliferatum*, *Epicoccum nigrum*, *Exserohilum rostratum*, and *Bionectria ochroleuca*, isolated from the Brazilian biodiversity, particularly from the mangrove and Caatinga biomes. The nanoparticles were coded as AgNP-AT, AgNP-Asp, AgNP-CPP, AgNP-FP, AgNP-EN, AgNP-ER, and AgNP-BO and characterized using spectrophotometry (UV-Vis), dynamic light scattering (DLS), zeta potential, transmission electron microcopy (TEM), and Fourier-transform infrared (FTIR) spectroscopy. All the AgNPs presented homogeneous size in the range from 43.4 to 120.6 nm (DLS) and from 21.8 to 35.8 nm (TEM), pH from 4.5 to 7.5, negative charge, and presence of protein coating on their surface. The antifungal activity of the AgNPs was evaluated on clinical strains of *Candida albicans*, and on the non-albicans species, *Candida krusei*, *Candida glabrata*, *Candida parapsilosis*, *Candida tropicalis*, and *Candida guilliermondii*, common in hospital infections, and against the phytopathogens *Fusarium oxysporum*, *Fusarium phaseoli*, *Fusarium sacchari*, *Fusarium subglutinans*, *Fusarium verticillioides*, and *Curvularia lunata*, which are species responsible for serious damage to agriculture production. The AgNPs were effective against the yeasts with MICs ranging from 1.25 to 40 μM and on the phytopathogens with MICs from 4 to 250 μM, indicating the promising possibility of application of these AgNPs as antifungal agents. The results indicated that the physicochemical parameters of the AgNPs, including the functional groups present on their surface, interfered with their antifungal activity. Overall, the results indicate that there is no specificity of the AgNPs for the yeasts or for the phytopathogens, which can be an advantage, increasing the possibility of application in different areas.

Keywords: biogenic silver nanoparticles; phytopathogens; *Candida* sp.; antifungal activity

Citation: Ribeiro, L.G.; Roque, G.S.C.; Conrado, R.; De Souza, A.O. Antifungal Activity of Mycogenic Silver Nanoparticles on Clinical Yeasts and Phytopathogens. *Antibiotics* **2023**, *12*, 91. https://doi.org/10.3390/antibiotics12010091

Academic Editors: Sotiris K Hadjikakou and Christina N. Banti

Received: 28 November 2022
Revised: 30 December 2022
Accepted: 31 December 2022
Published: 5 January 2023

Copyright: © 2023 by the authors. Licensee MDPI, Basel, Switzerland. This article is an open access article distributed under the terms and conditions of the Creative Commons Attribution (CC BY) license (https://creativecommons.org/licenses/by/4.0/).

1. Introduction

The growth of fungal infections has become a major threat to public health, with more than 300 million people suffering from fungal diseases each year, resulting in over 1 million deaths worldwide [1,2]. In addition to infections caused by fungi in humans and animals, there are also problems caused by phytopathogens in several types of crops that are important for food production [3,4].

A recent study reported that at least 137 phytopathogens and pests are responsible for infections in potato, wheat, rice, maize, and soybean, causing massive crop losses [5]. The loss in the production is estimated to be 17.2% for potato, around 20% for soybean, wheat, and maize, and 30.0% for rice [5]. The food loss is important in all countries; however, the maximum loss has been observed in food-insecure regions with high populations, significantly increasing the problem of starvation [5].

The situation is worsened by the increased resistance of these pathogens to the currently available treatments, which are scarce, for infections both in humans [2] and animals [6] and for phytopathogens [7]. Microbial resistance is considered a global health problem, which compromises the effectiveness of antibiotics, making the treatment of common infections unfeasible [8,9].

Candida sp. species are a global threat to public health, with frequent outbreaks in hospitals and a high rate of mortality [10]. The genus Candida can be found in several ecosystems, such as in the microbiota of man and of some animals. *Candida* sp. species degrade proteins and carbohydrates to acquire carbon and nitrogen for their development and have an adaptive ability to multiply in aerobic and anaerobic conditions. These species are commensal and usually are present in the gastrointestinal tract, vagina, urethra, and lung microbiota. In case of an imbalance in the body's microbiota, *Candida* sp. can become pathogenic for the host, causing candidiasis or even a severe candidemia, being, due to that, considered opportunistic microorganisms [11–14].

Among 15 Candida species usually present in infections in humans, *Candida albicans*, *Candida glabrata*, *Candida tropicalis*, *Candida parapsilosis*, and *Candida krusei* are the most common, causing more than 90% of all the candidemia cases [15].

A study performed in Portugal, in 2016 and 2017, reported 117 cases of candidemia in 114 patients, and among them, 51.3% were caused by *C. albicans*, followed by 22.2%, 15.4%, 4.3%, and 2.6% caused by *C. glabrata*, *C. parapsilosis*, *C. tropicalis*, and *C. lusitaniae*, respectively [10].

The most common fungi species responsible for frequent and problematic contamination of foods and feeds due to mycotoxins production belongs to the genera *Aspergillus*, *Fusarium*, and *Penicillium* [16–18]. Under natural conditions, these species are able to produce mycotoxins, in pre- and post-harvest cultures, which can be extremely harmful to humans and animals. The exposure to these mycotoxins occurs primarily by ingestion but can also occur by the dermal and inhalation routes. The effects of some food-borne mycotoxins can be acute, with symptoms of severe illness appearing rapidly after the consumption of food products contaminated [16–19].

For the treatment and successful control of the serious threats caused by phytopathogenic and human pathogenic fungi, new strategies against the shortage of antifungal drugs are necessary.

Nanomaterials of sizes between 1 and 100 nm have been successfully used in different medical and pharmaceutical applications [20]. In general, nanoscale materials have excellent antimicrobial activity due to the high surface area to volume ratio and physical chemical properties [20]. Several studies have demonstrated the antifungal activity of metallic nanoparticles, and among them, silver nanoparticles (AgNPs) represent a potential option as antimicrobial agents [20–25].

AgNPs can be obtained using physical, chemical, or biological methods [20,23–25]. However, biological synthesis stands out as an ecologically sustainable process, economically viable and able to be scaled up to industrial production without a big challenge. In the biological route, factors such as pH, luminosity, chemical characteristics, and concentration of the substrate or reducing agents are factors that can interfere with the nucleation, growth, agglomeration, and stability of the nanoparticles [22,23,25].

Through the bioreduction of the metal precursor (i.e., silver, gold), reducing or stabilizing agents produced by plants, algae, bacteria, yeasts, or fungi, such as ether, thiol, carbonyl, hydroxyl, amine, polyamine, proteins, and peptides groups, can also perform a coat on the metallic nanoparticles surface [22,23,25]. These biological coating groups, usually from proteins [26], can improve the uptake and functionality of the nanoparticles, reducing their toxicity and improving their potential as candidates to treat diseases in humans, animals, and in agriculture [20,21,27,28].

The mycogenic biosynthesis of AgNPs has been reported with the application of several species of fungi [22,23,25,29] and presence of a protein coating surrounding the metal ion, which provides high stability, avoiding aggregation of the nanoparticles [22,23,25]. This property is an advantage that improves the interaction of the nanoparticles with biological targets and their biological effect [27,28].

Among the possible organisms capable of biosynthesizing AgNPs, fungi offer advantages because they are easy to manipulate, allowing a large-scale culture and biomass production that can be used for metal nanoparticles biosynthesis, which own a protein coat that can contribute to an improved biological activity [22,23,25].

Considering the high demand for new antifungal agents, due to the emergence of microbial resistance to the existing treatments and the potential of AgNPs as an antifungal alternative, in this study seven AgNPs were obtained by green synthesis, using seven species of fungi isolated from the Brazilian biodiversity, particularly from the mangrove and Caatinga biomes. The mangroves in the Brazilian coast are anaerobic and saline, and the Caatinga is an arid environment with high temperature and hot soil through all the year. Both these environments harbor extremophiles microorganisms, and the fungi *Aspergillus* spp., *Cladosporium pini-ponderosae*, *Fusarium proliferatum*, *Epicoccum nigrum*, *Aspergillus tubingensis*, and *Bionectria ochroleuca* applied in this study were previously isolated from mangrove, while *Exserohilum rostratum* was isolated from the Caatinga biome.

To the best of our knowledge, this is the first report on the biosynthesis of AgNPs using the species of *F. proliferatum*, *C. pini-ponderosae*, and *E. rostratum*. The obtained AgNPs were characterized according to their physicochemical properties, using spectrophotometry (UV-Vis), Fourier-transform infrared (FTIR) spectroscopy, transmission electron microscopy (TEM), dynamic light scattering (DLS), and zeta potential. The antifungal activity of the AgNPs was evaluated in *C. albicans*, *C. krusei*, *C. glabrata*, *C. parapsilosis*, *C. tropicalis*, and *Candida guilliermondii*, that are important pathogenic yeasts present in hospital infection [10], and on the phytopathogens *Fusarium oxysporum*, *Fusarium phaseoli*, *Fusarium sacchari*, *Fusarium subglutinans*, *Fusarium verticillioides*, and *Curvularia lunata*, which are species responsible for serious damage in the agricultural production.

2. Results and Discussion

2.1. Physicochemical Characterization of the AgNPs

The AgNPs obtained using the fungi *Aspergillus* spp., *E. nigrum*, *F. proliferatum*, *C. pini-ponderosae*, *E. rostratum*, *A. tubingensis*, and *B. ochroleuca* were coded by the initial of the respective fungus species applied in their biosynthesis, being AgNP-Asp, AgNP-EN, AgNP-FP, AgNP-CPP, AgNP-ER, AgNP-AT, and AgNP-BO, respectively.

The formation of the AgNPs was observed by the reaction of the extracellular cell free aqueous extract (AE) obtained from the culture of the seven fungi species with $AgNO_3$ as represented in Figure 1. As can be observed in Figure 2, there was a color change of the reactional mixtures from colorless to brownish, indicating the formation of AgNPs in the first hours of the reaction.

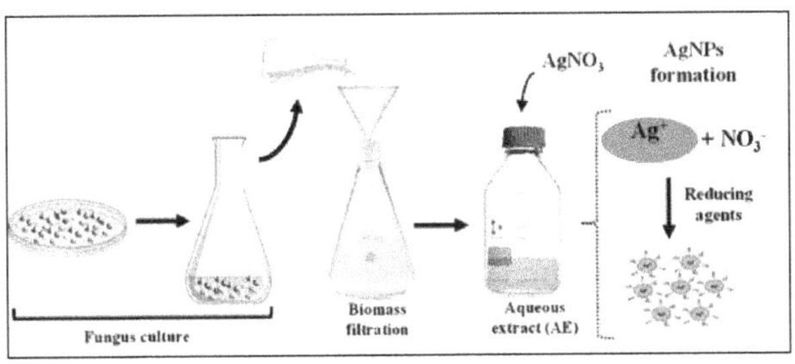

Figure 1. Representation of the silver nanoparticles biosynthesis by the reaction of the extracellular cell free aqueous extract (AE) obtained from the fungi culture with $AgNO_3$.

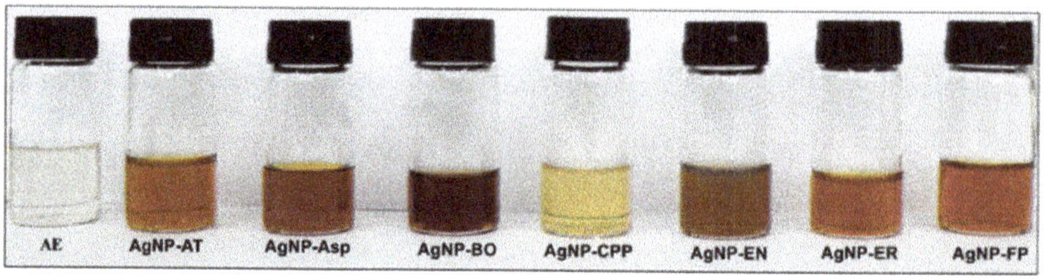

Figure 2. Color of the AgNP-AT, AgNP-Asp, AgNP-BO, AgNP-CPP, AgNP-EN, AgNP-ER, and AgNP-FP formed by the reaction of the extracellular cell free aqueous extract (AE) obtained using the fungi *A. tubingensis*, *Aspergillus* spp., *B. ochroleuca*, *C. pini- ponderosae*, *E. nigrum*, *E. rostratum*, and *F. proliferatum* with $AgNO_3$. Pictures were obtained 96 h after the beginning of the reaction.

The resonant oscillation of electrons density present on the surface of the AgNPs resulted in the formation of the surface plasmon resonance observed at 420 nm for all AgNPs (Figure 3A), which is the main characteristic of AgNPs formation [30].

It is possible to observe a difference in the intensity of the brownish color of the AgNPs (Figure 2) and in the absorbance (Figure 3A), which was higher for the AgNP-EN, followed by AgNP-ER, AgNP-Asp, AgNP-FP, and AgNP-CPP. The color intensity is related to the concentration of AgNPs formed after 96 h of reaction. The data allow inferring that the AgNP-EN was formed faster than the other AgNPs. The hydrodynamic sizes of the AgNPs measured by DLS were 44.9 ± 4.1, 43.4 ± 3.3, 120.6 ± 3.5, 87.1 ± 3.4, 71.2 ± 6.7, 86.4 ± 6.4, 59.6 ± 2.1 nm for the AgNP-Asp, AgNP-AT, AgNP-BO, AgNP-CPP, AgNP-EN, AgNP-ER, and AgNP-FP, respectively (Table 1). The sizes of the AgNPs measured by TEM were, respectively, 33.3 ± 2.7, 25.0 ± 6.5, 21.8 ± 4.1, 35.8 ± 5.16, 28.0 ± 6.31, 22.1 ± 2.9, 26.7 ± 5.3 nm. As expected, the sizes measured by TEM were smaller than those measured by DLS. Through the DLS, the protein corona surrounding the metal ion in the nanoparticles is also measured, and consequently, the size of the nanoparticles is higher. This data is interesting since the protein corona can interfere with the interaction of the nanoparticles with the target [27,28].

Table 1. Physicochemical properties of the AgNP-Asp, AgNP-AT, AgNP-BO, AgNP-CCP, AgNP-EN, AgNP-ER, and AgNP-FP: size (nm) by DLS and TEM, polydispersity index (PDI), zeta potential, and pH values.

AgNPs	Size (DLS) *	Size (TEM) *	PDI	Zeta Potential	pH
AgNP-Asp	44.9 ± 4.1	33.3 ± 2.7	0.343	−1.04	6.0
AgNP-AT	43.4 ± 3.3	25.0 ± 6.5	0.080	−22.26	7.0
AgNP-BO	120.6 ± 3.5	21.8 ± 4.1	0.257	−2.19	7.5
AgNP-CPP	87.1 ± 3.4	35.8 ± 5.16	0.334	−1.59	5.0
AgNP-EN	71.2 ± 6.7	28.0 ± 6.31	0.404	−33.28	5.0
AgNP-ER	86.4 ± 6.4	22.1 ± 2.9	0.230	−14.40	4.5
AgNP-FP	59.6 ± 2.1	26.7 ± 5.3	0.199	−15.33	6.0

* Mean ± standard deviation (n = 6).

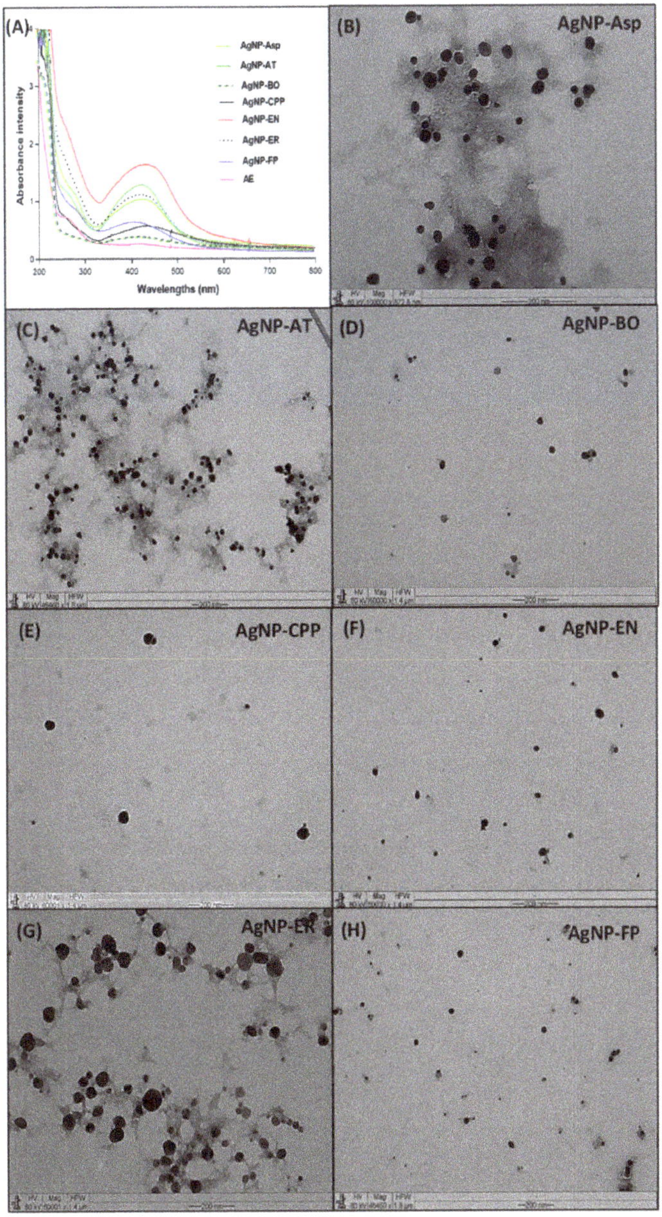

Figure 3. AgNPs characterization by (**A**) UV-Vis profile from 200 to 800 nm with presence of surface plasmon resonance around 420 nm for the AgNP-Asp, AgNP-AT, AgNP-BO, AgNP-CCP, AgNP-EN, AgNP-ER, and AgNP-FP, obtained using the extracellular cell free aqueous extract (AE) of the culture of *Aspergillus* spp., *A. tubingensis*, *B. ochroleuca*, *C. pini-ponderosae*, *E. nigrum*, *E. rostratum*, and *F. proliferatum*, respectively. Transmission electron microscopy for (**B**) AgNP-Asp (magnification 100,000×), (**C**) AgNP-AT (magnification 46,640×), (**D**) AgNP-BO (magnification 60,000×), (**E**) AgNP-CPP (magnification 60,001×), (F) AgNP-EN (magnification 50,000×), (**G**) AgNP-ER (magnification 60,001×), and (**H**) AgNP-FP (magnification 46,640×). The extracellular cell free aqueous extracts (AE) obtained from all fungi culture were used as control for the UV-Vis analysis and showed the same profile.

The polydispersity index (PDI) is related to the distribution of the AgNPs sizes. The smaller the PI values, the more homogeneous is the size of the AgNPs. The AgNP-AT presented the lowest PI, followed by AgNP-FP, AgNP-ER, AgNP-BO, AgNP-CPP, AgNP-Asp, and AgNP-EN. These data can be confirmed by the size and morphology observed for each AgNPs through the analysis by using TEM (Figure 3B–H). It can be also observed that all AgNPs are predominantly spherical, with a few or any aggregate formations (Figure 3B–F).

Literature describes that smaller AgNPs can be more cytotoxic than larger ones, due to intracellular uptake and greater surface area, which can promote more effective interaction with cells or their intracellular components [27,28,31]. The surface charge of AgNPs is defined by the presence of negatively or positively charged molecules on their coat, which is a relevant property for the nanoparticles' stability. The negative charge generates repulsiveness between particles and a low constant of force between then, preventing the formation of aggregates [32]. All seven AgNPs obtained in this study showed negative zeta potential values (Table 1) and, as it is shown in Figure 3B–F, there was no aggregation formation for almost all the AgNPs.

The pH value is an interesting parameter due to its influence on the size, shape, dispersion, and protein coating of AgNPs [33]. In the present study, the AgNPs with larger sizes showed lower pH values. Similar results were previously demonstrated in a study performed by Prakash and Soni [33], in which the size and shape of the AgNPs obtained using the fungi *Chrysosporium tropicum* and *F. oxysporum* were according to the pH and temperature applied for the synthesis. The pH of biogenic AgNPs is related to its protein coat, and both can interfere with the nanoparticles function and activity in biological systems, according to the possibility of nano–bio interaction [27,28,31].

The FTIR spectroscopy of the AgNPs showed the presence of functional groups attributed to aromatic rings, ether, carboxylic, and amide groups in their composition, which are from molecules present in the AE. These molecules acted as reducing agents for the formation of the AgNPs, or as capping agents promoting the protein corona formation and stability of the AgNPs (Table 2 and Figures 4 and 5). Additional functional groups were detected and attributed to amino acid residues present in the side chains of proteins presents in the AE (Table 2, and Figures 4 and 5).

The functional groups identified by FTIR for AgNP-Asp, AgNP-CCP, AgNP-EN, AgNP-ER, and AgNP-FP, and shown in Table 2, indicate the presence of several molecules around the nanoparticles' surface, as those previously reported for AgNP-AT and AgNP-BO [26,34]. The presence of a band in the region of 3200 cm^{-1} for all AgNPs was attributed to OH (chelate) groups and indicate the chelation of the metal ion (Ag^{+}) by molecules from the AE used for the biosynthesis of the AgNPs (Table 2 and Figures 4 and 5), as it was also observed for the AgNPs obtained using other microorganisms [35]. The results indicate that the secondary structure of proteins present in the AE used for the AgNPs biosynthesis was not modified by the oxidation–reduction reaction, or binding to the metallic core [26,34].

The effectiveness of AgNPs against specific targets is related to several aspects such as the morphology, size, PI, zeta potential, and pH values. In general, the AgNPs obtained in this study showed different functional groups on their surface forming the protein corona and interesting physicochemical properties, which are suitable for antimicrobial application. These groups were identified by FTIR spectroscopy and come from the AE used for the biosynthesis of the AgNPs (Table 2 and Figures 4 and 5). It is known that small and homogeneous size and absence of aggregates are interesting for the interaction of nanoparticles with biological systems [27,28,31], and in this study, in general the AgNPs showed this profile in their properties.

Table 2. Profile of vibrational frequencies (cm^{-1}) and functional groups present in the AgNP-Asp, AgNP-AT, AgNP-BO, AgNP-CPP, AgNP-EN, AgNP-ER, and AgNP-FP. Data obtained using Fourier-transform infrared (FTIR) spectroscopy.

AgNPs	≈Vibrational Frequencies (cm^{-1})	Functional Groups
AgNP-Asp	825–860	Aromatic ring (2 adjacent H)
	1020–1170	C-O (ether)
	1320	C-N (aromatic)
	1560–1490	N-H
	1790	C=O of acyl chloride
	2925	Aliphatic C-H
	3200	O-H (chelate)
AgNP-AT	840–775	R_2C=CHR (C-H out of plane)
	1165–1040	C-O (ether)
	1300	C-O (carboxyl)
	1590–1500	C=C (aromatic)
	2927	Aliphatic C-H
	3062	C-H (aromatic)
	3200	O-H (chelate)
AgNP-BO	825	Aromatic ring (2 adjacent H)
	1030–1018	C-O (ether)
	1350–1300	C-O (ester)
	1648	C=O (amides)
	2929–2900	Aliphatic C-H
	3400–3263	O-H (chelate)
AgNP-CPP	840–790	R_2C=CHR (C-H out of plane)
	1160–1021	C-O (ether)
	1289	C-N (aromatic)
	1490	N-H
	1625	C=C (aromatic)
	2936	O-H (chelate)
	3330	Free NH (secondary amine)
AgNP-EN	945	RCH=CH_2
	1190–1026	C-O (ethers)
	1336	SO_2 (sulfone)
	1500	C=C (aromatic)
	2360	CO_2
	2934	O-H (chelate)
	3343	Free NH (secondary amines)
AgNP-ER	820–798	R_2C=CHR
	1170–1025	C-O (ether)
	1312	SO_2 (sulfone)
	1493	N-H
	1621	C=C (aromatic)
	2936	Aliphatic C-H
	3230	O-H (chelate)
AgNP-FP	835–820	Aromatic ring
	1168–1020	C-O (ether)
	1300	C-O (ester)
	1490	N-H
	1550	NH_2
	1630	C=O (amides)
	1790	C=O (acyl chloride)
	2925	C-H (aliphatic)
	3055	C-H (alkene)
	3200	O-H (chelate)

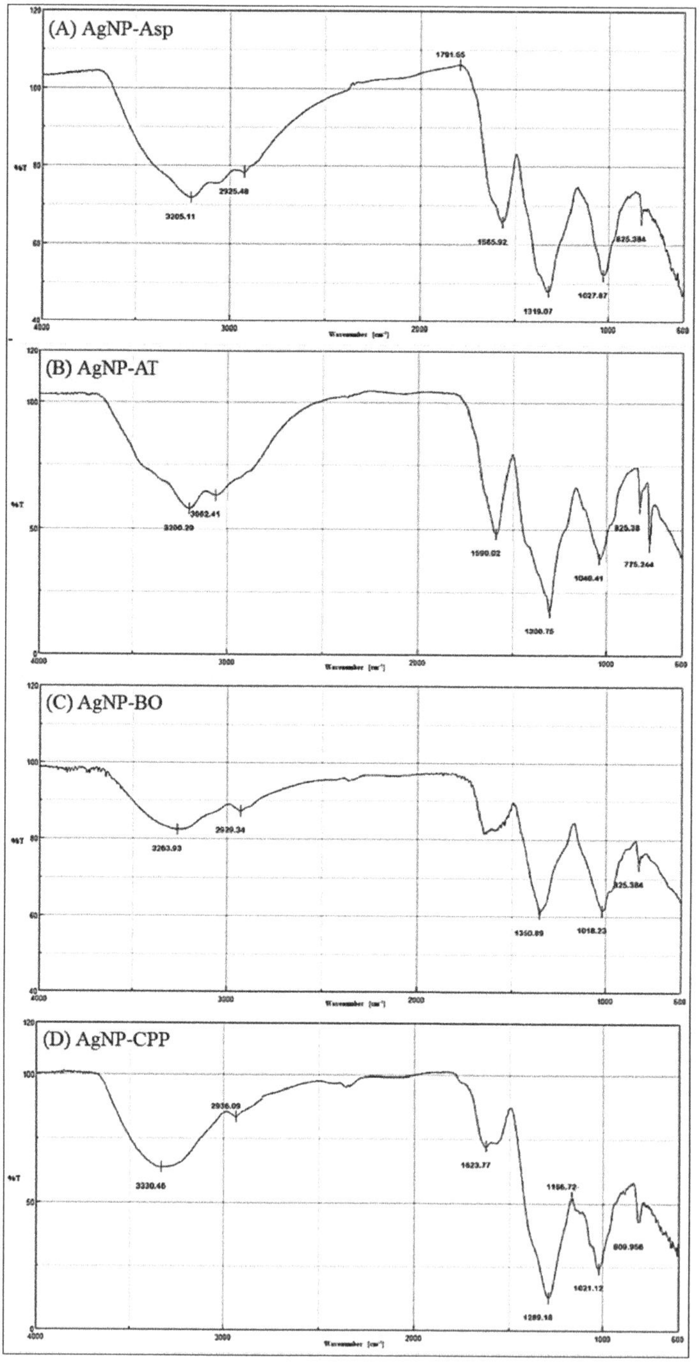

Figure 4. Fourier-transform infrared (FTIR) spectroscopy profile for the (**A**) AgNP-Asp, (**B**) AgNP-AT (**C**) AgNP-BO, and (**D**) AgNP-CPP.

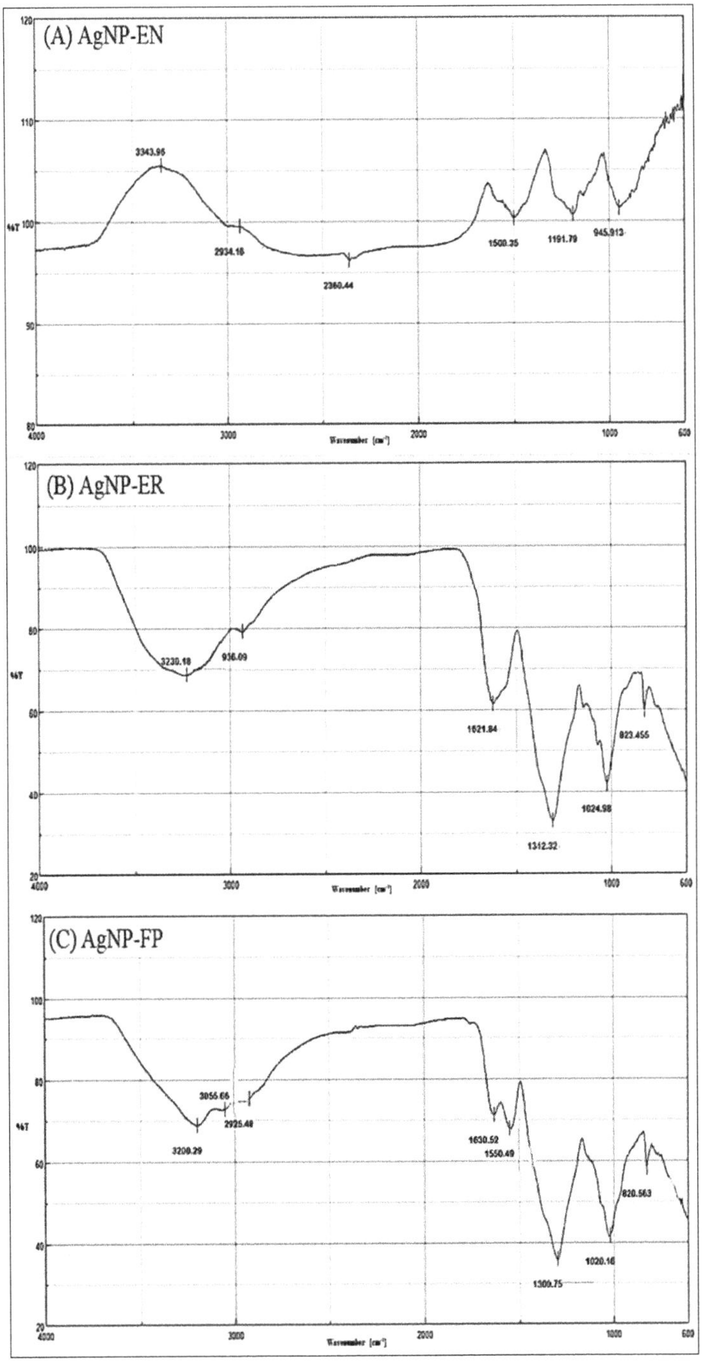

Figure 5. Fourier-transform infrared (FTIR) spectroscopy profile for the (**A**) AgNP-EN, (**B**) AgNP-ER, and (**C**) AgNP-FP.

2.2. Antifungal Activity

AgNPs have a broad spectrum of antimicrobial activity [20–24,34–37], including the ability to inhibit biofilm formation [36], and, due to that, are a promising option as antifungal agents. In this study, the antifungal activity of the AgNPs was evaluated against different species of *Candida* sp., common in hospital infections and against different relevant phytopathogens, capable of causing agricultural damage. The AgNPs showed antifungal activity against clinical strains of *C. albicans*, *C. krusei*, *C. glabrata*, *C. guillermondii*, *C. parapsilosis*, and *C. tropicalis*. The AgNPs were more effective than amphotericin B (AMB), which was used as the positive control, and was not effective until 32 μM in all the *Candida* sp. species. The AgNPs showed MIC in the range from 1.25 to 40 μM, and *C. guillermondii*, *C. krusei*, and *C. glabrata* were the most sensitive to the action of all AgNPs that showed MIC from 1.25 to 5 μM (Table 3). The only exception was the AgNP-FP, that until 40 μM was not effective on *C. glabrata*. It is noteworthy that the AgNP-CPP was not effective on any of the four *C. albicans* strains evaluated, but it was effective on non-albicans species with MICs of 1.25 to 5 μM (Table 3).

Table 3. Antifungal activity of the AgNP-Asp, AgNP-AT, AgNP-BO, AgNP-CPP, AgNP-EN, AgNP-ER, and AgNP-FP by minimal inhibitory concentration (MIC—μM) against an ATCC strain and clinical species of *Candida* sp.

Yeasts	AgNP-Asp	AgNP-AT	AgNP-BO	AgNP-CPP	AgNP-EN	AgNP-ER	AgNP-FP	AMB
C. albicans ATCC 36802	2.5	20	5	>40	1.25	20	10	32
C. albicans IOC 4556	10	40	20	>40	20	20	20	>32
C. albicans IOC 4525	2.5	20	20	>40	10	20	20	>32
C. albicans IOC 4558	10	40	20	>40	20	10	10	>32
C. glabrata IOC 4565	5	5	2.5	1.25	2.5	5	>40	>32
C. krusei IOC 4559	1.25	2.5	1.25	1.25	1.25	5	1.25	>32
C. guillermondii IOC 4557	1.25	2.5	1.25	1.25	1.25	2.5	1.25	32
C. parapsilosis IOC 4564	40	20	10	2.5	10	10	>40	>32
C. tropicalis IOC 4560	>40	20	20	5	20	10	5	>32

MIC = minimal inhibitory concentration refers to at least 90% of the yeast growth inhibition. AMB= amphotericin B; ATCC = American Type Culture Collection, IOC = Instituto Oswaldo Cruz Collection.

The AgNPs were effective on *C. parapsilosis* and *C. tropicalis*, but with MIC from 2.5 to 40 μM. Until 40 μM, the AgNP-Asp was not effective on *C. tropicalis* and AgNP-FP on *C. parapsilosis*.

The AgNPs showed antifungal activity against the phytopathogens *C. lunata*, *F. subglutinans*, *F. verticillioides*, and on one strain of *F. oxysporum*. The AgNPs showed higher MIC on the *F. oxysporum* WLA-FP07 that was less susceptible to their activity. These phytopathogens cause diseases in sugarcane and in crops such as in rice, maize, and beans with a great impact on agriculture and economy [38–40]. Synthetic fungicides are used in the management of these diseases, and their systematic and uncontrolled use has often led to fungal resistance [41].

Among the seven AgNPs, until 250 μM, none prevented the growth of *F. sacchari* (Table 4). The phytopathogens *C. lunata* and *F. verticillioides* were the most susceptible to the AgNPs, and even at low concentrations, all the AgNPs prevented the growth of these fungi (Table 4). These phytopathogens cause diseases in different cultivar species, and their impact on the economy is very relevant [42].

Table 4. Antifungal activity of the AgNP-Asp, AgNP-AT, AgNP-BO, AgNP-CPP, AgNP-EN, AgNP-ER, and AgNP-FP by minimal inhibitory concentration (MIC—μM) on the phytopathogens *C. lunata*, *F. sacchari*, *F. subglutinans*, *F. oxysporum* (WLA-FP07), *F. oxysporum* (WLA-FP25), and *F. verticillioides*.

Phytopathogens	AgNP-Asp		AgNP-AT		AgNP-BO		AgNP-CPP		AgNP-EN		AgNP-FP		AgNP-ER		AMB	
	MIC	FC	MIC	FC	MIC	FC	MIC	FC	MIC	FC	MIC	FC	MIC	FC	MIC	FC
C. lunata	8	8	4	4	16	16	4	4	4	4	4	4	4	4	4	4
F. sacchari	>250	>250	>250	>250	>250	>250	>250	>250	>250	>250	>250	>250	>250	>250	16	16
F. subglutinans	60	60	60	60	>250	>250	120	120	120	120	60	60	120	120	16	16
F. oxysporum (WLA-FP07)	120	120	120	120	>250	>250	250	250	120	120	120	120	120	120	>32	>32
F. oxysporum (WLA-FP25)	30	30	30	30	30	30	120	120	60	60	30	30	30	30	16	16
F. verticillioides	16	16	8	8	16	16	8	8	60	60	16	16	8	8	16	16

MIC = minimal inhibitory concentration refers to at least 90% of the fungi growth inhibition; AMB = amphotericin B; FC = fungicide concentration.

On the phytopathogens, the AgNPs showed the same values for MICs and fungicide concentration (FC), indicating that the activity of AgNPs is more fungicide than fungistatic.

In this study, the AgNPs presented sizes in the range from 43.4 to 120.6 nm (DLS) and from 21.8 to 35.8 nm (TEM), spherical and regular forms, with negative zeta potential and pH ranging from slightly acidic to neutral. The participation of biological molecules in the biosynthesis of AgNPs confers interesting advantages to the AgNPs, improving the interaction with biological targets and the biocompatibility [27,28,31]. The influence of these factors can be related to the antifungal activity.

The antifungal activity of biological AgNPs was previously described by several authors [20–24,34–37,43]. In the current study, the higher antifungal activity was observed for the AgNP-Asp on the different Candida species, including the non-albicans. However, the AgNP-CPP was the less effective.

An analysis of the physicochemical parameters (Table 3) for the AgNP-Asp and AgNP-CPP indicates that the size was important for the antifungal activity. The first one has a size of 44.9 and the last one of 87.1 nm. The lower size contributes to a better activity. Between both nanoparticles, there was not a high difference in pH, zeta potential, and PDI. On the yeasts, the range of MICs values obtained for the different AgNPs was from 1.25 to 40 μM. The MICs were different on the different strains of *C. albicans*, and that from ATCC was more susceptible to the action of AgNPs, while clinical isolates were more resistant. These differences were expected due to a possible genetic variability, and therefore it is important to test the susceptibility of different strains, especially of clinical isolates to the AgNPs activity.

The FTIR analysis showed the presence of aliphatic (C-H) functional group for AgNP-Asp, AgNP-AT, AgNP-BO, AgNP-ER, and AgNP-FP, and these nanoparticles were the most effective in killing the phytopathogens and yeasts, with lower MICs values. Interestingly, the AgNP-CPP and AgNP-EN did not present this functional group and were the less effective AgNPs on the yeasts and on phytopathogens with higher MICs value. This functional group has already been described as a potent antifungal agent against *Saccharomyces cerevisiae*, where primary aliphatic alkanols were tested and showed minimum fungicidal concentration (MFC) of 25 μg/mL (0.14 mM) [44].

The comparison between the physicochemical parameters of the AgNP-Asp and AgNP-FP shows a difference in their size and zeta potential. AgNP-FP has size and zeta potential of 56.9 ± 2.1 nm and −15.33, while for the AgNP-Asp these values are 44.9 ± 4.1 nm, and −1.04. Both nanoparticles have pH of 6.0 and are surrounded by molecules containing the functional aliphatic group. The AgNP-Asp is smaller with higher zeta potential value and was more effective then AgNP-FP as an antifungal agent on the yeasts. As previously reported [27,28,31], these data indicate that all the parameters are related and can interfere with the antifungal activity.

The results showed that the AgNPs were very effective against clinical yeasts and phytopathogens. The AgNPs were effective against phytopathogens, showing MICs from 4 to 120 μM, and against yeasts with MICs range of 1.25 to 40 μM, indicating the promising possibility of application of these AgNPs as an antifungal agent.

3. Materials and Methods

3.1. *Fungal Strains*

Fungi species employed in the AgNPs biosynthesis were previously isolated from the Brazilian biodiversity and deposited at the microorganisms collection from Instituto Oswaldo Cruz (IOC, Rio de Janeiro, RJ, Brazil), Instituto Adolfo Lutz (IAL, São Paulo, SP, Brazil), or in the "Collection of Microorganisms for Biocontrol of Phytopathogens and Weeds" at Embrapa Genetic Resources and Biotechnology (CENARGEN, Brasília, DF, Brazil). The species *Aspergillus* spp. is an endophyte and *C. pini-ponderosae* (IAL 7248), *F. proliferatum* (IOC 4682/IAL 7246), and *E. nigrum* (IAL 7249) are epiphytes from the leaf of the mangrove plant *Rhizophora mangle*, respectively. The *E. rostratum* (IAL 7247) was isolated as an endophyte from the plant *Croton blanchetianus*, collected in the Brazilian biome called Caatinga.

The pathogen *C. albicans* (ATCC 36802/IOC 3704) is a strain from the American Type Culture Collection (ATCC), and *C. albicans* (IOC 4525, IOC 4556, IOC 4558), *C. krusei* (IOC 4559), *C. glabrata* (IOC 4565), *C. parapsilosis* (IOC 4564), *C. tropicalis* (IOC 4560), and *C. guilliermondii* (IOC 4557) are clinical species from the IOC collection. The Fusarium species were isolated from the *Rhizosphere* of sugar cane [45] by the group of Prof. Welington L. Araújo, from Institute of Biomedical Sciences of the University of São Paulo, that kindly provided the phytopathogens from his collection (WLA) for this study. The phytopathogens were *C. lunata* (WLA-FP06), *F. sacchari* (WLA-FP04), *F. subglutinans* (WLA-FP21), *F. oxysporum* (WLA-FP07), *F. oxysporum* (WLA-FP25), and *F. verticillioides* (WLA-FP05).

3.2. *Biosynthesis and Physicochemical Analysis of the AgNPs*

The biosynthesis of the AgNPs was performed according to the protocol previously described by our group [34]. The fungi *A. tubingensis*, *Aspergillus* spp., *B. ochroleuca*, *C. pini-ponderosae*, *F. proliferatum*, *E. nigrum*, and *E. rostratum* were cultivated in potato dextrose agar (PDA, Himedia #M096) for a week at 28 °C and sub-cultivated in an Erlenmeyer flask of 500 mL with 150 mL of potato dextrose broth (PDB) (Himedia #M096) for 72 h at 28 °C and 150 rpm, in an orbital shaker (Marconi MA-420, Piracicaba, SP, Brazil). After filtration through a polypropylene membrane, the biomass was washed with sterile deionized water to remove any residue of the culture medium. The biomass was incubated with sterile deionized water (10 g/100 mL) at 28 °C and 150 rpm for 72 h. Subsequently, following a filtration and biomass discard, the supernatant was sterilized using a 0.22 μm polyethersulfone (PES) membrane, resulting in the aqueous extracellular cell free extract (AE). For the AgNPs biosynthesis, silver nitrate ($AgNO_3$) was added to the sterile AE for the final concentration of 1 mM. The reactional mixture was protected from light, and the formation of AgNPs was monitored by reading the absorbance from 200 to 800 nm (UV-Vis, Agilent 8453), showing the presence of a surface plasmon resonance around 420 nm.

The obtained AgNPs were coded according to the initials of the fungi species used for their biosynthesis being AgNP-AT, AgNP-Asp, AgNP-BO, AgNP-CPP, AgNP-FP, AgNP-EN, and AgNP-ER. The AgNPs were characterized by size (Dynamic Light Scattering, DLS, Nanoplus–Particulate Systems, Norcross, GA, USA), polydispersity index (PDI), zeta potential, pH, Fourier-transform infrared (FTIR) spectroscopy, and morphology by transmission electron microscopy (TEM, Zeiss LEO 906 E, de 120Kv, Freiburg, Germany), as previously described [26,34]. The size of the AgNPs was also measured by TEM for at least seven particles of each AgNPs and expressed by the mean ± the standard deviation.

3.3. Antifungal Activity Assay

The antifungal activity was evaluated on four strains of *C. albicans* (ATCC 36802/IOC 3704, IOC 4525, IOC 4556, IOC 4558), *C. krusei* (IOC 4559), *C. glabrata* (IOC 4565), *C. parapsilosis* (IOC 4564), *C. tropicalis* (IOC 4560), and *C. guilliermondii* (IOC 4557). The phytopathogens utilized were *C. Lunata*, *F. phaseoli*, *F. sacchari*, *F. subglutinans*, *F. oxysporum* (WLA-FP07 and WLA-FP25), and *F. verticillioides*.

The yeasts and phytopathogens were cultured on PDA and sub-cultured in PDB. The yeast suspensions were prepared in RPMI 1640 (Gibco #23400-013) culture media at 0.5–2.5 $\times$ 10^3 CFU/mL, according to standard curves previously established in our laboratory. For the phytopathogens, the spores were prepared in a saline solution at 0.9% and after being counted using a Neubauer chamber, the suspensions were prepared in RPMI 1640 with 1 $\times$ 10^6 spores/mL.

The antifungal assay was evaluated by the microdilution assay in a 96-well plate [46–48]. The AgNP-AT, AgNP-Asp, AgNP-BO, AgNP-CPP, AgNP-EN, AgNP-ER, and AgNP-FP were serially diluted in RPMI 1640, and each well received 100 µL of the yeasts or phytopathogens suspensions and 25 µL of resazurin dye at 0.02% in saline solution. The AgNPs were assayed at final concentrations of 1.25, 2.5, 5, 10, 20, and 40 µM for the yeasts, and of 4, 8, 16, 30, 60, 120, and 250 µM for the phytopathogens. Untreated culture that received only RPMI 1640 was used as the negative control and amphotericin B (AMB) at 16 and 32 µM was used as positive control. The plates were incubated at 30 °C for 24 h, and the minimal inhibitory concentration (MIC90) was defined as the lowest concentration in which the color of the dye resazurin was kept in blue due to the inhibition of at least 90% of the microorganism's growth [46]. For the yeasts, the MICs were expressed by the mean of three independent experiments, in duplicate for each AgNPs concentration, and for the phytopathogens by the mean of two assays in triplicate. The fungicidal concentration (FC) for phytopathogens was determined from the MICs. For that, 50 µL of two concentrations before and after the MICs were collected from the plate and incubated in Petri dishes containing PDA at 30 °C for 120 h. The absence of phytopathogens growth was defined as the FC.

4. Conclusions

In this study, seven biogenic AgNPs were obtained using the fungi species *A. tubingensis*, *Aspergillus* spp., *B. ochroleuca*, *C. pini-ponderosae*, *F. proliferatum*, *E. nigrum*, and *E. rostratum* isolated from the Brazilian biodiversity. Among them, this is the first report of *E. rostratum*, *F. proliferatum*, and *C. pini-ponderosae* application on the biosynthesis of AgNPs. The nanoparticles showed spherical morphology, with pH from 4.5 to 7.5, and size in the range from 43.4 to 120.6 nm (DLS) and from 21.8 to 35.8 nm (TEM). The functional groups from the biomolecules surrounding the AgNPs were analyzed by FTIR.

The AgNPs showed antifungal activity against clinical strains of *C. albicans*, *C. krusei*, *C. glabrata*, *C. parapsilosis*, *C. tropicalis*, and *C. guilliermondii*, common in hospital infections and against phytopathogens, responsible for serious damages in agricultural production. As expected, the results indicated that the physicochemical parameters of the AgNPs including the functional groups present on their surface interfere on their antifungal activity. Overall, the results indicate that there is no specificity of the AgNPs for yeasts or phytopathogens, which can be an advantage, increasing the possibility of application in different areas.

Author Contributions: L.G.R. and A.O.D.S. contributed to the study conception and design. Experimental assays and data analysis were performed by L.G.R., R.C., G.S.C.R. and A.O.D.S. The first draft of the manuscript was written by L.G.R. and G.S.C.R. and reviewed by A.O.D.S. All authors commented on previous versions of the manuscript. All authors have read and agreed to the published version of the manuscript.

Funding: The study was supported by grants #2010/50186-5, 2020/04799-7 and #2020/03883-4 from São Paulo Research Foundation (FAPESP), Brazil.

Institutional Review Board Statement: Not applicable.

Informed Consent Statement: Not applicable.

Data Availability Statement: Not applicable.

Acknowledgments: Authors thank the transmission electron microscopy facilities from Butantan Institute for technical support, and Welington L. Araújo for kindly providing the phytopathogens.

Conflicts of Interest: The authors declare no conflict of interest.

References

1. Denning, D.W.; Perlin, D.S.; Muldoon, E.G.; Colombo, A.L.; Chakrabarti, A.; Richardson, M.D.; Sorrell, T.C. Delivering on antimicrobial resistance agenda not possible without improving fungal diagnostic capabilities. *Emerg. Infect. Dis.* **2017**, *23*, 177–183. [CrossRef]
2. Cui, X.; Wang, L.; Lü, Y.; Yue, C. Development and research progress of anti-drug resistant fungal drugs. *J. Infect. Public Health* **2022**, *15*, 986–1000. [CrossRef] [PubMed]
3. Omran, B.A.; Baek, K.H. Control of phytopathogens using sustainable biogenic nanomaterials: Recent perspectives, ecological safety, and challenging gaps. *J. Clean. Prod.* **2022**, *372*, 133–729. [CrossRef]
4. Pradhan, A.; Ghosh, S.; Sahoo, D.; Jha, G. Fungal effectors, the double edge sword of phytopathogens. *Curr. Genet.* **2020**, *67*, 27–40. [CrossRef]
5. Savary, S.; Willocquet, L.; Pethybridge, S.J.; Esker, P.; McRoberts, N.; Nelson, A. The global burden of pathogens and pests on major food crops. *Nat. Ecol. Evol.* **2019**, *3*, 430–439. [CrossRef]
6. Guillot, J.; Bond, R. Malassezia yeasts in veterinary dermatology, An updated overview. *Front. Cell. Infect. Microbiol.* **2020**, *10*, 79. [CrossRef]
7. Wang, Y.; Pruitt, R.N.; Nürnberger, T.; Wang, Y. Evasion of plant immunity by microbial pathogens. *Nat. Rev. Microbiol.* **2022**, *20*, 449–464. [CrossRef] [PubMed]
8. Amin, A.; Zahra, T.; Raja, H.; Amin, M.; Dilshad, E.; Naveed, M.; Ahmed, I. Major natural sinks for harboring microorganisms with altered antibiotic resistance versus major human contributing sources of antibiotic resistance: A detailed insight. In *Advances in Environmental Pollution Research Series, Antibiotics and Antimicrobial Resistance Genes in the Environment*; Elsevier: Amsterdam, The Netherlands, 2020; Volume 1, pp. 70–98. [CrossRef]
9. Lee, Y.; Puumala, E.; Robbins, N.; Cowen, L.E. Antifungal drug resistance: Molecular mechanisms in *Candida albicans* and beyond. *Chem. Rev.* **2021**, *121*, 3390–3411. [CrossRef] [PubMed]
10. Pinto-Magalhães, S.; Martins, A.; Lacerda, S.; Filipe, R.; Prista-Leão, B.; Pinheiro, D.; Silva-Pinto, A.; Santos, L. Candidemia in a Portuguese tertiary care hospital: Analysis of a 2-year period. *J. Mycol. Med.* **2019**, *29*, 320–324. [CrossRef] [PubMed]
11. Barbosa, A.; Araújo, D.; Ribeiro, E.; Henriques, M.; Silva, S. *Candida albicans* adaptation on simulated human body fluids under different pH. *Microorganisms* **2020**, *8*, 511. [CrossRef]
12. Cortés, J.A.; Ruiz, J.F.; Melgarejo-Moreno, L.N.; Lemos, E.V. Candidemia en Colombia. *Biomédica* **2020**, *40*, 195–207. [CrossRef]
13. Du, H.; Bing, J.; Hu, T.; Ennis, C.L.; Nobile, C.J.; Huang, G. *Candida auris*: Epidemiology, biology, antifungal resistance, and virulence. *PLoS Pathog.* **2020**, *16*, e1008921. [CrossRef]
14. Miceli, M.H.; Díaz, J.A.; Lee, S.A. Emerging opportunistic yeast infections. *Lancet Infect. Dis.* **2011**, *11*, 142–151. [CrossRef] [PubMed]
15. Pappas, P.G.; Kauffman, C.A.; Andes, D.R.; Clancy, C.J.; Marr, K.A.; Ostrosky-Zeichne, L.; Reboli, A.C.; Schuster, M.G.; Vasquez, J.A.; Walsh, T.J.; et al. Clinical practice guideline for the management of candidiasis: 2016 update by the infectious diseases society of America. *Clin. Infect. Dis.* **2016**, *62*, 1–50. [CrossRef] [PubMed]
16. Sweeney, M.J.; Dobson, D.W. Mycotoxin production by *Aspergillus, Fusarium* and *Penicillium* species. *Int. J. Food Microbiol.* **1998**, *43*, 141–158. [CrossRef] [PubMed]
17. Marin, S.; Ramos, A.J.; Cano-Sancho, G.; Sanchis, V. Mycotoxins: Occurrence, toxicology, and exposure assessment. *Food Chem. Toxicol.* **2013**, *60*, 218–237. [CrossRef]
18. Lanubile, A.; Giorni, P.; Bertuzz, T.; Marocco, A.; Battilani, P. *Fusarium verticillioides* and *Aspergillus flavus* co-occurrence influences plant and fungal transcriptional profiles in maize kernels and in vitro. *Toxins* **2021**, *13*, 680. [CrossRef]
19. Alshannaq, A.; Yu, J.H. Occurrence, toxicity, and analysis of major mycotoxins in food. *Int. J. Environ. Res. Public Health* **2017**, *14*, 632. [CrossRef]
20. Mabrouk, M.; Das, D.B.; Salem, Z.A.; Beherei, H.H. Nanomaterials for Biomedical Applications: Production, Characterisations, Recent Trends and Difficulties. *Molecules* **2021**, *26*, 1077. [CrossRef]
21. Cruz-Luna, A.R.; Cruz-Martínez, H.; Vásquez-López, A.; Medina, D.I. Metal nanoparticles as novel antifungal agents for sustainable agriculture: Current advances and future directions. *J. Fungi* **2021**, *7*, 1033. [CrossRef]

22. Zafar, N.; Madni, A.; Khalid, A.; Khan, T.; Kousar, R.; Naz, S.S.; Wahid, F. Pharmaceutical and biomedical applications of green synthesized metal and metal oxide nanoparticles. *Curr. Pharm. Des.* **2020**, *26*, 5844–5865. [CrossRef]
23. Rai, M.; Ingle, A.P.; Trzcińska-Wencel, J.; Wypij, M.; Bonde, S.; Yadav, A.; Kratošová, G.; Golińska, P. Biogenic silver nanoparticles: What we know and what do we need to know? *Nanomaterails* **2021**, *11*, 2901. [CrossRef]
24. Lin, Y.; Betts, H.; Keller, S.; Cariou, K.; Gasser, G. Recent developments of metal-based compounds against fungal pathogens. *Chem. Soc. Rev.* **2021**, *50*, 10346–10402. [CrossRef] [PubMed]
25. Guilger-Casagrande, M.; Lima, R. Synthesis of silver nanoparticles mediated by fungi: A review. *Front. Bioeng. Biotechnol.* **2019**, *7*, 287. [CrossRef]
26. Ballottin, D.; Fulaz, S.; Souza, M.L.; Corio, P.; Rodrigues, A.G.; Souza, A.O.; Gaspari, P.M.; Gomes, A.F.; Gozzo, F.; Tasic, L. Elucidating protein involvement in the stabilization of the biogenic silver nanoparticles. *Nano Res. Lett.* **2016**, *11*, 1–9. [CrossRef]
27. De Souza, A.O. Overview of nanomaterials and cellular interactions. *Biointerface Res. Appl. Chem.* **2022**, *13*, 367. [CrossRef]
28. Jackson, T.C.; Patani, B.O.; Israel, M.B. Nanomaterials and cell interactions: A review. *J. Biomater. Nanobiotechnol.* **2017**, *8*, 220–228. [CrossRef]
29. Barabadi, H.; Tajani, B.; Moradi, M.; Damavandi, K.K.; Meena, R.; Honary, S.; Mahjoub, M.A.; Saravanan, M. Penicillium family as emerging nanofactory for biosynthesis of green nanomaterials, A journey into the world of microorganisms. *J. Clust. Sci.* **2019**, *30*, 843–856. [CrossRef]
30. Evanoff, D.D., Jr.; Chumanov, G. Synthesis and optical properties of silver nanoparticles and arrays. *Chemphyschem* **2005**, *6*, 1221–1231. [CrossRef]
31. Kim, T.H.; Kim, M.; Park, H.S.; Shin, U.S.; Gong, M.S.; Kim, H.W. Size-dependent cellular toxicity of silver nanoparticles. *J. Biomed. Mater. Res. A* **2012**, *100*, 1033–1043. [CrossRef] [PubMed]
32. Leite, F.L.; Bueno, C.C.; Da Róz, A.L.; Ziemath, E.C.; Oliveira, O.N. Theoretical models for surface forces and adhesion and their measurement using atomic force microscopy. *Int. J. Mol. Sci.* **2012**, *13*, 12773–12856. [CrossRef]
33. Prakash, S.; Soni, N. Factors affecting the geometry of silver nanoparticles synthesis in *Chrysosporium Tropicum* and *Fusarium Oxysporum*. *Am. J. Nanotech.* **2011**, *2*, 112–121. [CrossRef]
34. Rodrigues, A.G.; Ping, L.Y.; Marcato, P.D.; Alves, O.L.; Silva, M.C.P.; Ruiz, R.C.; Melo, I.S.; Tasic, L.; De Souza, A.O. Biogenic antimicrobial silver nanoparticles produced by fungi. *Appl. Microbiol. Biotechnol.* **2013**, *97*, 775–782. [CrossRef]
35. Shariq, A.M.; Soundhararajan, R.; Akther, T.; Kashif, M.; Khan, J.; Waseem, M.; Srinivasan, H. Biogenic AgNPs synthesized via endophytic bacteria and its biological applications. *Environ. Sci. Pollut. Res. Int.* **2019**, *26*, 26939–26946. [CrossRef] [PubMed]
36. Rodrigues, A.G.; Ruiz, R.C.; Selari, P.J.R.G.; Araújo, W.L.; De Souza, A.O. Anti-Biofilm action of biological silver nanoparticles produced by *Aspergillus tubingensis* and antimicrobial activity of fabrics carrying it. *Biointerface Res. Appl. Chem.* **2021**, *11*, 14764–14774. [CrossRef]
37. Rodrigues, A.G.; Romano, O.G.P.J.; Ottoni, C.A.; Ruiz, R.C.; Morgano, M.A.; Araújo, W.L.; Melo, I.S.; De Souza, A.O. Functional textiles impregnated with biogenic silver nanoparticles from *Bionectria ochroleuca* and its antimicrobial activity. *Biomed. Microdevices* **2019**, *21*, 1–10. [CrossRef] [PubMed]
38. Majeed, R.A.; Shahid, A.A.; Ashfaq, M.; Saleem, M.Z.; Haider, M.S. First report of *Curvularia lunata* causing brown leaf spots of rice in Punjab, Pakistan. *Plant Dis.* **2015**, *100*, 219. [CrossRef]
39. Bansal, Y.; Chander, J.; Kaistha, N.; Singla, N.; Sood, S.; Van Diepeningen, A.D. *Fusarium sacchari*, a cause of mycotic keratitis among sugarcane farmers—A series of four cases from North India. *Mycoses* **2016**, *59*, 705–709. [CrossRef]
40. Bleackley, M.R.; Samuel, M.; Garcia-Ceron, D.; McKenna, J.A.; Lowe, R.G.T.; Pathan, M.; Zhao, K.; Ang, C.-S.; Mathivanan, S.; Anderson, M.A. Extracellular vesicles from the cotton pathogen *Fusarium oxysporum f. sp. vasinfectum* induce a phytotoxic response in plants. *Front. Plant Sci.* **2020**, *10*, 1610. [CrossRef] [PubMed]
41. Kim, K.; Lee, Y.; Ha, A.; Kim, J.I.; Park, A.R.; Yu, N.H.; Son, H.; Choi, G.J.; Park, H.W.; Lee, C.W.; et al. Chemosensitization of *Fusarium graminearum* to chemical fungicides using cyclic lipopeptides produced by *Bacillus amyloliquefaciens* strain JCK-12. *Front. Plant Sci.* **2017**, *8*, 2010. [CrossRef] [PubMed]
42. Jia, H.; Huang, J.; Luo, X.; Yan, J. Agriculture: Science and technology safeguard sustainability. *Nat. Sci. Rev.* **2019**, *6*, 595–600. [CrossRef] [PubMed]
43. Zwar, I.P.; Trotta, C.V.; Ziotti, A.B.S.; Lima, N.M.; Araújo, W.L.; De Melo, I.S.; Ottoni, C.A.; De Souza, A.O. Biosynthesis of silver nanoparticles using actinomycetes, phytotoxicity on rice seeds, and potential application in the biocontrol of phytopathogens. *J. Basic Microbiol.* **2023**, *6*, 64–74. [CrossRef] [PubMed]
44. Kubo, I.; Cespedes, C.L. Antifungal activity of alkanols: Inhibition of growth of spoilage yeasts. *Phytochem. Rev.* **2013**, *12*, 961–977. [CrossRef]
45. Stuart, R.M.; Romão, A.S.; Pizzirani-Kleiner, A.A.; Azevedo, J.L.; Araújo, W.L. Culturable endophytic filamentous fungi from leaves of transgenic imidazolinone-tolerant sugarcane and its non-transgenic isolines. *Arch. Microbiol.* **2010**, *192*, 307–313. [CrossRef]
46. Palomino, J.C.; Martin, A.; Camacho, M.; Guerra, H.; Swings, J.; Portaels, F. Resazurin microtiter assay plate: Simple and inexpensive method for detection of drug resistance in *Mycobacterium tuberculosis*. *Antimicrob. Agent. Chemother.* **2002**, *46*, 2720–2722. [CrossRef] [PubMed]

47. Clinical and Laboratory Standards Institute (CLSI). *Reference Method for Broth Dilution Antifungal Susceptibility Testing of Filamentous Fungi*, 3rd ed.; CLSI standard M38. CLSI: Wayne, PA, USA, 2017. Available online: https://clsi.org/media/1894/m38ed3_sample.pdf (accessed on 27 November 2022).
48. Clinical and Laboratory Standards Institute (CLSI). *Method for Antifungal Disk Diffusion Susceptibility Testing of Yeasts*, 3rd ed.; CLSI Guideline M44. CLSI: Wayne, PA, USA, 2018. Available online: https://clsi.org/standards/products/microbiology/documents/m44/ (accessed on 27 November 2022).

Disclaimer/Publisher's Note: The statements, opinions and data contained in all publications are solely those of the individual author(s) and contributor(s) and not of MDPI and/or the editor(s). MDPI and/or the editor(s) disclaim responsibility for any injury to people or property resulting from any ideas, methods, instructions or products referred to in the content.

antibiotics

MDPI

Article

Development of 3D-Printed Collagen Scaffolds with In-Situ Synthesis of Silver Nanoparticles

Sofia Municoy [1], Pablo Edmundo Antezana [1], Martín Gonzalo Bellino [2] and Martín Federico Desimone [1,*]

1 Universidad de Buenos Aires, Facultad de Farmacia y Bioquímica, Instituto de Química y Metabolismo del Fármaco (IQUIMEFA), Consejo Nacional de Investigaciones Científicas y Técnicas (CONICET), Junín 956, Buenos Aires 1113, Argentina

2 Instituto de Nanociencia y Nanotecnología, Comisión Nacional de Energía Atómica, Consejo Nacional de Investigaciones Científicas y Técnicas (CONICET), San Martín 1650, Argentina

* Correspondence: desimone@ffyb.uba.ar

Abstract: UV-irradiation method has grown as an alternative approach to in situ synthetize silver nanoparticles (AgNPs) for avoiding the use of toxic reducing agents. In this work, an antimicrobial material by in situ synthesizing AgNPs within 3D-printed collagen-based scaffolds (Col-Ag) was developed. By modifying the concentration of $AgNO_3$ (0.05 and 0.1 M) and UV irradiation time (2 h, 4 h, and 6 h), the morphology and size of the in situ prepared AgNPs could be controlled. As a result, star-like silver particles of around 23 ± 4 µm and spherical AgNPs of 220 ± 42 nm were obtained for Ag 0.05 M, while for Ag 0.1 M cubic particles from 0.3 to 1.0 µm and round silver precipitates of 3.0 ± 0.4 µm were formed in the surface of the scaffolds at different UV irradiation times. However, inside the material AgNPs of 10–28 nm were obtained. The DSC thermal analysis showed that a higher concentration of Ag stabilizes the 3D-printed collagen-based scaffolds, while a longer UV irradiation interval produces a decrease in the denaturation temperature of collagen. The enzymatic degradation assay also revealed that the in situ formed AgNPs act as stabilizing and reinforcement agent which also improve the swelling capacity of collagen-based material. Finally, antimicrobial activity of Col-Ag was studied, showing high bactericidal efficiency against Gram-negative (*Escherichia coli*) and Gram-positive (*Staphylococcus aureus*) bacteria. These results showed that the UV irradiation method was really attractive to modulate the size and shape of in situ synthesized AgNPs to develop antimicrobial 3D-printed collagen scaffolds with different thermal, swelling and degradation properties.

Keywords: 3D-printing; UV irradiation method; AgNPs; collagen; antimicrobial scaffolds; post-printing

Citation: Municoy, S.; Antezana, P.E.; Bellino, M.G.; Desimone, M.F. Development of 3D-Printed Collagen Scaffolds with In-Situ Synthesis of Silver Nanoparticles. *Antibiotics* **2023**, *12*, 16. https://doi.org/10.3390/antibiotics12010016

Academic Editors: Sotiris K Hadjikakou and Christina N. Banti

Received: 30 November 2022
Revised: 17 December 2022
Accepted: 20 December 2022
Published: 22 December 2022

Copyright: © 2022 by the authors. Licensee MDPI, Basel, Switzerland. This article is an open access article distributed under the terms and conditions of the Creative Commons Attribution (CC BY) license (https://creativecommons.org/licenses/by/4.0/).

1. Introduction

The recent advances in 3D printing technology and its marriage with the biomaterials field allowed the development of 3D-printed scaffolds for biomedical applications [1]. This technology leads to the manufacture of custom-made 3D structures with high resolution and controlled architecture [2]. Among the different methods used to create 3D-materials, fuse deposition modelling (FDM) is highly used, and it is based on the material extrusion [3]. In this sense, extrusion-based printing is an adaptation from the FDM with the purpose of printing biocompatible materials and cells. This method used pneumatic pressure or mechanical force to extrude the ink out of the nozzle in an uninterrupted line [4].

There are several materials that could be used to produce 3D-printed scaffolds, such as collagen, chitosan, cellulose, hyaluronic acid, and alginate. Among them, collagen has gained popularity because of its interesting properties. Collagen is the main protein of the extracellular matrix of animals, almost 30% in vertebrates. Maintaining the biological and structural integrity of the extracellular matrix is the main role of collagen [5]. Collagen type I has important biocompatible properties and forms strong and stable fibers through its self-assembly and cross-linking that could lead to the development of useful scaffolds [6,7]. On the other hand, it is interesting to improve the properties of the scaffolds by adding other functionalities to the 3D-printed material. In this sense, the incorporation of antimicrobial agents is a relevant improvement. In this line, Ag nanoparticles (AgNPs) has gained popularity as a metal with great antimicrobial activity [8,9]. This metal has an important antibacterial capacity against both Gram-negative and Gram-positive bacteria, even the drug-resistant ones [10,11]. AgNPs has several mechanisms to achieve the antimicrobial activity: anchoring and penetrating the bacterial cell wall, leading to structural changes in cell membrane; binding to proteins; interacting with DNA bases leading to irreversible damages; production of free radicals; and inhibition of cell division and reproduction [12–14]. The presence of these different mechanisms of action, hinders the development of microbial resistance. It is known that the main responsible of the AgNPs antimicrobial activity is the silver ions release (Ag^+) [15].

There are several physical and chemical methods available to synthesize AgNPs, such as electrochemical changes, and chemical and photochemical reduction [16–19]. Among them, the chemical reduction is the most frequently used in order to reduce Ag^+ ions to obtain AgNPs. In this sense, AgNPs are usually synthesized by reduction of a silver salt with a reducing agent, such as sodium borohydride in the presence of a colloidal stabilizer. The most commonly used colloidal stabilizers are polyvinyl alcohol [20], poly(vinylpyrrolidone) [21], bovine serum albumin [22], and citrate [23]. In another work, polyethyleneimine was used to avoid the aggregation of AgNPs prepared on the surface of commercial cellulose filter papers by reduction with polydopamine [24]. Supramolecular agents, such as cyclophanes, were also used to stabilized AgNPs [25]. During the process of synthesizing the nanoparticles the different parameters, such as the interaction kinetics of the metal ions with the reducing agent and adsorption of the stabilizer agent, among others, can influence the morphology, stability, and physicochemical properties of AgNPs [26,27]. However, the reduced agent used during these chemical methods, such as sodium borohydride, hydrazine, and N,N-dimethylformamide could lead to environmental concerns due to their inherent toxicity [28]. To overcome this disadvantages, it is gaining attention the use of physical methods, such as photoreduction methods, like ultrasonic waves, laser, and ultraviolet light (UV) irradiation [29–32]. In particular, UV light has become a promising method to in situ synthesize AgNPs from silver solutions without using any reducing or stabilizer agent [33]. This leads to a method which is ecofriendly, with increased energy efficiency and reduced costs [34].

Since it is promising the development of materials that could be printed with 3D technology, the aim of this work was to develop a new 3D-printed material based on collagen, with the in situ synthesis of AgNPs in order to endow the biomaterial with the antimicrobial activity of silver.

2. Results and Discussion

2.1. Preparation of Antimicrobial 3D-Printed Collagen-Based Scaffolds and In situ Reduction in Ag

The first step for the manufacture of the antimicrobial 3D-printed AgNPs-loaded Collagen gels (Col-Ag) involved the preparation of suitable inks. In order to obtain a collagen-based ink with good printability, a high concentration of collagen was needed to enhance the shape fidelity. For this reason, a collagen solution in acetic acid was lyophilized and then resuspended in distilled water or $AgNO_3$ solutions to obtain a final concentration of collagen of 80 mg/mL. After 24 h of hydration and mixing, homogeneous inks were obtained, and used to print the 3D scaffolds.

After printing, 3D square grid collagen-based scaffolds (6 mm × 6 mm × 2 mm) were obtained as test samples. To induce in situ reduction in Ag, 3D-printed Col-Ag grids were exposed to UV light during different time intervals. Once jellified, 3D-printed collagen-based composites were finally obtained from the different $AgNO_3$ 0.05 and 0.1 M solutions, labeled as Col-Ag0.05 and Col-Ag0.1, respectively. As can be seen in Figure 1A, colorless 3D-printed Col-Ag scaffolds were obtained before UV irradiation, but after 2 h, 4 h, and 6 h of exposure to UV light, a gradual increase in brown color was observed as a result of the reduction in silver and a possible formation of silver particles within the collagen network. In addition, the brown color was more intense in Col-Ag0.1 than in Col-Ag0.05, due to a greater concentration of silver in the ink.

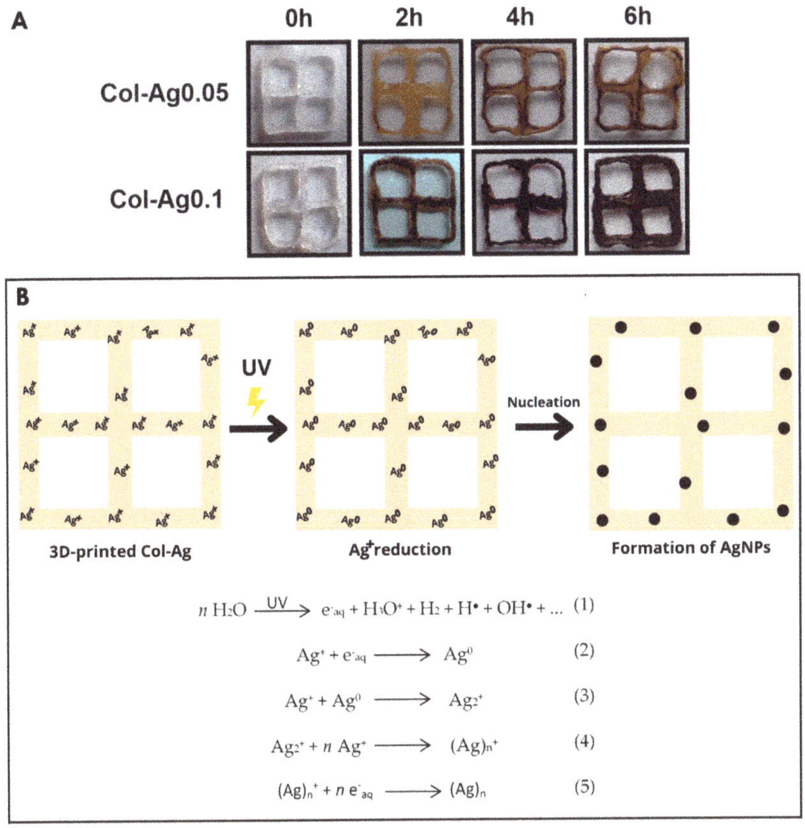

Figure 1. (**A**) Images of dry 3D-printed collagen-based scaffolds after in situ reduction of Ag by different UV irradiation intervals. Dimensions of 3D-printed collagen-based scaffolds are 6 mm × 6 mm × 2 mm. (**B**) Schematic illustration of the mechanism of in situ AgNPs formation by using the UV irradiation method.

The mechanism proposed by which silver is reduced within collagen suggests the formation of hydrated electrons ($e^{-}{}_{aq}$) by photolysis of aqueous solution during UV irradiation (Figure 1B-1) and the following capture of these electrons by Ag^{+} ions (Figure 1B-2). The neutral Ag^{0} atoms bind to other Ag^{+} ions to give $Ag_{n}{}^{+}$ species (Figure 1B-4), which in presence of the hydrated electrons leads to the formation of silver particles (Figure 1B-5) [32]. It was found that the high rate constant of the reaction between $e^{-}{}_{aq}$ and Ag^{+} in a solution of a biopolymer indicates that silver ions are favorably reduced to Ag^{0} by UV irradiation [35]. At the same time, the amino acids glycine, proline, and hydroxy proline present in collagen could be involved in the Ag^{+} ions reduction [36,37].

2.2. *Characterization of Antimicrobial 3D-Printed Col-Ag*

2.2.1. SEM Imaging

To confirm the formation of silver particles within collagen hydrogels after UV irradiation, the 3D-printed materials were analyzed by SEM. Figure 2 shows SEM images obtained for 3D-printed named Col-Ag0.05-2, Col-Ag0.05-4, Col-Ag0.05-6, Col-Ag0.1-2, Col-Ag0.1-4, and Col-Ag0.1-6, where the second tags denote the UV irradiation time. Silver particles with different morphology and sizes were formed on the surface of the scaffolds depending on the initial concentration of silver used to prepare the inks and on the UV irradiation time. In case of Col-Ag0.05-2, star-like micro-sized particles [38] of around 23 ± 4 μm were obtained (Figure 2A), but as UV irradiation interval increased to 4 h and 6 h (Figure 2B,C, respectively), a mixture of free spherical AgNPs [20] of 220 ± 42 nm and elongated clusters of around 4.0 ± 0.6 μm formed by the agglomeration of the AgNPs were observed. Col-Ag0.1-2 (Figure 2D) shows few round silver precipitates of around 3.0 ± 0.4 μm, whereas after 4 h and 6 h of UV irradiation cubic silver particles [39,40] from 0.3 to 1.0 μm were mainly observed in 3D-printed Col-Ag0.1-4 and Col-Ag0.1-6 (Figure 2E,F, respectively). Furthermore, at a longer time of UV radiation, the formation of a larger number of particles was observed. The analysis of the obtained nanomaterials by energy-dispersive analysis further confirms the presence of the silver in nanoparticles (Figure 2G).

From these results, it is evident that varying the condition of synthesis (concentration of $AgNO_3$ and UV irradiation time), the formation process of silver particles changes, leading to particles of different size and morphology [41]. The shape of the Ag particles obtained depends on the interaction between Ag^+ ions with collagen and electrons formed during the UV exposure, and the reaction time and the crystallinity of the initial nucleus. These factors could be controlled by modifying the concentration of $AgNO_3$ used for the preparation of the ink and the molar ratio between collagen and $AgNO_3$. In this sense, it was possible to tune the shape and size of in situ synthesized particles by a more environmentally friendly and rapid method, avoiding the use of toxic chemicals.

To further characterize the in situ prepared AgNPs, the material was completely degraded by a collagenase solution and the obtained suspension was studied by TEM (Figure 3). Interestingly, the particles formed inside the collagen have a smaller size (Table 1) than those formed on the surface of the material (Figure 2), possibly due to a stabilizing action of the collagen [42] and due to the different degree of UV light penetration among with limited space for particles growth (Figure 3G). This is attractive since the size and morphology are not only controlled through the different conditions of synthesis studied, but they are also given according to the place of the material they are formed, whether on the surface or inside the 3D-printed scaffolds.

Table 1. Size of AgNPs formed in situ, inside the 3D-printed collagen-based scaffolds.

Sample	AgNPs Size (nm)
Col-Ag0.05-2	26.17 ± 5.53
Col-Ag0.05-4	27.70 ± 7.45
Col-Ag0.05-6	24.39 ± 7.50
Col-Ag0.1-2	12.02 ± 3.34
Col-Ag0.1-4	9.74 ± 2.23
Col-Ag0.1-6	10.18 ± 2.45

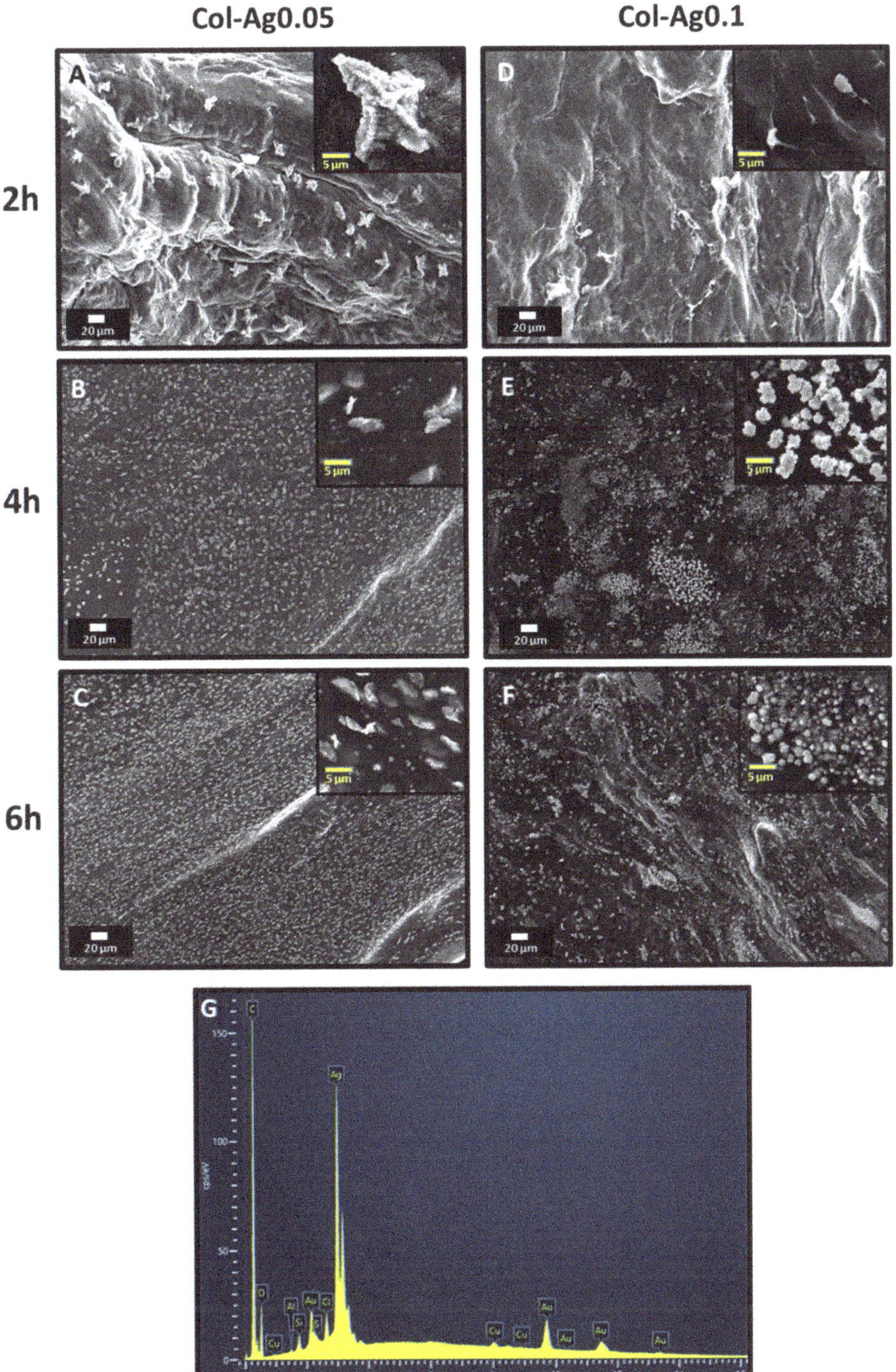

Figure 2. SEM images of (**A**): Col-Ag0.05-2; (**B**): Col-Ag0.05-4; (**C**): Col-Ag0.05-6; (**D**): Col-Ag0.1-2; (**E**): Col-Ag0.1-4; (**F**): Col-Ag0.1-6. (**G**): Representative EDS analysis of Col-Ag.

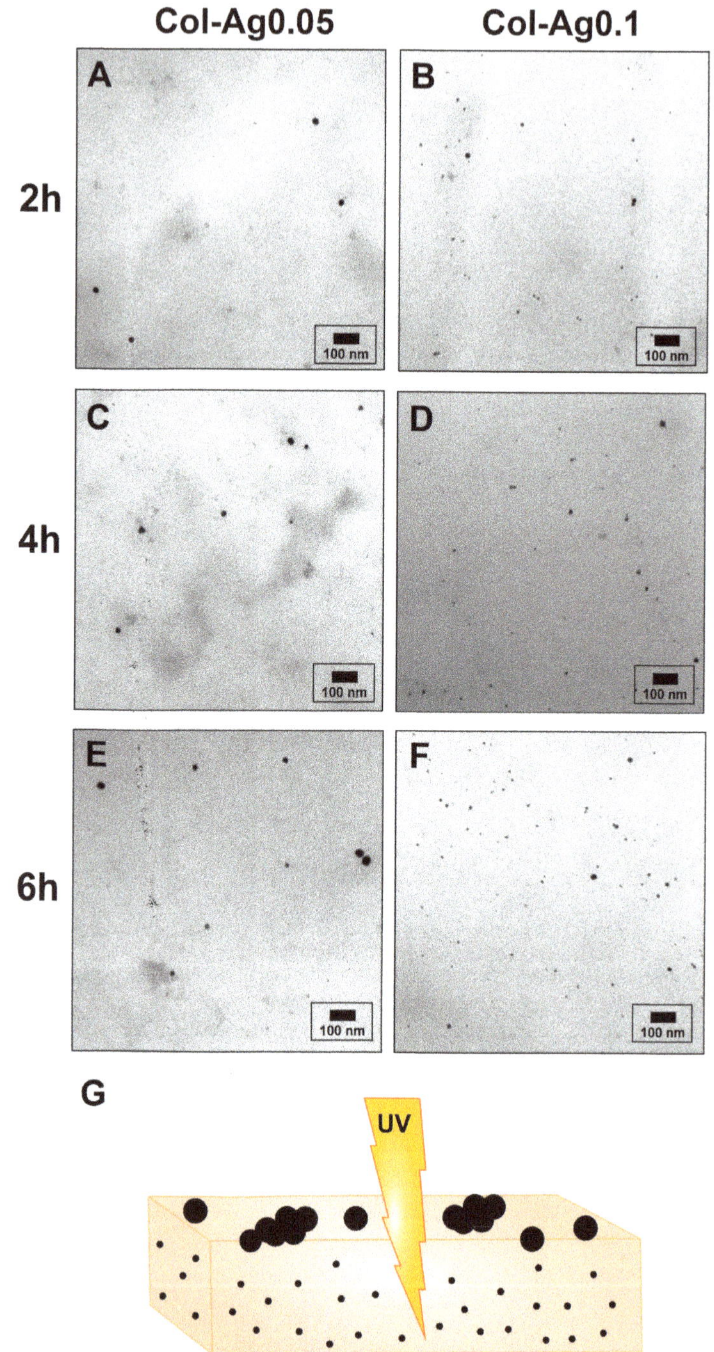

Figure 3. TEM images of the AgNPs formed inside the 3D-printed collagen-based scaffolds. (**A**): Col-Ag0.05-2; (**B**): Col-Ag0.05-4; (**C**): Col-Ag0.05-6; (**D**): Col-Ag0.1-2; (**E**): Col-Ag0.1-4; (**F**): Col-Ag0.1-6. (**G**): schematic illustration of the variation of the size of the silver particles according to the place of the material where they are formed.

2.2.2. FTIR Analyses

FTIR spectra of 3D-printed Col (in the absence of AgNPs), Col-Ag0.05 and Col-Ag0.1 scaffolds are shown in Figure 4. FTIR analysis was performed to study whether there was an interaction between collagen and in situ synthesized Ag particles. When comparing FTIR spectra of Col, Col-Ag0.05 (Figure 4A) and Col-Ag0.1 (Figure 4B) representative peaks were found. However, a close analysis of the spectra indicates that a more intense band at around 1340 and 1400 cm^{-1} was observed for 3D-printed Col-Ag0.05 and more clearly in Col-A0.1 scaffolds. These vibration signals, located at 1340 and 1400 cm^{-1}, respectively, could indicate the possible interaction between silver particles and collagen (antiparallel-βsheet) [43]. Furthermore, a decrease in the intensity of the bands at 1635 and 1556 cm^{-1} corresponding to C=O of amide I and NH groups of amine II of collagen was observed for these scaffolds. This may also indicate that these groups are involved in the reduction and stabilization of silver particles in collagen [36,44].

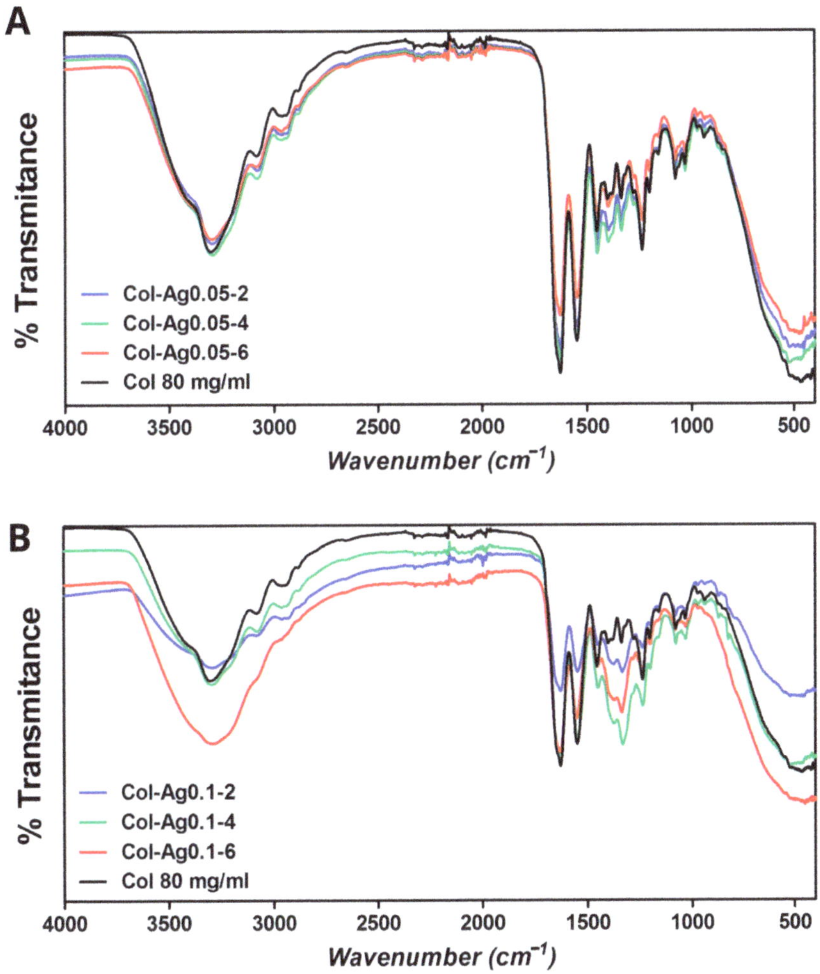

Figure 4. FTIR spectra of Col, Col-Ag0.05 (**A**), and Col-Ag0.1 (**B**) after different time of UV light exposure (0 h, 2 h, 4 h, and 6 h).

2.2.3. Silver Content

Silver content in the 3D-printed Col-Ag materials was quantified by Atomic Absorption technique. Figure 5 shows the final concentration of silver present in Col-Ag0.05 and Col-Ag0.1 after the different UV irradiation times.

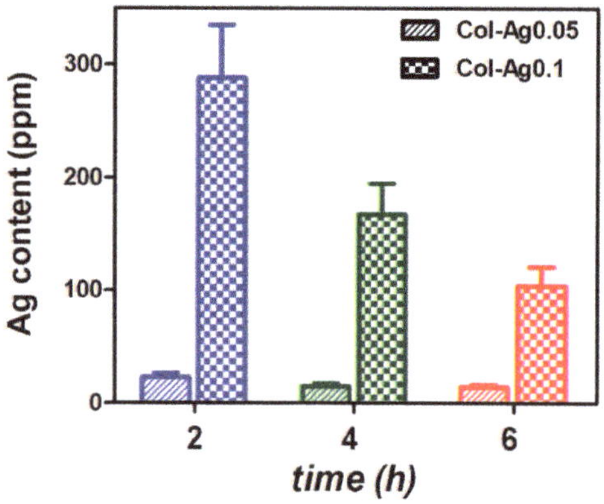

Sample code	Ag concentration (ppm)
Col-Ag0.05-2	23 ± 3
Col-Ag0.05-4	15 ± 3
Col-Ag0.05-6	14 ± 2
Col-Ag0.1-2	288 ± 47
Col-Ag0.1-4	167 ± 27
Col-Ag0.1-6	104 ± 17

Figure 5. Silver content in 3D-printed Col-Ag0.05 and Col-Ag0.1 scaffolds, after 2 h, 4 h, and 6 h of UV irradiation, measured by Atomic Absorption spectroscopy.

As expected, Ag concentration of Col-Ag0.1 is higher than in Col-Ag0.05. The reduction in Ag concentration from the initial amount of $AgNO_3$ used for the preparation of the inks is related with the successive washes of the materials after the gelation with NH_3. Interestingly, a longer exposure to UV radiation light produces a decrease in the final silver content in the 3D-printed collagen materials. This could be attributed to the higher amount of Ag particles in situ formed at longer exposure time to UV light (Figure 2). It has been probed that interaction between proteins and silver ions is stronger than with Ag^0 formulations [45], thus, after successive washes, a higher percentage of Ag particles are prone to be released, remaining less Ag in contact with the collagen.

2.2.4. DSC Thermal Analysis

To determine the changes in the morphological integrity and stability of the 3D-printed collagen scaffolds after in situ synthesized Ag particles, the denaturation temperature of the collagen-based scaffolds was analyzed using DSC. Figure 6 shows the DSC curves for Col, Col-Ag0.05, and Col-Ag0.1 exposed to 2 h, 4 h, and 6 h of UV light.

DSC curves of 3D-printed Col scaffolds (Figure 6A) exposed at different UV times show the typical denaturation temperature at ca. 52.0 ± 0.9 °C found in fibrillar collagen [46]. Interestingly, for Col-Ag0.05 and Col-Ag0.1, opposite effects were found. In case of Col-Ag0.05 (Figure 6B), a shifting of DSC curves to lower temperatures is observed. This indicates that the addition of 0.05 M concentration of Ag produces a destabilization of collagen fibrils by interfering in the self-assembling processes of collagen molecules. This effect can be attributed to the binding of silver ions with collagen, mainly sulfhydryl groups of the fibrillar protein, which finally leads to a decrease in the denaturation temperature of the collagen [47,48]. However, when Col-Ag0.1 (Figure 6C) was analyzed, DSC curves were obtained at higher temperatures than Col. This suggests that under this conditions collagen is stabilized by a higher concentration of Ag (0.1 M) through other chemical interactions [49]. In fact, FTIR spectra indicate that the interactions between Ag and amides groups of collagen are stronger in Col-Ag0.1 than in Col-Ag0.05. This effect is in good agreement with previous data showing that silicic acid influenced the self-assembly of collagen triple helices and, consequently, their denaturation temperature. On the one hand, at low silicification rates, a slightly diminution in the denaturation temperature was attributed to the charges provided by the inorganic component which hinders the correct aggregation of collagen. On the other hand, a significant increase in collagen thermal stability was reported in the presence of higher silicate content due to the coverage of collagen fibril with the inorganic component that stabilized its structure [6,50].

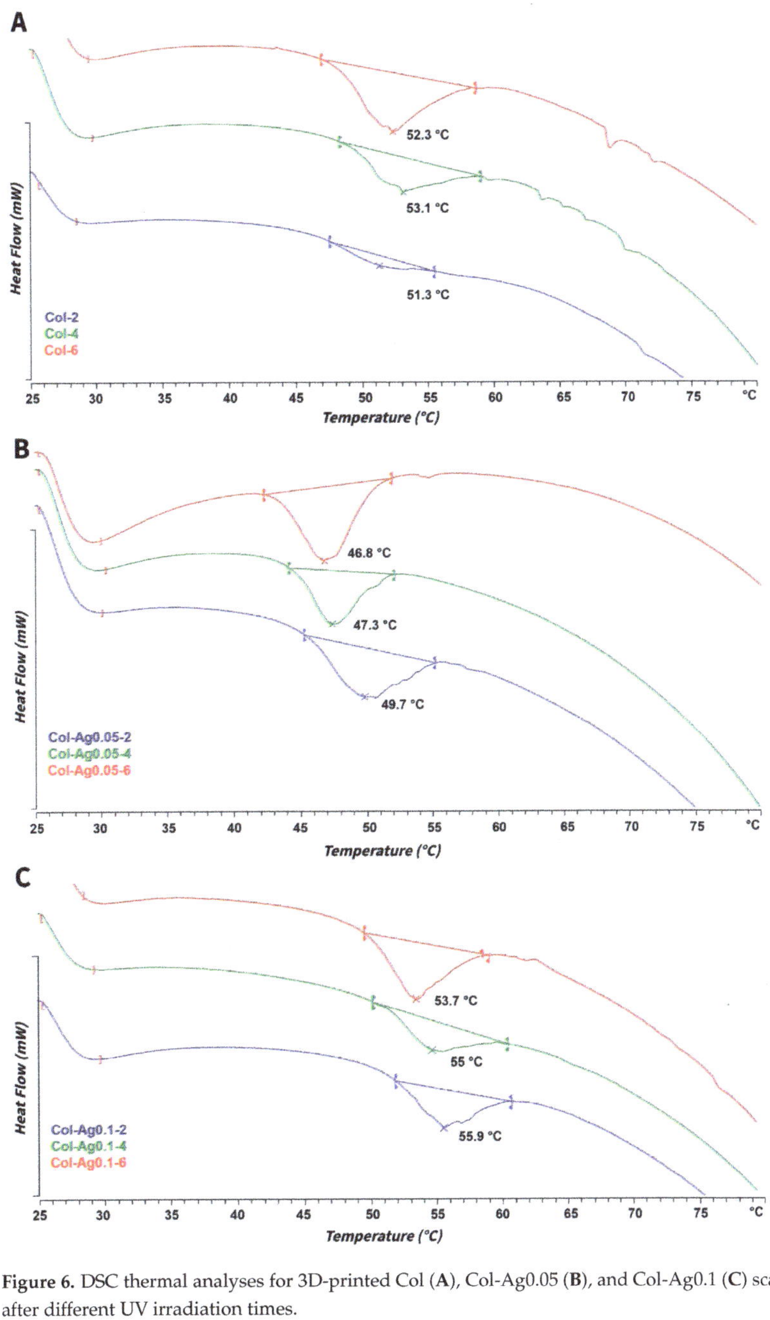

Figure 6. DSC thermal analyses for 3D-printed Col (**A**), Col-Ag0.05 (**B**), and Col-Ag0.1 (**C**) scaffolds after different UV irradiation times.

Furthermore, the longer the UV irradiation time, the lower the denaturation temperature of collagen. This behavior was observed for both Col-Ag0.05 and Col-Ag0.1, demonstrating that an increase in the irradiation dose damages collagen molecules and produces conformational changes, such as a loss of triple helical content of the protein [51]. In addition, longer irradiation times lead to the formation of more particles and, consequently, less Ag^+ ions are available to interact with collagen. From these results, it could be assumed that stability of collagen matrices depends on Ag concentration in the ink and the UV radiation dose.

2.2.5. Enzymatic Degradation

Stability of the 3D-printed collagen-based hydrogels was also evaluated by collagenase degradation test. For this, Col, Col-Ag0.05 and Col-Ag0.1 were incubated with a solution of collagenase 20 U/mL at 37 °C and their weight was measured over time. As a result, after 24 h Col without Ag was completely degraded and Col-Ag0.05 preserved around 10% of its initial weight (Figure 7A). In contrast, 3D-printed Col-Ag0.1 scaffolds only lost about 50% of their initial mass (Figure 7B) indicating, and the thermal analysis, that the in situ formation of silver particles from a higher Ag concentration improves stabilization of 3D-printed fibrillar collagen. According to previous works, silver particles act as reinforcement of collagen-based material enhancing its stability and slowing down collagenase degradation by a physical filling effect and chemical interactions [20,52].

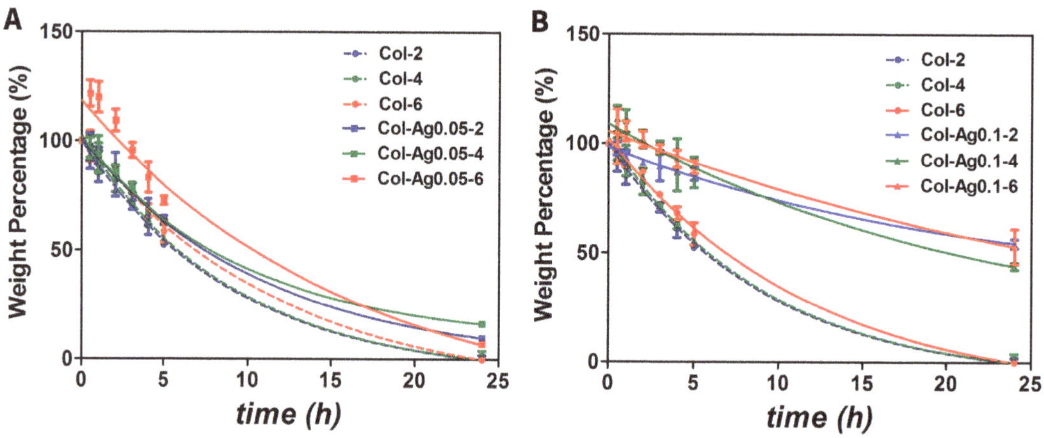

Figure 7. Collagenase degradation test of 3D-printed collagen-based scaffolds. Weight percentage is shown as a function of time comparing Col with Col-Ag0.05 (**A**), and Col with Col-0.1 (**B**). Col-2(–•–blue), Col-4(–•–green), Col-6 (–•–red), Col-Ag0.05-2 (-■-blue), Col-Ag0.05-4 (-■-green), Col-Ag0.05-6 (-■-red), Col-Ag0.05-2 (-▲-blue), Col-Ag0.05-4 (-▲-green), and Col-Ag0.05-6 (-▲-red). Results are expressed as mean ± SD from triplicate experiments.

2.2.6. Swelling Capacity

The swelling response of the 3D-printed Col-Ag materials was also investigated as it is related to the exchange of substances, such as Ag ions and Ag particles, and subsequently, to their antimicrobial activity [53]. Figure 8 shows that Col, Col-Ag0.05, and Col-Ag0.1 reach the maximum water uptake after 6 h, and no further change was observed after 24 h. Col-Ag0.01 (Figure 8B) exhibited a higher capacity to take up water compared to Col-Ag0.05 (Figure 8A), and both showed a higher water absorbency than Col. As the number of fixed charges in collagen increases in presence of a higher Ag concentration, the electrostatic repulsions between collagen chains also increases. As a result, hydrophilicity of the collagen network is increased, leading to a greater swelling capacity [54]. Furthermore, it was observed that at longer UV irradiation intervals the swelling capacity decreases, more significantly in 3D-printed Col scaffolds. According to DSC results, UV radiation causes a possible collapse of the structure and damage of collagen [55,56].

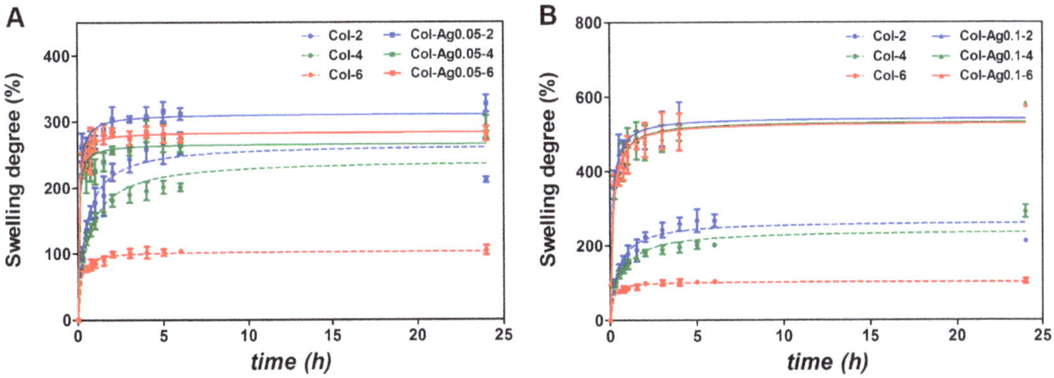

Figure 8. Swelling capacity of Col-Ag0.05 (**A**) and Col-Ag0.1 (**B**). Col-2(–•–blue), Col-4(–•–green), Col-6 (–•–red), Col-Ag0.05-2 (-■-blue), Col-Ag0.05-4 (-■-green), Col-Ag0.05-6 (-■-red), Col-Ag0.05-2 (-▲-blue), Col-Ag0.05-4 (-▲-green), and Col-Ag0.05-6 (-▲-red). Results are expressed as mean ± SD from triplicate experiments.

2.3. Antimicrobial Activity of 3D-printed Col-Ag Scaffolds

Antimicrobial activity of 3D-printed Col-Ag was studied against two bacterial strains, *E. coli* and *S. aureus*, using two different methods, the solid and the dilution technique. As it is shown in Figure 9A, when the antimicrobial activity was evaluated against *E. coli*, there were clear inhibition zones both around the samples and between the gaps for Col-Ag0.05-2, Col-Ag0.05-4, Col-Ag0.05-6, Col-Ag0.1-2, Col-Ag0.1-4, and Col-Ag0.1-6. When inhibition zones with commercial nanoparticles were studied, it was observed that the inhibition zone for the Ag concentration of 288 ppm was (1.3 ± 0.05 cm) similar to the inhibition zone obtained for Col-Ag0.05-6 (1.0 ± 0.06 cm). This result shows that collagen material with Ag presented a higher antimicrobial activity compared to Ag alone. This may be due to the fact that collagen's own hydration improves the silver diffusion phenomenon [57]. In addition, when the antimicrobial activity was evaluated by the dilution method, 100% inhibition of bacterial growth was observed for Col-Ag0.05-2, Col-Ag0.05-4, Col-Ag0.05-6, Col-Ag0.1-2, Col-Ag0.1-4, and Col-Ag0.1-6 (Figure 9B). These results clearly demonstrate that 3D-printed Col-Ag scaffolds present antimicrobial activity against Gram-negative bacteria. In addition, the MIC for Gram-negative bacteria was 4.66 ppm.

When the antimicrobial activity against *S. aureus* was studied by the solid method, well-defined inhibition zones were also observed around the samples and between the gaps for Col-Ag0.05-2, Col-Ag0.05-4, Col-Ag0.05-6, Col-Ag0.1-2, Col-Ag0.1-4, and Col-Ag0.1-6 (Figure 10A). When inhibition zones with commercial nanoparticles was studied, it was observed that the inhibition zone for the Ag concentration of 167 ppm was (1.2 ± 0.07 cm) equal to the inhibition zone obtained for Col-Ag0.05-6 (1.2 ± 0.05 cm). This result shows that collagen material with Ag presented a higher antimicrobial activity compared to Ag alone. This may be due to the fact that collagen's own hydration improves the silver diffusion phenomenon. Moreover, as it is shown in Figure 10B, when evaluated the antimicrobial activity using the dilution method, it was found a percentage of inhibition of bacterial growth greater than 90% for Col-Ag0.05-2, Col-Ag0.05-6, Col-Ag0.1-2, Col-Ag0.1-4, and Col-Ag0.1-6, except for Col-Ag0.05-4. These results suggest that 3D-printed Col-Ag materials also present antimicrobial activity against Gram-positive bacteria. In addition, the MIC for Gram-negative bacteria was 5.83 ppm. However, the slight difference between the percentage of inhibition of bacterial growth among both bacteria strains could be explained by the structural difference in the bacterial cell walls between Gram-negative and Gram-positive bacteria [58]. The antimicrobial activity observed against both bacterial strains is attributed to the reported well-known antibacterial capacity of the AgNPs thanks to the release of Ag^+ ions [20,52]. In fact, when the materials were incubated for 24 h in liquid medium, it was found that Col-Ag0.05-2, Col-Ag0.05-4, Col-Ag0.05-6, Col-Ag0.1-2, Col-Ag0.1-4, and Col-Ag0.1-6 released 32.09 ± 0.37, 29.47 ± 0.38, 27.30 ± 1.86, 55.21 ± 0.40, 55.42 ± 0.31, and 56.44 ± 0.37 ppb of Ag^+, respectively.

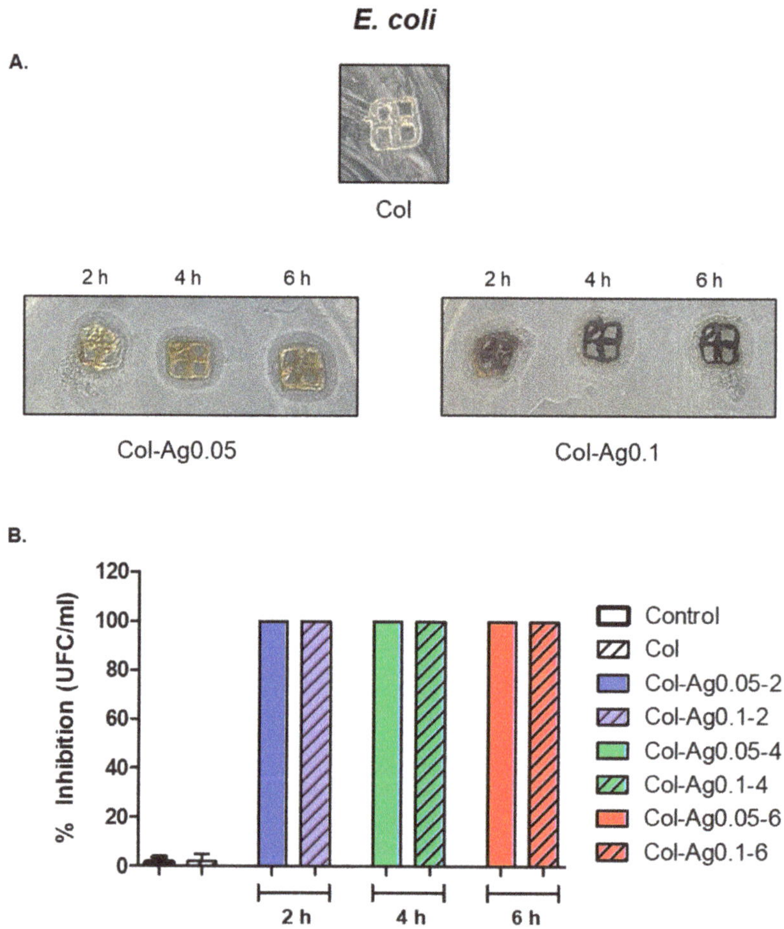

Figure 9. Antimicrobial activity of 3D-printed Col and Col-Ag gels against *E. coli*. (**A**) Disk diffusion method. (**B**) Percentage of growth inhibition, evaluated by the dilution method.

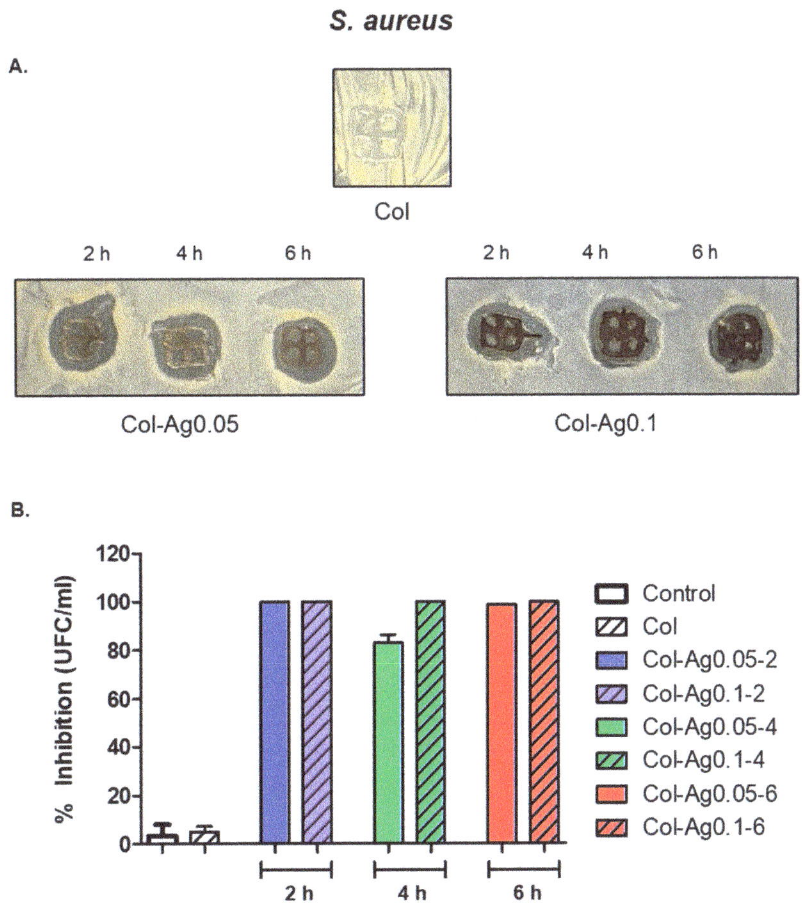

Figure 10. Antimicrobial activity of silver in 3D-printed Col and Col-Ag gels against *S. aureus*. (**A**) Disk diffusion method. (**B**) Percentage of growth inhibition, evaluated by the dilution method.

3. Materials and Methods

3.1. Preparation of Inks

Collagen (type I) was obtained from rat tails tendons and dissolved in 17.5 mM acetic acid. The collagen solution was lyophilized, then the dry powder was resuspended with the necessary volume of distilled water or a solution of $AgNO_3$ in order to obtain a final collagen concentration of 80 mg/mL. Based on previous works, a concentration of 0.05 M [20,52] of $AgNO_3$ was used to in situ prepared AgNPs, and for comparison a higher concentration 0.1 M was also evaluated. The three prepared inks were allowed to hydrate at 4 °C in darkness, and, after 24 h, were transferred to a 1 mL syringe.

3.2. Printing of Collagen-Based Scaffolds

A square grid design was chosen as model structure to study 3D-printed Col-Ag material properties. The design was created with the software Onshape® (2013–Present) and saved as stereolithography (.stl) file, which was then converted to a G-code file by Repetier software. This G-code file was then used to dictate the scaffold dimensions and printing coordinates to the bioprinter. Scaffolds were printed using an extrusion-based Life SI 3D-bioprinter (Life SI, Humanizing Technology, Córdoba, Argentina). For this, the optimal printing conditions were first determined (25 °C temperature, delay of 40 μseg, and 1 nL/4 μm for the amount of printed material), then 6 mm × 6 mm × 2 mm square grids were printed by extruding the inks through a 21G needle with an internal diameter of 514 μm.

3.3. In situ Reduction in Ag by Using the UV Irradiation Method

To induce in situ reduction in silver nitrate, the UV irradiation method was used. The 3D scaffolds printed from the inks prepared with collagen dissolved in 0.05 or 0.1 M of $AgNO_3$ were irradiated utilizing an ultraviolet light (UV) chamber with a UV lamp at λ: 365 nm. UV light exposure was applied during 2 h, 4 h, or 6 h at room temperature. As control, 3D-printed scaffolds obtained from collagen inks without $AgNO_3$ were also irradiated with UV light.

3.4. Gelation of 3D-Printed Scaffolds

After UV irradiation, the in situ Ag particles-synthesized 3D-printed Col scaffolds were exposed to ammonia vapors to produce gelation of collagen. After 12 h, the 3D-printed gels were removed from the basic atmosphere, and finally washed with water until a neutral pH was reached. The hydrogels were defined as Col-2, Col-4, Col-6, Col-Ag0.05-2, Col-Ag0.05-4, Col-Ag0.05-6, Col-Ag0.1-2, Col-Ag0.1-4, and Col-Ag0.1-6, respectively, according to the silver concentration (0.05 or 0.1 M) and UV irradiation time (2 h, 4 h, and 6 h).

3.5. Characterization of 3D-Printed Col-Ag Scaffolds

3.5.1. SEM Imaging

Col-Ag gels were characterized by scanning electron microscopy (SEM) using a Carl Zeiss-EVO 10 HV W microscope. For this, the 3D-printed Col-Ag gels were fixed with a 2.5% glutaraldehyde solution and lyophilized for 48 h. The freeze-dried samples were placed on carbon double-sided tape supported in aluminum SEM sample holder and sputter-coated with gold.

3.5.2. TEM Imaging

The morphology and size of the in situ formed AgNPs were studied by transmission electron microscopy (TEM) using a Zeiss EM109T electron microscope. For this, the 3D-printed Col-Ag scaffolds were completely degraded by incubation with type I collagenase solution of 100 U/mL (Gibco® (Waltham, MA, USA), 340 U/mg) at 37 °C for 12 h. Finally, a drop of each obtained suspension was added to carbon copper grids and let dry for a few minutes before introducing them to the microscope.

3.5.3. FTIR Analysis

Fourier transform infrared (FTIR) spectra of 3D-printed collagen gels (Col and Col-Ag) were performed using an FTIR-Raman Nicolet iS50 (Thermo Scientific, Waltham, MA, USA). Briefly, the three-dimensional samples were dried under a nitrogen flow, then a small slice of each one was placed on the attenuated total reflection accessory of the spectrometer. The spectra were recorded over the range of 4000–500 cm^{-1}. The SpectraGryph software (v.1.2.16) was used to normalize FTIR data.

3.5.4. DSC

Differential Scanning Calorimetry (DSC) analysis of 3D-printed collagen gels (Col and Col-Ag) were performed using a DSC Calorimeter (DSC 822, Mettler Toledo, Switzerland). For this purpose, a known amount (5–10 mg) of each sample was placed in an aluminum crucible and the DSC curves were obtained with a heat flow of 10 °C min^{-1} from 25 to 80 °C, under a nitrogen atmosphere. The collagen denaturation temperature was determined at the maximum of the melting peak after baseline subtraction.

3.5.5. Swelling Capacity

To determine the amount of liquid that 3D-printed collagen-based gels can absorb, Col and Col-Ag gels were dried and weighed. Then, they were re-hydrated with 500 μL of distilled water and incubated at room temperature. The change of the weight was measured over time for 24 h. For each time, the 3D-printed hydrogels were removed and wiped dry. In order to quantify the swelling capacity of each sample, the following Equation (1) was used:

$$\%\ \text{Swelling} = \frac{W_{wet} - W_{dried}}{W_{dried}} \times 100 \quad (1)$$

3.5.6. Enzymatic Degradation

The degradation of the 3D-printed collagen-based gels with and without Ag was evaluated by collagenase digestion. Briefly, the Col and Col-Ag gels were first weighed and placed in a 24-well plate. Then, 500 μL of a 50 U/mL solution of type I collagenase enzyme (Gibco®, 340 U/mg) was added to each of the wells containing the 3D-printed scaffolds and incubated at 37 °C. The weight of the Col and Col-Ag gels was measured after 0.5 h, 1 h, 2 h, 3 h, 4 h, 5 h, and 24 h. For this, the material was removed from the enzyme solution, the excess of liquid was dried and finally weighted. The changes of weight of Col and Col-Ag gels were analyzed over time.

3.5.7. Silver Content in 3D-Printed Collagen-Based Gels

To determine the final concentration of silver in 3D-printed Col-Ag gels, Col-Ag0.05-2, Col-Ag0.05-4, Col-Ag0.05-6, Col-Ag0.1-2, Col-Ag0.1-4, and Col-Ag0.1-6 were digested with 5 mL of an acid mixture of HNO_3 (65%)/H_2O_2 (10 vol.) 9:1 for 24 h, when a complete degradation of collagen-based gels was reached. The final silver concentration was determined by Atomic Absorption Spectrometry in a 210 VGP spectrometer (Buck Scientific, Norwalk, CA, USA) by the electrothermal atomization method using pyrolytic graphite tubes.

3.5.8. Silver Release

To study the release of Ag+ from the 3D-printed collagen scaffolds containing in situ prepared AgNPs, the materials were submerged in 500 μL of H_2O and incubated at room temperature for 24 h. The concentration of silver in the supernatant was measured by Atomic Absorption Spectrometry in a 210 VGP spectrometer (Buck Scientific) by the electrothermal atomization method using pyrolytic graphite tubes.

3.6. *Antimicrobial Activity*

To study antimicrobial activity of silver in 3D-printed Col-Ag gels, Col-Ag0.05-2, Col-Ag0.05-4, Col-Ag0.05-6, Col-Ag0.1-2, Col-Ag0.1-4, and Col-Ag0.1-6 Gram-positive (*Staphylococcus aureus*, ATCC 29213) and Gram-negative (*Escherichia coli*, ATCC 9637) bacteria were used. The antimicrobial activity was evaluated using two methods: the solid and the dilution method. Both, Gram-negative and positive bacteria were grown in Luria Bertani (LB) medium (yeast extract, 5 g/L; NaCl, 10 g/L, and tryptone, 10 g/L) at 37 °C and diluted up to 1×10^6 Colony Forming Units (CFU)/mL. Briefly, for the disk diffusion method, we spread 50 μL of a bacterial suspension 1:1000 on an agar Petri dish. Then, the 3D-printed Col-Ag gels, Col-Ag0.05-2, Col-Ag0.05-4, Col-Ag0.05-6, Col-Ag0.1-2, Col-Ag0.1-4, and Col-Ag0.1-6 were placed on the agar surface and incubated them at 37 °C. Finally, the diameter of the bacteria-free areas that surround the 3D-printed Col-Ag materials to determine the antimicrobial activity was measure. In addition, the mentioned groups were compared to the antimicrobial activity of Ag (Nanotek S.A., 20–40 nm, colloidal suspension of 1% *w/v*, (nanArgen®, Buenos Aires, Argentina), CAS no. 7440–22-4) in different concentrations. On the other hand, for the dilution method, 100 μL of the bacterial suspension 1:1000 was mixed with 3D-printed Col-Ag gels, Col-Ag0.05-2, Col-Ag0.05-4, Col-Ag0.05-6, Col-Ag0.1-2, Col-Ag0.1-4, and Col-Ag0.1-6 with 900 μL of LB medium, and were incubated at 37 °C for 24 h. Then, 20 μL of the dilutions was seeded on agar Petri dishes and incubated at 37 °C for 24 h. Finally, CFU was determined by counting manually the colonies on the plate. To determine the minimum inhibitory concentration (MIC) of the material, Col-Ag0.05-6 was selected, since it is the condition with fewer concentration of Ag. Different amounts of the material were placed in the LB medium with *S. aureus* or *E. coli* and incubated for 24 h at 37 °C. Finally, the MIC was determined as the lowest concentration of the material at which there was no visible growth of the organism. The growth was evaluated by turbidimetry. Results are expressed as mean ± SD from triplicate experiments.

4. Conclusions

In this work, antimicrobial 3D-printed collagen-based scaffolds through in situ synthesizing AgNPs by the UV irradiation method were developed. First, lyophilized collagen was resuspended in $AgNO_3$ (0.05 and 0.1 M) to prepare suitable inks for 3D printing. Once the scaffolds were 3D-printed, they were irradiated during different UV intervals (2, 4, 6 h). As a result, AgNPs with different morphologies and sizes were obtained within the collagen scaffolds. This is very interesting since it was possible to grow and control the shape and size of the Ag nanoparticles through an environmentally friendly and simple method, without using toxic reagents. DSC analysis showed that the concentration of $AgNO_3$ and UV time exposure influence the thermal stability of collagen. This was also verified by the enzymatic degradation assay, which showed that 3D-printed Col-Ag scaffolds prepared from a higher concentration of $AgNO_3$ were more resistant to collagenase action, indicating that a higher amount of silver particles acts as better reinforcement of the 3D-printed material. In addition to this, swelling capacity of Col-Ag also varied according to the preparation conditions of the antimicrobial 3D-printed materials. Finally, each and every Col-Ag scaffold showed bactericidal activity against Gram-positive and Gram-negative bacteria.

In summary, in this work the UV irradiation method and 3D printing technique were combined to in situ prepare AgNPs, and thus develop broad spectrum antimicrobial collagen-based scaffolds. This is attractive to design and prepare new bactericidal materials by nontoxic and cost-effective methodologies.

Author Contributions: S.M.: methodology, formal analysis, investigation, writing—original draft preparation; P.E.A.: methodology, formal analysis, investigation, writing—original draft preparation; M.G.B.: supervision, review and editing; M.F.D.: conceptualization, writing—review and editing, supervision, project administration, funding acquisition. All authors have read and agreed to the published version of the manuscript.

Funding: The authors would like to acknowledge grants from the Universidad de Buenos Aires, UBACYT 20020150100056BA and PIDAE 2022 (Martín F. Desimone), and PIBAA 28720210100962CO (Sofia Municoy), which supported this work.

Institutional Review Board Statement: Not applicable.

Informed Consent Statement: Not applicable.

Data Availability Statement: Not applicable.

Acknowledgments: Authors acknowledge to J. Infatino of Centro de Ingeniería en Materiales-Instituto Tecnológico de Buenos Aires for SEM images and M. A. Rosasco of FFYB for help in DSC measurements. S.M. acknowledges CONICET and UBA.

Conflicts of Interest: The authors declare no conflict of interest.

References

1. Antezana, P.E.; Municoy, S.; Álvarez-Echazú, M.I.; Santo-Orihuela, P.L.; Catalano, P.N.; Al-Tel, T.H.; Kadumudi, F.B.; Dolatshahi-Pirouz, A.; Orive, G.; Desimone, M.F. The 3D Bioprinted Scaffolds for Wound Healing. *Pharmaceutics* **2022**, *14*, 464. [CrossRef] [PubMed]
2. Antezana, P.E.; Municoy, S.; Orive, G.; Desimone, M.F. Design of a New 3D Gelatin—Alginate Scaffold Loaded with Cannabis sativa Oil. *Polymers* **2022**, *14*, 4506. [CrossRef] [PubMed]
3. Chua, C.K.; Leong, K.F. *3D Printing and Additive Manufacturing*; World Scientific: Singapore, 2016; ISBN 978-981-314-675-4.
4. Vikram Singh, A.; Hasan Dad Ansari, M.; Wang, S.; Laux, P.; Luch, A.; Kumar, A.; Patil, R.; Nussberger, S. The Adoption of Three-Dimensional Additive Manufacturing from Biomedical Material Design to 3D Organ Printing. *Appl. Sci.* **2019**, *9*, 811. [CrossRef]

5. David, G.; Bargan, A.I.; Drobota, M.; Bele, A.; Rosca, I. Comparative Investigation of Collagen-Based Hybrid 3D Structures for Potential Biomedical Applications. *Materials* **2021**, *14*, 3313. [CrossRef] [PubMed]
6. Desimone, M.F.; Hélary, C.; Rietveld, I.B.; Bataille, I.; Mosser, G.; Giraud-Guille, M.M.; Livage, J.; Coradin, T. Silica-collagen bionanocomposites as three-dimensional scaffolds for fibroblast immobilization. *Acta Biomater.* **2010**, *6*, 3998–4004. [CrossRef]
7. Desimone, M.F.; Hélary, C.; Quignard, S.; Rietveld, I.B.; Bataille, I.; Copello, G.J.; Mosser, G.; Giraud-Guille, M.M.; Livage, J.; Meddahi-Pellé, A.; et al. In vitro studies and preliminary in vivo evaluation of silicified concentrated collagen hydrogels. *ACS Appl. Mater. Interfaces* **2011**, *3*, 3831–3838. [CrossRef]
8. Mateo, E.M.; Jiménez, M. Silver Nanoparticle-Based Therapy: Can It Be Useful to Combat Multi-Drug Resistant Bacteria? *Antibiotics* **2022**, *11*, 1205. [CrossRef]
9. Alotaibi, A.M.; Alsaleh, N.B.; Aljasham, A.T.; Tawfik, E.A.; Almutairi, M.M.; Assiri, M.A.; Alkholief, M.; Almutairi, M.M. Silver Nanoparticle-Based Combinations with Antimicrobial Agents against Antimicrobial-Resistant Clinical Isolates. *Antibiotics* **2022**, *11*, 1219. [CrossRef]
10. Fiore, M.; Bruschi, A.; Giannini, C.; Morante, L.; Rondinella, C.; Filippini, M.; Sambri, A.; De Paolis, M. Is Silver the New Gold? A Systematic Review of the Preclinical Evidence of Its Use in Bone Substitutes as Antiseptic. *Antibiotics* **2022**, *11*, 995. [CrossRef]
11. Sim, W.; Barnard, R.T.; Blaskovich, M.A.T.; Ziora, Z.M. Antimicrobial Silver in Medicinal and Consumer Applications: A Patent Review of the Past Decade (2007–2017). *Antibiotic.* **2018**, *7*, 93. [CrossRef]
12. Sondi, I.; Salopek-Sondi, B. Silver nanoparticles as antimicrobial agent: A case study on *E. coli* as a model for Gram-negative bacteria. *J. Colloid Interface Sci.* **2004**, *275*, 177–182. [CrossRef] [PubMed]
13. Hovhannisyan, Z.; Timotina, M.; Manoyan, J.; Gabrielyan, L.; Petrosyan, M.; Kusznierewicz, B.; Bartoszek, A.; Jacob, C.; Ginovyan, M.; Trchounian, K.; et al. Characteristics of Ribes nigrum L. extract-mediated silver nanoparticles and their action mechanisms. *Antibiotics* **2022**, *11*, 1415. [CrossRef] [PubMed]
14. Mikhailova, E.O. Silver Nanoparticles: Mechanism of Action and Probable Bio-Application. *J. Funct. Biomater.* **2020**, *11*, 84. [CrossRef] [PubMed]
15. Durán, N.; Durán, M.; de Jesus, M.B.; Seabra, A.B.; Fávaro, W.J.; Nakazato, G. Silver nanoparticles: A new view on mechanistic aspects on antimicrobial activity. *Nanomed. Nanotechnol. Biol. Med.* **2016**, *12*, 789–799. [CrossRef]
16. Iravani, S. Bacteria in Nanoparticle Synthesis: Current Status and Future Prospects. *Int. Sch. Res. Not.* **2014**, *2014*, 359316. [CrossRef]
17. Geethalakshmi, R.; Sarada, D.V.L. Gold and silver nanoparticles from Trianthema decandra: Synthesis, characterization, and antimicrobial properties. *Int. J. Nanomed.* **2012**, *7*, 5375–5384. [CrossRef]
18. Zhou, P.; Xia, Z.; Qi, C.; He, M.; Yu, T.; Shi, L. Construction of chitosan/Ag nanocomposite sponges and their properties. *Int. J. Biol. Macromol.* **2021**, *192*, 272–277. [CrossRef]
19. Wu, Z.; Zhou, W.; Deng, W.; Xu, C.; Cai, Y.; Wang, X. Antibacterial and Hemostatic Thiol-Modified Chitosan-Immobilized AgNPs Composite Sponges. *ACS Appl. Mater. Interfaces* **2020**, *12*, 20307–20320. [CrossRef]
20. Municoy, S.; Antezana, P.E.; Pérez, C.J.; Bellino, M.G.; Desimone, M.F. Tuning the antimicrobial activity of collagen biomaterials through a liposomal approach. *J. Appl. Polym. Sci.* **2021**, *138*, 50330. [CrossRef]
21. Rónavári, A.; Bélteky, P.; Boka, E.; Zakupszky, D.; Igaz, N.; Szerencsés, B.; Pfeiffer, I.; Kónya, Z.; Kiricsi, M. Polyvinyl-Pyrrolidone-Coated Silver Nanoparticles—The Colloidal, Chemical, and Biological Consequences of Steric Stabilization under Biorelevant Conditions. *Int. J. Mol. Sci.* **2021**, *22*, 8673. [CrossRef]
22. Dasgupta, N.; Ranjan, S.; Patra, D.; Srivastava, P.; Kumar, A.; Ramalingam, C. Bovine serum albumin interacts with silver nanoparticles with a "side-on" or "end on" conformation. *Chem. Biol. Interact.* **2016**, *253*, 100–111. [CrossRef] [PubMed]
23. Boehmler, D.J.; O'Dell, Z.J.; Chung, C.; Riley, K.R. Bovine Serum Albumin Enhances Silver Nanoparticle Dissolution Kinetics in a Size- and Concentration-Dependent Manner. *Langmuir* **2020**, *36*, 1053–1061. [CrossRef] [PubMed]
24. Chien, H.-W.; Tsai, M.-Y.; Kuo, C.-J.; Lin, C.-L. Well-Dispersed Silver Nanoparticles on Cellulose Filter Paper for Bacterial Removal. *Nanomaterials* **2021**, *11*, 595. [CrossRef] [PubMed]
25. Padnya, P.; Gorbachuk, V.; Stoikov, I. The Role of Calix[n]arenes and Pillar[n]arenes in the Design of Silver Nanoparticles: Self-Assembly and Application. *Int. J. Mol. Sci.* **2020**, *21*, 1425. [CrossRef]
26. Shenashen, M.A.; El-Safty, S.A.; Elshehy, E.A. Synthesis, Morphological Control, and Properties of Silver Nanoparticles in Potential Applications. *Part. Part. Syst. Charact.* **2014**, *31*, 293–316. [CrossRef]
27. Nene, A.; Galluzzi, M.; Hongrong, L.; Somani, P.; Ramakrishna, S.; Yu, X.-F. Synthetic preparations and atomic scale engineering of silver nanoparticles for biomedical applications. *Nanoscale* **2021**, *13*, 13923–13942. [CrossRef] [PubMed]
28. Jorge de Souza, T.A.; Rosa Souza, L.R.; Franchi, L.P. Silver nanoparticles: An integrated view of green synthesis methods, transformation in the environment, and toxicity. *Ecotoxicol. Environ. Saf.* **2019**, *171*, 691–700. [CrossRef] [PubMed]
29. Wagener, P.; Ibrahimkutty, S.; Menzel, A.; Plech, A.; Barcikowski, S. Dynamics of silver nanoparticle formation and agglomeration inside the cavitation bubble after pulsed laser ablation in liquid. *Phys. Chem. Chem. Phys.* **2013**, *15*, 3068–3074. [CrossRef] [PubMed]
30. Peng, H.; Yang, A.; Xiong, J. Green, microwave-assisted synthesis of silver nanoparticles using bamboo hemicelluloses and glucose in an aqueous medium. *Carbohydr. Polym.* **2013**, *91*, 348–355. [CrossRef]
31. Wani, I.A.; Ganguly, A.; Ahmed, J.; Ahmad, T. Silver nanoparticles: Ultrasonic wave assisted synthesis, optical characterization and surface area studies. *Mater. Lett.* **2011**, *65*, 520–522. [CrossRef]

32. Calamak, S.; Ermis, M. In situ silver nanoparticle synthesis on 3D-printed polylactic acid scaffolds for biomedical applications. *J. Mater. Res.* **2021**, *36*, 166–175. [CrossRef]
33. Catalano, P.N.; Pezzoni, M.; Costa, C.; Soler-Illia, G.J.D.A.A.; Bellino, M.G.; Desimone, M.F. Optically transparent silver-loaded mesoporous thin film coating with long-lasting antibacterial activity. *Microporous Mesoporous Mater.* **2016**, *236*. [CrossRef]
34. Shameli, K.; Ahmad, M.B.; Yunus, W.M.Z.W.; Rustaiyan, A.; Ibrahim, N.A.; Zargar, M.; Abdollahi, Y. Green synthesis of silver/montmorillonite/chitosan bionanocomposites using the UV irradiation method and evaluation of antibacterial activity. *Int. J. Nanomed.* **2010**, *5*, 875–887. [CrossRef] [PubMed]
35. Huang, L.; Zhai, M.L.; Long, D.W.; Peng, J.; Xu, L.; Wu, G.Z.; Li, J.Q.; Wei, G.S. UV-induced synthesis, characterization and formation mechanism of silver nanoparticles in alkalic carboxymethylated chitosan solution. *J. Nanopart. Res.* **2008**, *10*, 1193–1202. [CrossRef]
36. Yorseng, K.; Siengchin, S.; Ashok, B.; Rajulu, A.V. Nanocomposite egg shell powder with in situ generated silver nanoparticles using inherent collagen as reducing agent. *J. Bioresour. Bioprod.* **2020**, *5*, 101–107. [CrossRef]
37. Alarcon, E.I.; Udekwu, K.; Skog, M.; Pacioni, N.L.; Stamplecoskie, K.G.; González-Béjar, M.; Polisetti, N.; Wickham, A.; Richter-Dahlfors, A.; Griffith, M.; et al. The biocompatibility and antibacterial properties of collagen-stabilized, photochemically prepared silver nanoparticles. *Biomaterials* **2012**, *33*, 4947–4956. [CrossRef]
38. Pourjavadi, A.; Soleyman, R. Novel silver nano-wedges for killing microorganisms. *Mater. Res. Bull.* **2011**, *46*, 1860–1865. [CrossRef]
39. Skrabalak, S.E.; Au, L.; Li, X.; Xia, Y. Facile synthesis of Ag nanocubes and Au nanocages. *Nat. Protoc.* **2007**, *2*, 2182–2190. [CrossRef]
40. Wiley, B.; Sun, Y.; Mayers, B.; Xia, Y. Shape-controlled synthesis of metal nanostructures: The case of silver. *Chem.-A Eur. J.* **2005**, *11*, 454–463. [CrossRef]
41. Khodashenas, B.; Ghorbani, H.R. Synthesis of silver nanoparticles with different shapes. *Arab. J. Chem.* **2019**, *12*, 1823–1838. [CrossRef]
42. Nogueira, S.S.; de Araujo-Nobre, A.R.; Mafud, A.C.; Guimarães, M.A.; Alves, M.M.M.; Plácido, A.; Carvalho, F.A.A.; Arcanjo, D.D.R.; Mascarenhas, Y.; Costa, F.G.; et al. Silver nanoparticle stabilized by hydrolyzed collagen and natural polymers: Synthesis, characterization and antibacterial-antifungal evaluation. *Int. J. Biol. Macromol.* **2019**, *135*, 808–814. [CrossRef] [PubMed]
43. Drobota, M.; Grierosu, I.; Radu, I.; Vasilescu, D.S. The effect of silver nanoparticles on the collagen secondary structure. *Key Eng. Mater.* **2014**, *638*, 8–13. [CrossRef]
44. Cardoso, V.S.; Quelemes, P.V.; Amorin, A.; Primo, F.L.; Gobo, G.G.; Tedesco, A.C.; Mafud, A.C.; Mascarenhas, Y.P.; Corrêa, J.R.; Kuckelhaus, S.A.S.; et al. Collagen-based silver nanoparticles for biological applications: Synthesis and characterization. *J. Nanobiotechnol.* **2014**, *12*, 1–9. [CrossRef] [PubMed]
45. Kędziora, A.; Wieczorek, R.; Speruda, M.; Matolínová, I.; Goszczyński, T.M.; Litwin, I.; Matolín, V.; Bugla-Płoskońska, G. Comparison of Antibacterial Mode of Action of Silver Ions and Silver Nanoformulations With Different Physico-Chemical Properties: Experimental and Computational Studies. *Front. Microbiol.* **2021**, *12*, 1707. [CrossRef]
46. Badea, E.; Della Gatta, G.; Usacheva, T. Effects of temperature and relative humidity on fibrillar collagen in parchment: A micro differential scanning calorimetry (micro DSC) study. *Polym. Degrad. Stab.* **2012**, *97*, 346–353. [CrossRef]
47. Walker, M.; Parsons, D. The biological fate of silver ions following the use of silver-containing wound care products–A review. *Int. Wound J.* **2014**, *11*, 496–504. [CrossRef]
48. Sayed, M.; Hiraishi, N.; Matin, K.; Abdou, A.; Burrow, M.F.; Tagami, J. Effect of silver-containing agents on the ultra-structural morphology of dentinal collagen. *Dent. Mater.* **2020**, *36*, 936–944. [CrossRef]
49. González-Masís, J.; Cubero-Sesin, J.M.; Corrales-Ureña, Y.R.; González-Camacho, S.; Mora-Ugalde, N.; Baizán-Rojas, M.; Loaiza, R.; Vega-Baudrit, J.R.; Gonzalez-Paz, R.J. Increased fibroblast metabolic activity of collagen scaffolds via the addition of propolis nanoparticles. *Materials* **2020**, *13*, 3118. [CrossRef]
50. Desimone, M.F.; Hélary, C.; Mosser, G.; Giraud-Guille, M.-M.; Livage, J.; Coradin, T. Fibroblast encapsulation in hybrid silica-collagen hydrogels. *J. Mater. Chem.* **2010**, *20*. [CrossRef]
51. Demeter, M.; Călina, I.; Scărișoreanu, A.; Micutz, M.; Kaya, M.A. Correlations on the Structure and Properties of Collagen Hydrogels Produced by E-Beam Crosslinking. *Materials* **2022**, *15*, 7663. [CrossRef]
52. Antezana, P.E.; Municoy, S.; Perez, C.J.; Desimone, M.F. Collagen Hydrogels Loaded with Silver Nanoparticles and Cannabis Sativa Oil. *Antibiotics* **2021**, *10*, 1420. [CrossRef] [PubMed]
53. Holback, H.; Yeo, Y.; Park, K. Hydrogel swelling behavior and its biomedical applications. In *Biomedical Hydrogels*; Woodhead Publishing Limited: Soston, UK, 2011; pp. 3–24.
54. Peppas, N.A.; Bures, P.; Leobandung, W.; Ichikawa, H. Hydrogels in pharmaceutical formulations. *Eur. J. Pharm. Biopharm.* **2000**, *50*, 27–46. [CrossRef]
55. Jariashvili, K.; Madhan, B.; Brodsky, B.; Kuchava, A.; Namicheishvili, L.; Metreveli, N. Uv damage of collagen: Insights from model collagen peptides. *Biopolymers* **2012**, *97*, 189–198. [CrossRef] [PubMed]
56. Stellavato, A.; Pirozzi, A.V.A.; Donato, S.; Scognamiglio, I.; Reale, S.; DI Pardo, A.; Filosa, S.; Vassallo, V.; Bellia, G.; De Rosa, M.; et al. Positive Effects against UV-A Induced Damage and Oxidative Stress on an in Vitro Cell Model Using a Hyaluronic Acid Based Formulation Containing Amino Acids, Vitamins, and Minerals. *Biomed Res. Int.* **2018**, *2018*, 8481243. [CrossRef] [PubMed]

57. Cunliffe, A.J.; Askew, P.D.; Stephan, I.; Iredale, G.; Cosemans, P.; Simmons, L.M.; Verran, J.; Redfern, J. How Do We Determine the Efficacy of an Antibacterial Surface? A Review of Standardised Antibacterial Material Testing Methods. *Antibiotics* **2021**, *10*, 1069. [CrossRef] [PubMed]
58. Slavin, Y.N.; Asnis, J.; Häfeli, U.O.; Bach, H. Metal nanoparticles: Understanding the mechanisms behind antibacterial activity. *J. Nanobiotechnol.* **2017**, *15*, 65. [CrossRef] [PubMed]

Disclaimer/Publisher's Note: The statements, opinions and data contained in all publications are solely those of the individual author(s) and contributor(s) and not of MDPI and/or the editor(s). MDPI and/or the editor(s) disclaim responsibility for any injury to people or property resulting from any ideas, methods, instructions or products referred to in the content.

antibiotics

Article

A Promising Antifungal and Antiamoebic Effect of Silver Nanorings, a Novel Type of AgNP

Sara González-Fernández [1,2,†], Victor Lozano-Iturbe [1,2,†], Mª Fe Menéndez [3], Helena Ordiales [1,2], Iván Fernández-Vega [2,4,5], Jesús Merayo [2,5,6], Fernando Vazquez [1,2,5,7], Luis M. Quirós [1,2,5,*] and Carla Martín [1,2,5,*]

1 Department of Functional Biology, University of Oviedo, 33006 Oviedo, Spain; saragonzalezfernandez7@gmail.com (S.G.-F.); lozanoiturbevict.fuo@uniovi.es (V.L.-I.); helenaordiales@gmail.com (H.O.); fvazquez@uniovi.es (F.V.)
2 Instituto Universitario Fernández-Vega, University of Oviedo, Av. Drs Fernández Vega 34, 33012 Oviedo, Spain; fernandezvivan@uniovi.es (I.F.-V.); merayo@fio.as (J.M.)
3 Department of Photonics, ITMA Materials Technology, 33490 Avilés, Spain; marife.menendez@idonial.com
4 Department of Pathology, Hospital Universitario Central de Asturias, 33011 Oviedo, Spain
5 Instituto de Investigación Sanitaria del Principado de Asturias, Av. del Hospital Universitario, s/n, 33011 Oviedo, Spain
6 Department of Surgery, University of Oviedo, 33006 Oviedo, Spain
7 Department of Microbiology, Hospital Universitario Central de Asturias, 33011 Oviedo, Spain
* Correspondence: quirosluis@uniovi.es (L.M.Q.); cmartincueto@gmail.com (C.M.)
† These authors contributed equally to this work.

Abstract: Silver nanoparticles (AgNPs) play an important role in the medical field due to their potent antimicrobial activity. This, together with the constant emergence of resistance to antimicrobial drugs, means AgNPs are often investigated as an alternative to solve this problem. In this article, we analyzed the antifungal and antiamoebic effects of a recently described type of AgNP, silver nanorings (AgNRs), and compared them with other types of AgNPs. Tests of the activity of AgNPs against various fungal and amoebic species were carried out. In all cases, AgNPs showed a high biocidal effect, although with fungi this depended on the species involved. Antifungal activity was detected by the conditioning of culture media or water but this effect was not dependent on the release of Ag ions. On the other hand, the proliferation of *Acanthamoeba castellanii* trophozoites was reduced by silver nanorings (AgNRs) and silver nanowires (AgNWs), with AgNWs being capable of totally inhibiting the germination of *A. castellanii* cysts. AgNRs constitute a new type of AgNP with an antifungal and antiacanthamoebic activity. These results open the door to new and effective antimicrobial therapies as an alternative to the use of antifungals or antiamoebic drugs, thus avoiding the constant appearance of resistance and the difficulty of eradicating infections.

Keywords: silver nanoparticles; silver nanorings; silver nanospheres; silver nanowires; antifungals; antiamoebics

Citation: González-Fernández, S.; Lozano-Iturbe, V.; Menéndez, M.F.; Ordiales, H.; Fernández-Vega, I.; Merayo, J.; Vazquez, F.; Quirós, L.M.; Martín, C. A Promising Antifungal and Antiamoebic Effect of Silver Nanorings, a Novel Type of AgNP. *Antibiotics* **2022**, *11*, 1054. https://doi.org/10.3390/antibiotics11081054

Academic Editors: Sotiris K. Hadjikakou and Christina N. Banti

Received: 20 July 2022
Accepted: 2 August 2022
Published: 3 August 2022

Publisher's Note: MDPI stays neutral with regard to jurisdictional claims in published maps and institutional affiliations.

Copyright: © 2022 by the authors. Licensee MDPI, Basel, Switzerland. This article is an open access article distributed under the terms and conditions of the Creative Commons Attribution (CC BY) license (https://creativecommons.org/licenses/by/4.0/).

1. Introduction

In recent years, nanotechnology has been in continuous development and recent advances show that nanoparticles (NPs), specifically silver nanoparticles (AgNPs), are playing and will play an important role in the medical, biological and pharmaceutical fields. NPs are between 1–100 nm and the increase in the use of AgNPs compared with other metal NPs is due to their physical-chemical characteristics and the well-known toxic effect of silver, which gives them good antimicrobial capabilities [1–3]. Depending on how they are synthesized, their size, and their morphology, AgNPs have unique and different characteristics, which allows them to be applied in a wide range of functions such as biosensors and anticancer therapy, in bioimaging, wound healing, disease treatment and drug delivery, or as nutraceuticals [4,5]. To date, several studies described the potent antimicrobial activity of AgNPs against bacteria, considering them an effective alternative to dealing with the problem of bacterial multidrug resistance such as ampicillin-resistant *Escherichia coli*, erythromycin-resistant *Streptococcus pyogenes* and methicillin-resistant or vancomycin-resistant *Staphylococcus aureus* and the concomitant increase in the number of infections [3,6–8]. However, little is known about the effect of these particles on eukaryotic microorganisms such as fungi or parasites such as amoebas and, despite the strong antibacterial effect of AgNPs that has been widely described, the mechanisms of this action have not yet been completely elucidated. It has been observed that the use of AgNPs conjugated with multipurpose solutions for contact lenses significantly reduces the adherence of *Acanthamoeba* trophozoites to the surface of the contact lens, decreasing the risk of *Acanthamoeba* keratitis infection. It has also been seen that the conjugation of AgNPs with amphotericin B, means that nystatin has amoebicidal activity, being more effective against *Acanthamoeba castellanii* compared to drugs alone [9,10]. In relation to the size of AgNPs, it is known that the smaller the size, the higher the surface-to-volume ratio and the greater the effectiveness of their use as anti-infective agents. Small-sized AgNPs can interact with the surface of microorganisms altering numerous basic functions such as membrane permeability and respiration [11]. What is more, the silver ions released can interact with the negative charges of the membrane, increasing the toxic effect, and they can also interact with the sulfate or phosphate groups of DNA and proteins in the cytoplasm, causing irreversible damage and inhibiting their growth and replication [1,11]. Moreover, the production of reactive oxygen species (ROS), DNA fragmentation and apoptosis have also been reported [12]. On the other hand, there is evidence that AgNP activity depends not only on their size, but also on their concentration and shape [1], although little is known about the mechanism involved [12–17]. It is generally agreed that concentration seems to be key to their cytotoxic action, albeit being dependent on the cell type with which they interact; in the case of bacteria, the greater sensitivity of Gram-negatives to the action of AgNPs was observed [16]. In the case of fungi of the genus *Candida*, the correlation between the fungicidal effect and AgNP concentrations varies depending on the species [15,18].

In addition to this, the shape of the AgNP itself plays an important role and seems to influence its activity [14–17]. Using different synthesis methods, it is possible to create AgNPs that are spherical, rod-like or triangular in shape, and the latter demonstrate the greatest bactericidal and antifungal potential with respect to dental implants, this apparently being due to their larger active surface area [19]. A similar effect was observed in *Escherichia coli*, where triangular nanoparticles were found to be qualitatively more effective than those that were rod-like or spherical in shape [14]. A new ring-shaped AgNP morphology was recently described, and these NPs demonstrated greater effectiveness against a broad spectrum of bacteria such as *Enterococcus faecalis*, *Streptococcus pneumoniae*, *E. coli* and *Neisseria gonhorreae* in terms of biocidal activity when compared with spherical and wire-shaped NPs [20]. Since the 3 morphologies were obtained by the same procedure, the observed effect highlighted the influence of the shape on the antibacterial activity of AgNPs [20].

On the other hand, the formation of biofilms by a large number of bacteria and fungi is also a serious health problem due to their resistance to existing treatments. Numerous studies analyzed the ability of AgNPs to eradicate these structures, showing that their effectiveness depends on their shape, concentration, size and the species involved in the formation of said structures [1,21,22], although the mechanism underlying their action remains unclear, as in the planktonic cases.

Currently, the increased prevalence of fungal infections, the low availability of antifungal drugs that have low levels of side effects on host cells, and the increase in fungal drug resistance resulted in AgNPs being targeted as potential antifungal agents. AgNPs exhibited very good antifungal activity against *Candida* spp. and *Candida* biofilms, which are implicated in many infections, although their impact on other fungal pathogens is scarcely explored [23–26]. However, in the case of *Candida*, it was described that AgNPs damage the structure of the cell membrane, producing holes on their surface, increasing membrane permeability and the release of potassium ions, as well as inhibiting cellular processes that are involved in yeast budding through the disruption of the membrane integrity and producing reactive oxygen species that ultimately cause cell apoptosis [18,27]. On the other hand, infections by the free living amoeba *Acanthamoeba* sp. are one of the most difficult to eradicate due to the existence of two different cell forms in the environment, the trophozoite and the cyst. A lack of effective therapeutic agents led to the analysis of the use of NPs conjugated with available drugs for the treatment of this protist, resulting in the significant inhibition of trophozoite growth, along with the inhibition of the encystation and excystation processes [28–30].

Taking into account the currently limited knowledge of AgNPs in eukaryotic organisms and that the new ring-shaped NPs known as nanorings (AgNRs) have only been tested in bacteria, the aim of this article was to analyze and compare the antifungal and antiamoebic activity of three different AgNPs, namely nanowires (AgNW), nanospheres (AgNS), and the new AgNRs. All Nps were obtained through the same synthesis process and the tests were carried out using six different fungal species and *A. castellanii*.

2. Results

2.1. Antifungal Effect of AgNPs Depends on Their Structure and the Species with Which They Interact

The addition of AgNPs to the different species of fungi produced, in all cases, a biocidal effect although its extent was dependent on the microorganism and the type of AgNP used. (Figure 1). *Candida albicans* and *Candida glabrata* showed high inhibition values, of above 70%, from very short incubation periods, with no notable changes afterwards, and without great differences between the types of nanostructure used (Figure 1a,c). In the case of filamentous fungi *Fusarium solani* and *Scedosporium apiospermum*, a similar result was obtained (Figure 1e,f). In contrast, *Candida parapsilosis* showed differences in inhibition patterns, which were dependent on incubation time, particularly when AgNRs were used (Figure 1b), and *S. cerevisiae* treated with AgNWs also showed an effect clearly dependent on the incubation time (Figure 1d). With both these fungi, the inhibition observed in periods longer than 15 min of incubation were high, and similar for all three nanostructures used.

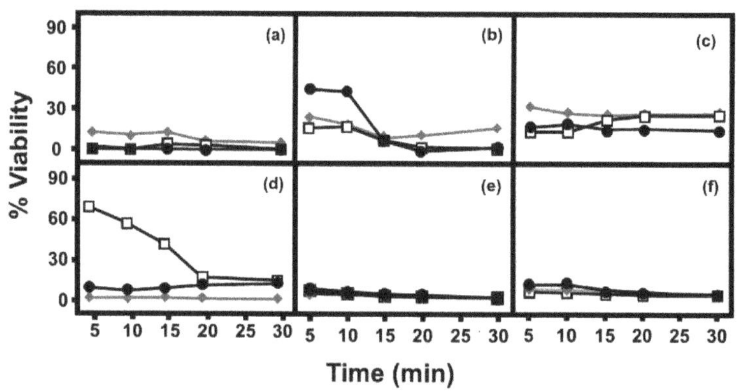

Figure 1. Antifungal effect of AgNPs. (**a**) *C. albicans*, (**b**) *C. parapsilosis*, (**c**) *C. glabrata*, (**d**) *S. cerevisiae*, (**e**) *F. solani*, (**f**) *S. angiospermum*. (♦) AgNSs, (□) AgNWs, (•) AgNRs.

2.2. Fungal Growth Is Influenced by the Presence of AgNPs and by Their Morphology

The effect of AgNPs on growth curves was also analyzed. The results showed that both AgNWs and AgNRs were able to completely inhibit growth in both yeast and filamentous fungi (Figure 2). In contrast, AgNSs showed a variable effect depending on the fungal species involved, that is, they were able to completely inhibit the growth of *C. albicans*, *C. parapsilosis* and *S. cerevisiae* but in the case of *C. glabrata* and *F. solani*, limited growth continued over an extended period, although this was greater in *F. solani* (Figure 2). AgNSs hardly produced any inhibition in the presence of *S. apiospermum* (Figure 2b).

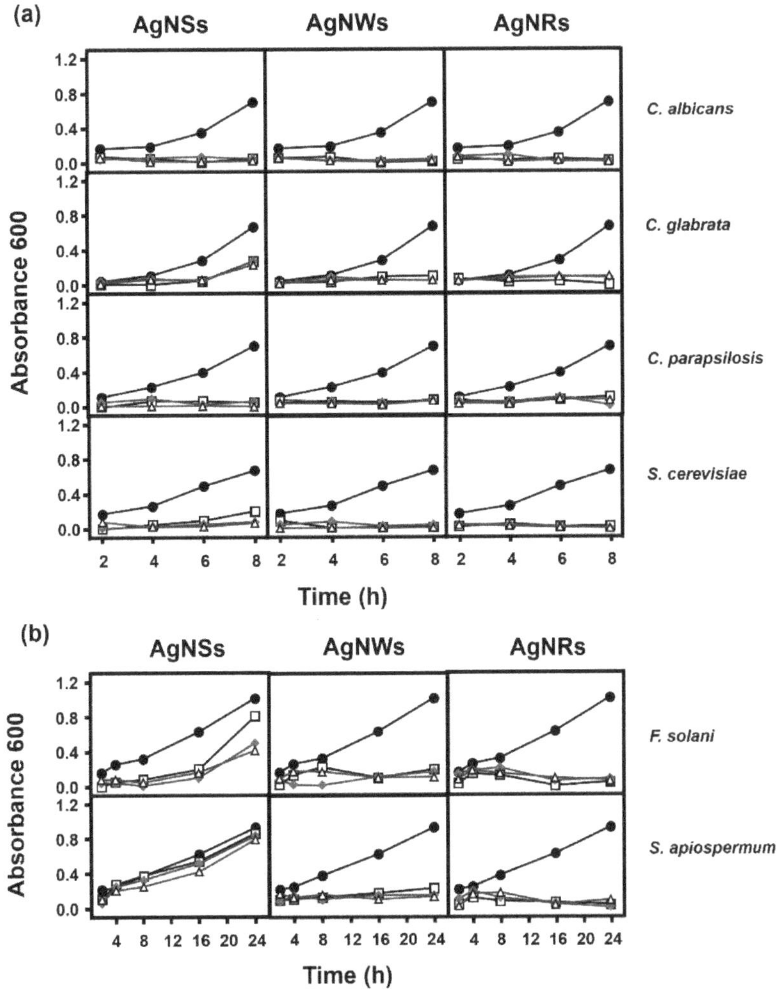

Figure 2. Influence of AgNPs on the growth of fungi species. (**a**) Ratio of yeast:AgNPs used: (•) 1:0, (□) 1:1, (♦) 1:2, (Δ) 1:3. (**b**) Ratio of filamentous fungi:AgNPs used: (•) 1:0, (□) 1:1, (♦) 1:2, (Δ) 1:3.

2.3. Culture Medium Conditioned by the Presence of AgNPs Shows an Antifungal Effect, although the Effect Is Not Dependent on the Release of Ag Ions

The observed antifungal effect could be due to the direct physical contact of the AgNPs with the microorganisms, or to the toxic conditioning of the culture medium, or both. Taking into account that any such toxic conditioning should be dependent on contact time, both the Sabouread medium and the deionized water were conditioned with AgNPs using different contact times, which was then followed, after the elimination of the nanoparticles, by immediate contact with *C. albicans*.

The results showed a variable effect that was highly dependent on whether conditioning was performed with deionized water or culture medium, as well as on the type of AgNP used and on the conditioning time of the medium (Figure 3a). In deionized water, AgNWs induced an intense antifungal effect after only 10 min of incubation, while AgNRs had a smaller effect that increased over time, reaching values close to 60% toxicity after 120 min of incubation. Interestingly, AgNSs had no effect, and an initial increase in the growth of the microorganism was observed (Figure 3a).

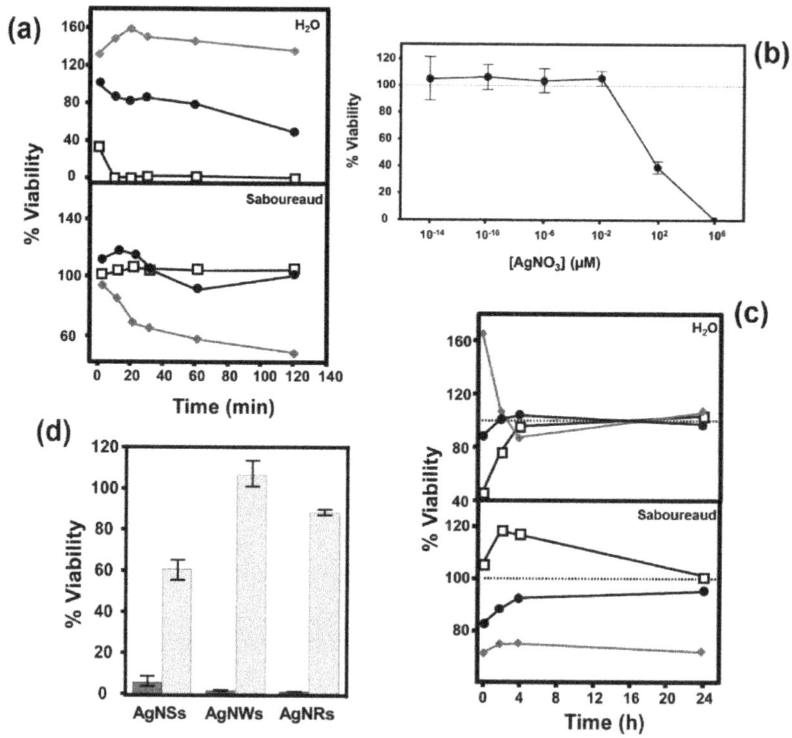

Figure 3. Antifungal effect of medium conditioned by contact with AgNPs. (**a**) Influence of contact time with AgNPs on the antifungal activity of deionized water and Sabouraud conditioned with (♦) AgNSs,(□) AgNWs, (•) AgNRs. (**b**) Effect of $AgNO_3$ concentration on the viability of *C. albicans*. (**c**) Temporal decay of the antibacterial activity of deionized water and Sabouraud from the moment conditioning with (♦) AgNSs, (□) AgNWs or (•) AgNRs was stopped. (**d**) Comparative antifungal effect of AgNPs (black bars) and Sabouraud conditioned (gray bars).

In contrast to the above, when conditioning was carried out using Sabouraud medium, AgNWs did not have any effect, AgNRs reduced viability to values below 20% in incubations of 60 min or longer, while AgNSs produced progressive toxic effects that reached more than 70% after 2 h of conditioning (Figure 3a).

To analyze the possibility that the release of Ag ions by the nanostructures could be responsible for the observed antifungal effect, the susceptibility of *C. albicans* to these ions was analyzed by determining its viability after keeping the fungus in a wide range of $AgNO_3$ concentrations for 30 min. The results showed that viability was reduced by 50% under these conditions at concentrations of around 10 μM (Figure 3b). Next, the release levels of Ag ions in conditioned medium without AgNPs were determined by ICP-AES, using various combinations of Sabouraud and deionized water with the different nanostructures, both in the presence and absence of *C. albicans*. The results obtained showed that in all cases the concentration of Ag was below 5 ppb (Table 1), which rules out the possibility of attributing the observed antifungal effect to the presence of a sufficient concentration of Ag ions in the conditioned medium.

Table 1. Presence of Ag ions in AgNP-free conditioned medium analyzed by inductively coupled plasma atomic emission spectroscopy (ICP-OES).

Sample	[Ag] (ppb)
H_2O milliQ	<5
H_2O milliQ + AgNSs	<5
H_2O milliQ + AgNWs	<5
H_2O milliQ + AgNRs	<5
H_2O milliQ + AgNSs + *C. albicans*	<5
H_2O milliQ + AgNWs + *C. albicans*	<5
H_2O milliQ + AgNRs + *C. albicans*	<5
Saboureaud	<5
Saboureaud + AgNSs	<5
Saboureaud + AgNWs	<5
Saboureaud + AgNRs	<5
Saboureaud + AgNSs + *C. albicans*	<5
Saboureaud + AgNWs + *C. albicans*	<5
Saboureaud + AgNRs + *C. albicans*	<5

Having established the existence of the antifungal capacity of the medium conditioned by AgNPs in certain circumstances, the analysis of its duration and stability seemed appropriate. For this, the toxic effect on *C. albicans* cells in Sabouread medium and deionized water previously conditioned with AgNPs for 1 h, and subsequently maintained for a variable interval after elimination of the nanostructures, was measured. In the case of deionized water, the observed effects dissipated in all cases in less than 4 h. Interestingly, the analysis of the conditioned Saboureaud medium showed that the effects observed with AgNSs, notably higher than those of the other nanostructures, only partially decreased in the first few hours, after which a decrease in viability of around 30% was maintained (Figure 3c).

The existence of antifungal activity in the conditioned medium made it interesting to compare it with the effect observed in the presence of AgNPs. To determine this, Saboureaud was conditioned for 30 min with the different AgNPs, and its effect on the viability of *C. albicans* was compared with yeast incubations in the presence of each AgNP, with incubation for a duration of 30 min in both situations. The results showed that in all cases the reduction in viability was clearly higher when there was physical contact with the AgNPs (>95% for AgNSs, 99.5% for AgNWs and 100% AgNRs), while the medium conditioned by AgNSs reduced viability by 41%, and by 14% when conditioned with AgNRs, while AgNWs had no effect (Figure 3d).

2.4. AgNWs and AgNRs Reduce the Proliferation of Acanthamoeba castellanii Trohozoites

Acanthamoeba spp. comprises free-living amoeba that can act as opportunistic pathogens. It constitutes a model of a non-fungal eukaryotic microorganisms, which makes it interesting to analyze the effect of AgNPs on its viability. As controls for the experiment, the trophozoites initially present (t = 0 h) and the additional trophozoites formed after 48 h+ of incubation in the absence of AgNPs were used. In all cases, there was some statistically significant proliferation, even in the presence of AgNPs, especially AgNSs (around 195%), although also, to a lesser extent, in the presence of AgNRs (47%) and AgnWs (23%). However, when the proliferation values were compared with those obtained in the absence of AgNPs no significant differences were observed for AgNSs, while AgNRs and AgNWs inhibited proliferation by around 71% and 86%, respectively (Figure 4).

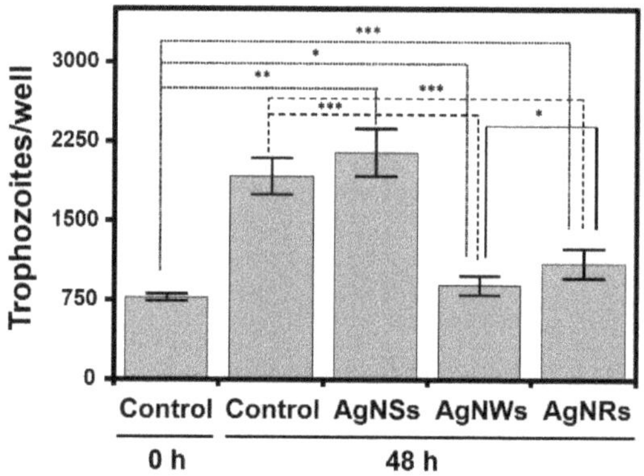

Figure 4. Influence of AgNPs on *A. castellanii* trophozoite proliferation. Statistically significant differences are denoted by ***, ** and * which indicate, respectively, $p < 0.001$, $p < 0.01$ and $p < 0.05$.

2.5. AgNWs Inhibits Germination of A. castellanii Cysts

The effect of AgNPs on the germination of *A. castellanii* cysts was studied by quantifying the appearance of trophozoites from these forms of resistance in the absence and the presence of nanoparticles. In all cases, it was found that AgNPs significantly inhibited germination, although the magnitude of the effect was highly dependent on AgNP morphology. Both AgNSs and AgNRs reduced trophozoite formation by around 40%, while AgNWs were far more effective, completely inhibiting their formation (Figure 5).

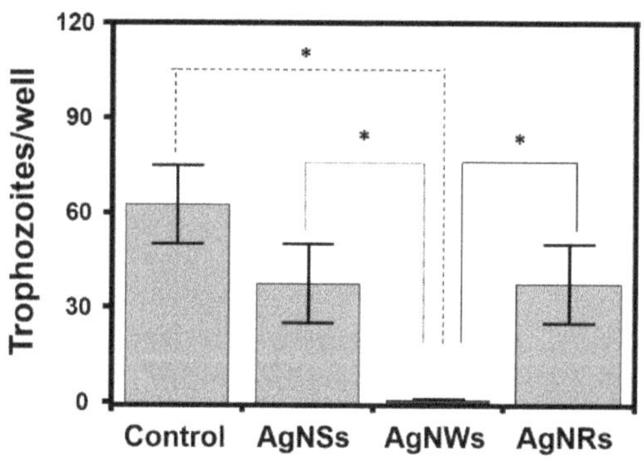

Figure 5. Influence of AgNPs on *A. castellanii* cyst germination. Statistically significant differences are denoted by * which indicate $p < 0.05$.

3. Discussion

The great cytotoxic potential of AgNPs has been widely described in the literature. Numerous studies analyzed their antibacterial capacity against a wide spectrum of microorganisms among which are some such as *S. aureus*, *Staphylococcus epidermidis*, *Bacillus subtilis*, *Klebsiella pneumoniae*, *E. coli*, *Salmonella typhi* [1,17]. However, the antifungal effects of AgNPs received only marginal attention and only a few studies have been published on this area, with most of them taking the genus *Candida*, and specifically *C. albicans*, as a model [23]. Even less is known about their activity against other parasitic eukaryotic microorganisms such as amoeba [31]. The shape, size and concentration of AgNPs are important aspects that influence the effects observed [15,32]. This work analyzed the biocidal capacity of a recently described morphology of AgNPs, namely AgNRs, which have a filament diameter of 80 nm and a ring diameter of between 12 and 18 μm. In addition, the effect of AgNRs was compared against two other types of AgNPs, namely AgNSs and AgNWs. The biocidal effects of AgNSs and AgNWs were described previously in various studies, unlike those of AgNRs, whose activity has only been described in a single study against a broad spectrum of bacteria such as *S. aureus*, *S. epidermidis*, *Streptococcus pyogenes*, *Streptococcus pneumoniae*, *E. faecalis*, *B. globisporus*, *E. coli*, *Serratia marcescens*, *Haemophilus influenzae*, *Klebsiella pneumoniae*, *Neisseria gonorrhoeae* and *Pseudomonas aeruginosa* [20]. Due to the fact that the synthesis of these particles can be carried out by a variety of physical and chemical processes, and that their properties may vary, the comparison of these three morphologies was carried out after obtaining them in parallel using the same methodology.

When the toxic effect of the different AgNPs on fungi was analyzed, a wide range of efficiency could be observed in most cases from the outset and without great differences being observed depending on the nanostructure employed, with the exception of *S. cerevisiae* and *C. parapsilosis*, where AgNWs and AgNRs showed a lower effect at the start of incubation but then achieved inhibition values similar to the other two nanoparticles over longer periods of time. In the case of filamentous fungi, the toxicity values reached were very high, were similar for all three types of AgNPs and did not show great differences with the results obtained in most yeasts. This differs from what has previously been described with bacteria, where differences between the bacteria analyzed were observed and, furthermore, the antibacterial effect of each AgNP varied depending on the microorganism tested and their gram nature [20]. Interestingly, in many cases this inhibitory effect was not as high as in fungi. In addition, AgNSs showed a greater antibacterial effect compared to AgNWs and AgNRs, corroborating the finding that the smaller the size of a nanoparticle, the greater its effectiveness [16,17]. These results, however, differ from what was obtained in this work, where the differences in the size and morphology of the nanoparticles did not seem to show uniform behavior in terms of the level of inhibition when they were in contact with the fungi for short periods of time.

Although the inhibition values obtained over short periods were greater than 70%, the growth of the small surviving fraction was analyzed in longer incubation periods. In this scenario, the efficiency of the AgNPs was different from that observed in the shorter periods. AgNRs and AgNWs displayed a strong inhibitory effect with all fungi analyzed although AgNSs reduced the rate of the fungus growth in a variable manner depending on the fungal species. AgNSs were able to completely inhibit the growth of all fungi with the exceptions of *C. glabrata* and *F. solani*, where limited growth appeared over longer periods, and *S. apiospermum*, where no effect was observed. The effect observed with AgNRs is similar to that obtained in the earlier experiment with bacteria, where they represented the nanoparticles with the greatest inhibitory effect regardless of the gram nature of the bacteria, unlike AgNSs, whose efficacy was lower than the other AgNPs with both bacteria and fungi, although the extent of their impact depended on the microorganism involved [20]. These data do not support the idea that the reduced size of the AgNSs favors their effect in long incubation periods [33], although they do reinforce the notion that the shape of the particle has an important role in terms of its effectiveness. In addition to the physical-chemical characteristics of AgNPs, a key factor in their toxicity is the physiological characteristics of the microorganism on which they act. Thus, the differential response to the action of AgNPs may be due in part to differences in the composition of the cell wall, metabolism or virulence, as occurs in other antimicrobial drug treatments [34]. AgNPs adhere to the cell wall or membrane in both bacteria and fungi, and can penetrate into the cells, leading to serious and varied alterations in cell physiology, causing changes to signal transduction pathways, the induction of oxidative stress and damage to the intracellular structure [16]. The effect of AgNPs on the genus *Candida* has been the subject of some studies and seems to lie in the accumulation of AgNPs on the outside of the fungal wall and the release of Ag ions. As a consequence, alterations in the physiological state of cells and the fluidity of their membranes occur, along with a reduction in the levels of ergosterol and fatty acids in them, inducing cell death [23,35]. This could also explain the differences observed in the viability of the fungal species tested here. On the other hand, some mechanisms do not always involve direct contact with the microorganism and it may even be that the release of Ag ions by the AgNPs is also involved in toxicity [16]. In the case of bacteria, it was described that a large effect occurs when a medium previously conditioned with AgNPs is used, although mortality rates are lower compared to when nanoparticles are present. In the present study, unlike with the bacteria, where a toxic effect was observed in both media and with all types of AgNPs, in *C. albicans* strong differences were observed between type of conditioned medium, deionized water or Saboureaud medium, as well as on the basis of the type of nanoparticle used. While AgNWs produced the highest mortality rate when deionized water was used, in the Saboureaud medium the most effective AgNPs were AgNSs. For their part, AgNRs produced their greatest effect in deionized water. The toxicity of Ag to fungi has been previously described [35] and the possibility that AgNPs release Ag ions has therefore been investigated, finding that the concentration of silver necessary to reduce the viability of *C. albicans* to 50% was 10 μM, a concentration 10 times higher than that required with bacteria [20]. However, the effect does not seem to be due to the release of Ag ions by AgNPs since concentrations of ions sufficient to account for the effect were not detected in either medium. These data suggest the formation of chemical species capable of producing, at least in part, a toxic effect on fungi that is highly dependent on the type of AgNP involved. On the other hand, the average duration of this effect depended on the type of medium and AgNP used. In deionized water, the toxic capacity of all AgNPs progressively dissipated after the first 5 min, similar to what was observed in bacteria. However, analysis of the conditioned Sabouraud medium showed that, in the case of AgNSs, the loss of toxicity over time was limited, unlike that observed in bacteria, where the antibacterial activity in BHI was lost more quickly. This could indicate that either the toxic reactive species released differ depending on the composition of the conditioned medium, or that the effect on fungi is different to in bacteria. Although an antifungal effect was observed in the conditioned medium, it is less than that observed after direct contact with AgNPs, which suggests that, although the mechanism of action is mixed, the AgNP–fungus interaction plays the most important role.

Another eukaryotic opportunistic pathogen is the genus *Acanthamoeba*. Due to the wide distribution of these amoebas in the environment and the variety of diseases that they can cause in humans, together with the difficulties in diagnosing them and the ineffectiveness of treatment, which is often toxic to human cells [28], analyzing the effect of AgNPs on their viability is of great value. To date, few studies have investigated the activity of AgNPs against *A. castellanii*. Cobalt NPs have been studied for their anti-amoebic potential, hexagonal microflakes showing better anti-*Acanthamoeba* results compared to nanoflakes and granular cobalt NPs [36]. In the case of AgNPs, it was observed that their use, alone or conjugated with different drugs, has anti-amoebic and amoebistatic effects, along with inhibited encystation and excystation; in the case of conjugation, this also improves their bioavailability, sustained release, intracellular permeability and efficacy against the eradication of the infection [9,37]. However, there are no data in the literature about the influence of AgNPs morphology on *Acanthamoeba* eradication. The present work was able to establish that the presence of AgNPs reduces the proliferation of trophozoites, albeit without achieving their complete inhibition, when compared to control conditions where AgNPs were not present. The results demonstrate that AgNSs did not have a significant effect, unlike AgNWs, which were the most effective, and AgNRs, which inhibited proliferation, though to a lesser degree. In addition, using AgNPs in the presence of *A. castellanii* cysts lowered the percentage of excystation and, consequently, germination was also notably reduced; therefore, AgNWs once again were identified as the most effective followed by AgNRs and AgNSs. Once again, these results are not consistent with previously described observations that a smaller AgNP size produces a greater effect, instead suggesting the existence of other important factors that influence the mechanism of action of these nanostructures. However, in this work, it seems to be that the shape of an AgNP has a strong influence on its activity against both trophozoites and cyst forms, in contrast to what was observed by other authors, who reported that the shape of an AgNP does not seem to be such an important factor in terms of its activity [38].

4. Material and Methods

4.1. Fungal Species, Amoeba and Culture Conditions

The fungal species used in this study were *C. albicans*, *C. parapsilosis*, *Candida glabrata*, *S. cerevisiae*, *F. solani* and *S. apiospermum*, with all of them being clinical isolates obtained from the Hospital Universitario Central de Asturias and identified at the species level by MALDI-TOF MS spectrometry (Bruker Daltonics, Bremen, Germany). All species were grown in Saboureaud medium at 37 °C for 48 h. In the case of *F. solani* and *S. angiospermum*, both of which are filamentous fungi, they were grown in a volumetric flask with glass balls at 30 °C for 5 days in a shaking incubator. The amoeba used was *A. castellanii*, in both its active form (throphozoite) and its dormant form (cyst), obtained from Fundación de Investigación Oftalmológica. *A. castellanii* was grown at 30 °C in Peptone yeast glucose medium (PYG) supplemented with 0.5 mM $CaCl_2$, 4 mM $MgSO_4$, 2.5 mM Na_2HP_4, 2.5 mM KH_2PO_4, 3.4 mM $C_6H_9Na_3O_9$, 0.05 M Fe $(NH_4)_2(SO_4)_2$ and penicillin G/streptomycin (5000 IU/mL, 5000 µg/mL).

4.2. Synthesis of AgNPs

The synthesis of AgNPs was carried out in the same way as described in Gonzalez-Fernandez et al. [20]. Briefly, this involves the reduction of $AgNO_3$ in ethylene glycol in the presence of polyvinylpyrrolidone 360 k. The solution obtained had a transparent to a pearly appearance, denoting the presence of silver nanostructures. The reaction was performed at 170 °C under magnetic stirring. When the reaction was complete, the solution was submerged in ice water until room temperature was reached. After this, the separation and purification steps were carried out. AgNPs were separated by centrifugation and the remaining solution was left to decant for 3 days, after which, from the supernatant obtained, a concentrated solution of AgNRs together with a small quantity of AgNWs was obtained. Finally, the majority of AgNWs were then obtained from the remaining solution. The AgNPs were suspended in water and in order to verify the initial concentrations a gravimetric method was used, the final data being as follows: 1.7×10^7 AgNSs/µL suspension; 6.0×10^4 AgNRs/µL suspension; 2.4×10^4 AgNWs/µL suspension. The structural features of the nanostructures obtained through this process were analyzed by FEG-SEM, and were previously published by our group [20]. These features are AgNSs, 40–60 nm in diameter; AgNRs, 80 nm wire diameter, 12–18 µm (average 14 µm) ring diameter; AgNWs, 200 nm in diameter and 50–100 µm in length. The chemical nature and purity of the silver nanostructures developed were verified by means of an FRX probe, coupled to the FEG-SEM equipment, resulting in a content of 99.9% silver.

4.3. Effect of AgNPs on Cell Viability in Fungal Cultures

To test the influence of different AgNPs on yeast and filamentous fungi viability, fungal cultures were grown at an A_{600} of 0.5 and kept at room temperature under agitation with the AgNPs for periods of between 5 and 30 min. The yeast:NP and filamentous fungi:NP ratio were 1:1 and 1:100, respectively. Next, the NPs were removed by centrifugation at 800 rpm for 2 min, and different dilutions of the supernatant were seeded on solid Saboureaud medium plates. After incubation of the plates at 37 °C overnight, the colonies obtained were quantified.

4.4. Effect of AgNPs on Cell Growth in Fungal Cultures

The influence of different AgNPs on yeast and filamentous fungi growth was analyzed by incubating the AgNPs with fungal cultures at an A_{600} of 0.02, and using fungi:NP proportions of 1:1, 1:2 and 1:3. The effect on yeast growth was quantified by absorbance after incubation periods of 2, 4, 6 and 8 h. In the case of filamentous fungi, absorbance was determined after incubation for 2, 4, 8, 16 and 24 h.

4.5. Determination of the Concentration of Silver Ions Released by AgNPs

Different samples of *C. albicans* in Saboureaud at an A600 of 0.5, sterile Saboureaud, suspensions of *C. albicans* in deionized water at an A_{600} of 0.5 or sterile deionized water were individually treated with an amount of different AgNPs equivalent to a yeast:NP ratio of 1:1 in all cases. After incubating for 30 min with stirring, the suspensions were centrifuged at 3000 rpm for 5 min to discard the AgNPs. The concentration of Ag in the particle-free supernatants was determined by inductively coupled plasma atomic emission spectroscopy (ICP-OES) using an Agilent 5110 ICP-OES Instrument (Agilent, Santa Clara, CA, USA).

4.6. Study of the Toxic Effect of Media Conditioned by AgNPs

To analyze the effect that conditioning through the presence of AgNPs might have on fungal viability, aliquots of deionized water or BHI were kept in contact with NPs at different concentrations (10^{10}, 2×10^{10}, 4×10^{10}, 8×10^{10}, 16×10^{10} Units/L) for a variable time (5, 10, 20, 30, 60 or 120 min, depending on the experiment) with stirring. Next, the AgNPs were removed by centrifugation at 3000 rpm for 5 min, and *C. albicans* was incubated in the conditioned media at an A_{600} of 0.5 for 30 min with stirring. The quantity of nanostructures used in the experiments was adjusted to a fungi:NP ratio of 1:1, equivalent to that used in the cell viability experiments in planktonic fungal cultures. Finally, different dilutions of these cultures were seeded onto solid Saboureaud plates and, after incubation overnight at 37 °C, the colonies obtained were quantified. In order to establish the temporary durability of the culture medium conditioning, aliquots of deionized water and BHI were kept in contact with AgNPs at the same concentrations described in the previous paragraph, after which the AgNPs were removed by centrifugation at 3000 rpm for 5 min, and the supernatant was kept at room temperature for 0, 2, 4 and 24 h. Each sample of conditioned medium was incubated with *C. albicans* at an A_{600} of 0.5 for 30 min with stirring, and the quantification was carried out by dilution in solid BHI medium as described above.

4.7. Effect of AgNPs on Cell Viability in Trophozoites and Cyst Form of A. castellanii

The influence of NPs on the trophozoite and cyst forms of *A. castellanii* was analyzed by culturing the amoeba in 24-well plates. The trophozoites were coincubated with the different NPs at 30 °C for two days in the trophozoite:NP 1:250 ratio, while the cysts were coincubated with the NPs at 30 °C for 8 days, using a cyst:NP ratio of 1:500. After this time, the trophozoites were quantified using a Neubauer chamber. The cysts were also analyzed following the same method, although the quantification was carried out after both 4 and 8 days of coincubation.

4.8. Statistical Analysis

The results were analyzed using a Kruskal–Wallis test in the Statistics for Windows program (Statsoft Inc.; Tulsa, OK, USA). The differences were considered significant when $p < 0.05$.

5. Conclusions

This work analyzed the antifungal and antiamoebic properties of AgNRs, a new type of AgNP that has recently been described, and compared the results with those of two other morphologies, namely AgNSs and AgNWs. AgNRs showed antifungal activity in all cases although the effectiveness varied depending on the fungus and incubation time involved. In addition, they were able to inhibit the growth of yeasts and in the case of filamentous fungi, the AgNRs showed great effectiveness compared to the other AgNPs. In addition, AgNPs were able to partially inhibit the proliferation of trophozoites and the germination of cysts of *A. castellanii*, the best results being seen with AgNWs. These results open the door to new and effective antimicrobial therapies as an alternative to the use of antifungals or antiamoebic drugs, thus avoiding the constant appearance of resistance and addressing the difficulty of eradicating infections.

Author Contributions: Conceptualization, L.M.Q. and C.M.; Methodology, S.G.-F., V.L.-I. and M.F.M.; Software, V.L.-I. and I.F.-V.; Validation, L.M.Q. and C.M.; Formal Analysis, S.G.-F.; Investigation, S.G.-F., V.L.-I. and H.O.; Resources, M.F.M. and F.V.; Data Curation, L.M.Q. and I.F.-V.; Writing—Original Draft Preparation, C.M.; Writing—Review and Editing, C.M.; Supervision, L.M.Q. and C.M.; Project Administration, C.M. and J.M.; Funding Acquisition, L.M.Q. and J.M. All authors have read and agreed to the published version of the manuscript.

Funding: This research was funded by Instituto de Desarrollo Económico del Principado de Asturias (IDEPA), Government of the Principado de Asturias (Spain), grant number IDE/2019/000366.

Institutional Review Board Statement: Not applicable.

Informed Consent Statement: Not applicable.

Data Availability Statement: The data presented in this study are contained within the article.

Conflicts of Interest: The authors declare no conflict of interest.

References

1. Franci, G.; Falanga, A.; Galdiero, S.; Palomba, L.; Rai, M.; Morelli, G.; Galdiero, M. Silver nanoparticles as potential antibacterial agents. *Molecules.* **2015**, *20*, 8856–8874. [CrossRef] [PubMed]
2. Ahmad, S.A.; Das, S.S.; Khatoon, A.; Ansari, T.H.; Afzal, M.; Hasnain, M.S.; Nayak, A.K. Bactericidal activity of silver nanoparticles: A mechanistic review. *Mater. Sci. Energy Technol.* **2020**, *3*, 756–769. [CrossRef]
3. Bruna, T.; Maldonado-Bravo, F.; Jara, P.; Caro, N. Silver Nanoparticles and Their Antibacterial Applications. *Int. J. Mol. Sci.* **2021**, *22*, 7202. [CrossRef] [PubMed]
4. Naganthran, A.; Verasoundarapandian, G.; Khalid, F.E.; Masarudin, M.J.; Zulkharnain, A.; Nawawi, N.M.; Karim, M.; Che Abdullah, C.A.; Ahmad, S.A. Synthesis, Characterization and Biomedical Application of Silver Nanoparticles. *Materials* **2022**, *15*, 427. [CrossRef]
5. Jeevanandam, J.; Krishnan, S.; Hii, Y.S.; Pan, S.; Chan, Y.S.; Acquah, C.; Danquah, M.K.; Rodrigues, J. Synthesis approach-dependent antiviral properties of silver nanoparticles and nanocomposites. *J. Nanostruct. Chem.* **2022**, 1–23. [CrossRef]
6. Ahmed, S.; Ahmad, M.; Swami, B.L.; Ikram, S. A review on plants extract mediated synthesis of silver nanoparticles for antimicrobial applications: A green expertise. *J. Adv. Res.* **2016**, *7*, 17–28. [CrossRef]
7. Rai, M.K.; Deshmukh, S.D.; Ingle, A.P.; Gade, A.K. Silver nanoparticles: The powerful nanoweapon against multidrug-resistant bacteria. *J. Appl. Microbiol.* **2012**, *112*, 841–852. [CrossRef]
8. Faiz, M.B.; Amal, R.; Marquis, C.P.; Harry, E.J.; Sotiriou, G.A.; Rice, S.A.; Gunawan, C. Nanosilver and the microbiological activity of the particulate solids versus the leached soluble silver. *Nanotoxicology* **2018**, *12*, 263–273. [CrossRef]
9. Anwar, A.; Siddiqui, R.; Hussain, M.A.; Ahmed, D.; Shah, M.R.; Khan, N.A. Silver nanoparticle conjugation affects antiacanthamoebic activities of amphotericin B, nystatin, and fluconazole. *Parasitol. Res.* **2018**, *117*, 265–271. [CrossRef]
10. Hendiger, E.B.; Padzik, M.; Sifaoui, I.; Reyes-Batlle, M.; López-Arencibia, A.; Zyskowska, D.; Grodzik, M.; Pietruczuk-Padzik, A.; Hendiger, J.; Olędzka, G.; et al. Silver Nanoparticles Conjugated with Contact Lens Solutions May Reduce the Risk of Acanthamoeba Keratitis. *Pathogens* **2021**, *10*, 583. [CrossRef]
11. Begum, T.; Follett, P.A.; Mahmud, J.; Moskovchenko, L.; Salmieri, S.; Allahdad, Z.; Lacroix, M. Silver nanoparticles-essential oils combined treatments to enhance the antibacterial and antifungal properties against foodborne pathogens and spoilage microorganisms. *Microb. Pathog.* **2022**, *164*, 105411. [CrossRef]
12. Hwang, I.S.; Lee, J.; Hwang, J.H.; Kim, K.J.; Lee, D.G. Silver nanoparticles induce apoptotic cell death in *Candida albicans* through the increase of hydroxyl radicals. *FEBS J.* **2012**, *279*, 1327–1338. [CrossRef]
13. Morones, J.R.; Elechiguerra, J.L.; Camacho, A.; Holt, K.; Kouri, J.B.; Ramírez, J.T.; Yacaman, M.J. The bactericidal effect of silver nanoparticles. *Nanotechnology* **2005**, *16*, 2346. [CrossRef]
14. Pal, S.; Tak, Y.K.; Song, J.M. Does the antibacterial activity of silver nanoparticles depend on the shape of the nanoparticle? A study of the gram-negative bacterium *Escherichia coli*. *Appl. Environ. Microbiol.* **2007**, *73*, 1712–1720. [CrossRef]
15. Raza, M.A.; Kanwal, Z.; Rauf, A.; Sabri, A.N.; Riaz, S.; Naseem, S. Size- and Shape-Dependent Antibacterial Studies of Silver Nanoparticles Synthesized by Wet Chemical Routes. *Nanomaterials* **2016**, *6*, 74. [CrossRef]
16. Dakal, T.C.; Kumar, A.; Majumdar, R.S.; Yadav, V. Mechanistic basis of antimicrobial actions of silver nanoparticles. *Front. Microbiol.* **2016**, *7*, 1831. [CrossRef]
17. Vimbela, G.V.; Ngo, S.M.; Fraze, C.; Yang, L.; Stout, D.A. Antibacterial properties and toxicity from metallic nanomaterials. *Int. J. Nanomed.* **2017**, *12*, 3941–3965. [CrossRef]
18. Rozhin, A.; Batasheva, S.; Kruychkova, M.; Cherednichenko, Y.; Rozhina, E.; Fakhrullin, R. Biogenic Silver Nanoparticles: Synthesis and Application as Antibacterial and Antifungal Agents. *Micromachines* **2021**, *12*, 1480. [CrossRef]

19. Kowalczyk, P.; Szymczak, M.; Maciejewska, M.; Laskowski, Ł.; Laskowska, M.; Ostaszewski, R.; Skiba, G.; Franiak-Pietryga, I. All That Glitters Is Not Silver—A New Look at Microbiological and Medical Applications of Silver Nanoparticles. *Int. J. Mol. Sci.* **2021**, *22*, 854. [CrossRef]
20. González-Fernández, S.; Lozano-Iturbe, V.; García, B.; Andrés, L.J.; Menéndez, M.F.; Rodríguez, D.; Vazquez, F.; Martín, C.; Quirós, L.M. Antibacterial effect of silver nanorings. *BMC Microbiol.* **2020**, *20*, 172. [CrossRef]
21. Markowska, K.; Grudniak, A.M.; Wolska, K.I. Silver nanoparticles as an alternative strategy against bacterial biofilms. *Acta Biochim. Pol.* **2013**, *60*, 523–530. [CrossRef] [PubMed]
22. Mohanty, S.; Mishra, S.; Jena, P.; Jacob, B.; Sarkar, B.; Sonawane, A. An investigation on the antibacterial, cytotoxic, and antibiofilm efficacy of starch-stabilized silver nanoparticles. *Nanomedicine* **2012**, *8*, 916–924. [CrossRef] [PubMed]
23. Vazquez-Muñoz, R.; Avalos-Borja, M.; Castro-Longoria, E. Ultrastructural analysis of *Candida albicans* when exposed to silver nanoparticles. *PLoS ONE* **2014**, *9*, e108876. [CrossRef] [PubMed]
24. Ayatollahi Mousavi, S.A.; Salari, S.; Hadizadeh, S. Evaluation of Antifungal Effect of Silver Nanoparticles Against *Microsporum canis*, *Trichophyton mentagrophytes* and *Microsporum gypseum*. *Iran J. Biotechnol.* **2015**, *13*, 38–42. [CrossRef] [PubMed]
25. Ahmad, A.; Wei, Y.; Syed, F.; Tahir, K.; Taj, R.; Khan, A.U.; Hameed, M.U.; Yuan, Q. Amphotericin B-conjugated biogenic silver nanoparticles as an innovative strategy for fungal infections. *Microb. Pathog.* **2016**, *99*, 271–281. [CrossRef] [PubMed]
26. Lara, H.H.; Ixtepan-Turrent, L.; Jose Yacaman, M.; Lopez-Ribot, J. Inhibition of *Candida auris* Biofilm Formation on Medical and Environmental Surfaces by Silver Nanoparticles. *ACS Appl. Mater. Interfaces* **2020**, *12*, 21183–21191. [CrossRef]
27. Kim, K.-J.; Sung, W.S.; Suh, B.K.; Moon, S.K.; Choi, J.S.; Kim, J.G.; Lee, D.G. Antifungal activity and mode of action of silver nano-particles on *Candida albicans*. *BioMetals* **2009**, *22*, 235–242. [CrossRef]
28. Sharma, G.; Kalra, S.K.; Tejan, N.; Ghoshal, U. Nanoparticles based therapeutic efficacy against *Acanthamoeba*: Updates and future prospect. *Exp. Parasitol.* **2020**, *218*, 108008. [CrossRef]
29. Anwar, A.; Soomaroo, A.; Anwar, A.; Siddiqui, R.; Khan, N.A. Metformin-coated silver nanoparticles exhibit anti-acanthamoebic activities against both trophozoite and cyst stages. *Exp. Parasitol.* **2020**, *215*, 107915. [CrossRef]
30. Anwar, A.; Ting, E.L.S.; Anwar, A.; Ain, N.U.; Faizi, S.; Shah, M.R.; Khan, N.A.; Siddiqui, R. Antiamoebic activity of plant-based natural products and their conjugated silver nanoparticles against *Acanthamoeba castellanii* (ATCC 50492). *AMB Express* **2020**, *10*, 24. [CrossRef]
31. Hendiger, E.B.; Padzik, M.; Sifaoui, I.; Reyes-Batlle, M.; López-Arencibia, A.; Rizo-Liendo, A.; Bethencourt-Estrella, C.J.; Nicolás-Hernández, D.S.; Chiboub, O.; Rodríguez-Expósito, R.L.; et al. Silver Nanoparticles as a Novel Potential Preventive Agent against *Acanthamoeba* Keratitis. *Pathogens* **2020**, *9*, 350. [CrossRef]
32. Bhattacharya, R.; Mukherjee, P. Biological properties of "naked" metal nanoparticles. *Adv. Drug Deliv. Rev.* **2008**, *60*, 1289–1306. [CrossRef]
33. Szerencsés, B.; Igaz, N.; Tóbiás, Á.; Prucsi, Z.; Rónavári, A.; Bélteky, P.; Madarász, D.; Papp, C.; Makra, I.; Vágvölgyi, C.; et al. Size-dependent activity of silver nanoparticles on the morphological switch and biofilm formation of opportunistic pathogenic yeasts. *BMC Microbiol.* **2020**, *20*, 176. [CrossRef]
34. Espinel-Ingroff, A. Mechanisms of resistance to antifungal agents: Yeasts and filamentous fungi. *Rev. Iberoam. Micol.* **2008**, *25*, 101–106. [CrossRef]
35. Radhakrishnan, V.S.; Reddy Mudiam, M.K.; Kumar, M.; Dwivedi, S.P.; Singh, S.P.; Prasad, T. Silver nanoparticles induced alterations in multiple cellular targets, which are critical for drug susceptibilities and pathogenicity in fungal pathogen (*Candida albicans*). *Int. J. Nanomed.* **2018**, *13*, 2647–2663. [CrossRef]
36. Anwar, A.; Chi Fung, L.; Anwar, A.; Jagadish, P.; Numan, A.; Khalid, M.; Shahabuddin, S.; Siddiqui, R.; Khan, N.A. Efects of shape and size of cobalt phosphate nanoparticles against *Acanthamoeba castellanii*. *Pathogens* **2019**, *8*, 260. [CrossRef]
37. Anwar, A.; Mungroo, M.R.; Anwar, A.; Sullivan, W.J.; Khan, N.A.; Siddiqui, R. Repositioning of Guanabenz in Conjugation with Gold and Silver Nanoparticles against Pathogenic Amoebae *Acanthamoeba castellanii* and *Naegleria fowleri*. *ACS Infect. Dis.* **2019**, *5*, 2039–2046. [CrossRef]
38. Hendiger, E.B.; Padzik, M.; Żochowska, A.; Baltaza, W.; Olędzka, G.; Zyskowska, D.; Bluszcz, J.; Jarzynka, S.; Chomicz, L.; Grodzik, M.; et al. Tannic acid-modified silver nanoparticles enhance the anti-*Acanthamoeba* activity of three multipurpose contact lens solutions without increasing their cytotoxicity. *Parasites Vectors* **2020**, *13*, 624. [CrossRef]

antibiotics

MDPI

Article

Multidrug-Resistant Bacterial Pathogens and Public Health: The Antimicrobial Effect of Cyanobacterial-Biosynthesized Silver Nanoparticles

Nermin A. El Semary [1,2,3,*] and Esam M. Bakir [4,5]

1 Al Bild Bank Scholarly Chair for Food Security in Saudi Arabia, Deanship of Scientific Research, Vice Presidency for Graduate Studies and Scientific Research, King Faisal University, Al-Ahsa 31982, Saudi Arabia
2 Biological Sciences Department, College of Science, King Faisal University, Al-Ahsa 31982, Saudi Arabia
3 Botany and Microbiology Department, Faculty of Science, Helwan University, Ain Helwan, Cairo 11795, Egypt
4 Chemistry Department, Faculty of Science, Ain Shams University, Al-Abassia, Cairo 11566, Egypt; ebakir@kfu.edu.sa
5 Chemistry Department, College of Science, King Faisal University, Al-Ahsa 31982, Saudi Arabia
* Correspondence: nelsemary@kfu.edu.sa

Abstract: Background: Cyanobacteria are considered as green nano-factories. Manipulation of the size of biogenic silver nanoparticles is needed to produce particles that suit the different applications such as the use as antibacterial agents. The present study attempts to manipulate the size of biosynthesized silver nanoparticles produced by cyanobacteria and to test the different-sized nanoparticles against pathogenic clinical bacteria. Methods: *Cyanothece*-like. coccoid unicellular cyanobacterium was tested for its ability to biosynthesize nanosilver particles of different sizes. A stock solution of silver nitrate was prepared from which three different concentrations were added to cyanobacterial culture. UV-visible spectroscopy and FTIR were conducted to characterize the silver nanoparticles produced in the cell free filtrate. Dynamic Light Scattering (DLS) was performed to determine the size of the nanoparticles produced at each concentration. The antimicrobial bioassays were conducted on broad host methicillin-resistant *Staphylococcus aureus* (MRSA), and *Streptococcus* sp., was conducted to detect the nanoparticle size that was most efficient as an antimicrobial agent. Results. The UV-Visible spectra showed excellent congruence of the plasmon peak characteristic of nanosilver at 450 nm for all three different concentrations, varying peak heights were recorded according to the concentration used. The FTIR of the three solutions revealed the absence of characteristic functional groups in the solution. All three concentrations showed spectra at 1636 and 2050–2290 nm indicating uniformity of composition. Moreover, DLS analysis revealed that the silver nanoparticles produced with lowest concentration of precursor $AgNO_3$ had smallest size followed by those resulting from the higher precursor concentration. The nanoparticles resulting from highest concentration of precursor $AgNO_3$ were the biggest in size and tending to agglomerate when their size was above 100 nm. The three types of differently-sized silver nanoparticles were used against two bacterial pathogenic strains with broad host range; MRSA-(Methicillin-resistant *Staphylococcus aureus*) and *Streptococcus* sp. The three types of nanoparticles showed antimicrobial effects with the smallest nanoparticles being the most efficient in inhibiting bacterial growth. Discussion: Nanosilver particles biosynthesized by *Cyanothece*-like cyanobacterium can serve as antibacterial agent against pathogens including multi-drug resistant strains. The most appropriate nanoparticle size for efficient antimicrobial activity had to be identified. Hence, size-manipulation experiment was conducted to find the most effective size of nanosilver particles. This size manipulation was achieved by controlling the amount of starting precursor. Excessive precursor material resulted in the agglomeration of the silver nanoparticles to a size greater than 100 nm. Thereby decreasing their ability to penetrate into the inner vicinity of microbial cells and consequently decreasing their antibacterial potency. Conclusion: Antibacterial nanosilver particles can be biosynthesized and their size manipulated by green synthesis. The use of biogenic nanosilver particles as small as possible is recommended to obtain effective antibacterial agents.

Citation: El Semary, N.A.; Bakir, E.M. Multidrug-Resistant Bacterial Pathogens and Public Health: The Antimicrobial Effect of Cyanobacterial-Biosynthesized Silver Nanoparticles. *Antibiotics* **2022**, *11*, 1003. https://doi.org/10.3390/antibiotics11081003

Academic Editors: Sotiris K Hadjikakou, Christina N. Banti and Anthony William Coleman

Received: 17 June 2022
Accepted: 15 July 2022
Published: 26 July 2022

Publisher's Note: MDPI stays neutral with regard to jurisdictional claims in published maps and institutional affiliations.

Copyright: © 2022 by the authors. Licensee MDPI, Basel, Switzerland. This article is an open access article distributed under the terms and conditions of the Creative Commons Attribution (CC BY) license (https://creativecommons.org/licenses/by/4.0/).

Keywords: cyanobacteria; DLS; FTIR; molecular characterization; silver nanoparticles; size manipulation

1. Introduction

The green synthesis of nanoparticles is ideal as it does not cause pollution and is of minimal cost. In addition, allows nanoparticles production in large quantities with no toxic by-products [1]. Inorganic nanoparticles of noble metals such as gold and silver nanoparticles are increasingly used in biology and medicine due to their distinctive characteristics such as ease of use, good functionality, biocompatibility and ability to target specific cells [2]. In regard to their production, cyanobacteria are considered an active source of nanomaterials [3,4]. Several cyanobacterial genera are reported to produce nanoparticles including; *Anabaena*, *Calothrix*, and *Leptolyngbya* which actively produced Au, Ag, Pd, and Pt nanoparticles. These particles are naturally released in the culture medium and stabilized by algal polysaccharides/peptides that enable easy recovery. The size of the recovered particles and yield depend on the cyanobacterial genus [5]. The mechanisms by which those nanoparticles are produced were recently reviewed [6] and their biosynthesis was classified into extracellular and intracellular. In intracellular biosynthesis, ions are reduced by electrons in the electron transport systems that are involved in photosynthesis and respiration. Enzymes such as NADH-dependent reductases are mostly involved in electron transport and redox reactions in the cytoplasm, thylakoid membranes, and plasma membrane [7,8]. Extracellular synthesis involves cellular exudates such as pigments, proteins, enzymes, hormones, and ions which play an important role in the reduction and capping process of nanoparticles [9,10]. Biomolecules such as NADH-reductases and sulfur-containing proteins in the cell-free supernatant are important in bio-reduction of nanoparticles [11–13]. Due to the antibacterial action of silver nanoparticles, they are incorporated in footwear, cosmetics, wound dressings and plastics [11]. The antimicrobial effect of silver nanoparticles, biosynthesized by algae against human bacterial pathogens has been previously reported [14,15]. The cyanobacterium under study is unicellular, photosynthetic prokaryotic microorganism. It can be grown easily with minimal growth requirements in the presence of light source. This ease of growth ensures a continuous supply of cells capable of biosynthesizing nanosilver particles. In addition, this cyanobacterium is Gram-negative which means it has a large outer lipid membrane made of fatty acids that are important functional molcules in the binding and possibly reduction of nanosilver ions from a solution [16]. The *Cyanothece* genus is a quite unique as it is able to undergo diurnal cycle of photosynthesis in the day and perform nitrogen fixation in the night. Both of the two are reducing processes which only indicates its strong reducing abilities [17]. Indeed, *Cyanothece* spp. showed unique reducing ability and ease of manipulation for the production of nanogold particles of different sizes [4]. Here we isolated and used *Cyanothece*-like cyanobacterial strain in an attempt to biosynthesize silver nanoparticles to be used as antimicrobial agents. Analysis by Ultra violet-visible (UV-vis.) spectroscopy and Fourier Transmission Infrared spectroscopy (FTIR) were carried out for all different Silver nanoparticles (AgNPs) samples. In the UV-vis. spectroscopy, surface plasmon resonance (SPR) was recorded and indicated the specific vibration modes of electrons limited by the size and shape of the nanoparticles. [18]. Fourier transmission Infrared spectroscopy will also be performed to detect any functional groups associated with silver nanoparticles [14] in the water-based cell-free filtrate. Moreover, the Dynamic Light Scattering will be used for accurate determination of the size of the three AgNPs samples [19]. The AgNPs samples were tested in antimicrobial bioassays against bacterial pathogens. It is noteworthy that the extent of antimicrobial activity of biosynthesized nanosilver particles against bacterial pathogens appeared to be linked to particles' size [20]. Unfortunately, the size manipulation of silver nanoparticles biosynthesized using cyanobacteria has been scarcely studied. On the other hand, size-manipulation of gold nanoparticles that were biologically synthesized was successfully achieved [4]. Here, we provide a simple protocol for the manipulation of the size of silver nanoparticles produced and we investigated the antibacterial impact of differently-sized nanosilver particles on two multi-drug resistant clinical pathogenic bacteria.

2. Methods and Materials

2.1. Cyanobacterial Culture Establishment and Identification

The *Cyanothece*-like coccoid unicellular cyanobacterium used in the experiment was originally isolated from a rice-field in Al Ahsa, Eastern Province, KSA. Water samples were taken from canals filled with irrigation water. The samples were centrifuged at 3000 rpm for 10 min to eliminate contaminating bacteria by discarding the supernatant under a 12:12 h (light:dark) cycle at ambient temperature. The biomass pellet was streaked on agar plates based on BG11 growth medium. Cultures were kept under 12:12 h (light:dark) cycle at ambient temperature. The green colonies were picked and re-streaked on agar plates and then examined by light microscopy for morphological description. Liquid cultures of BG11 were inoculated with pure colonies and left to grow as monoalgal cultures. The cultures were subjected to antibiotic treatment using Ampicillin (200 μL/L) and left in the dark for two days to kill heterotrophic bacteria and then brought back to light and centrifuged. The supernatant was discarded and the biomass was re-suspended in sterile water for washing. Centrifugation and washing steps were repeated then the biomass was inoculated into fresh sterile BG11 medium, left to grow and checked microscopically to ascertain that axenic cyanobacterial culture was established for use in experiments.

2.2. Preparation of Nanosilver Particles

A stock solution of 10 mM silver nitrate (Sigma, Aldrich) was prepared. Three different volumes of the same stock were prepared (2 mL, 1 mL and 0.2 mL) and added to a cyanobacterial biomass (0.5 g fresh weight in 5 mL BG 11 growth medium) then, completed with distilled water to obtain a 20 mL total volume. The solutions were left for three days, the colour change was observed from the onset of the experiment. The external solution containing nanosilver was purified from the cells through filtration using Millipore filters of diameter size of 0.2 μm. This tiny pore size excluded all aggregates, cells, and bulky cellular components. The cell-free filtrate was watery in nature, as the stock precursor $AgNO_3$ was prepared in distilled water and the cyanobacterium was cultured in a water-based mineral growth medium (BG11). Siver nanoparticles were synthesized in a mixture containing distilled water, +5 mL of the cyanobacterial culture, and +10 mM $AgNO_3$ (added in three different volumes; 0.2, 2, and 1 mL). We assessed the nanoparticles in the cell-free supernatant after filtration through Millipore filters of pore diameter size of 0.2 μm.

2.3. UV–Visible Spectroscopic (UV–Vis) Analysis

The initial characterization of the silver nanoparticles was performed using a UV–vis spectrophotometer (Genesys10S UV–visible double beam spectrophotometer). The scans were recorded at room temperature using 1 mL of each nanosilver concentration in the range of 200 nm to 700 nm.

2.4. FTIR

Infrared spectrometric analysis of silver nanoparticles was performed on cell-free supernatant containing nanosilver using Fourier-Transform infrared spectrometer (FT-IR, Agilent Cory 630, Agilent Technologies, Santa Clara, CA, USA). A control sample with no silver nitrate was also used.

2.5. Dynamic Light Scattering Analysis

The size of the silver nanoparticles was studied by Dynamic light scattering (DLS), using Dual Scattering Particle Size Analyzer (cilas, Nano DS). Each sample was analyzed in triplicates at 25 °C with scattering angle 60°. Deionized water was used as the dispersal medium. The integration time was 30 min and the algorithm used was cumulative. The samples were loaded into quartz microcuvettes, and replicate measurements were recorded, of which the mean was calculated.

2.6. Antibacterial Bioassay

The antibacterial activity of the three solutions of the nanosilver particles was assessed against freshly sub-cultured bacteria originally isolated from clinical samples at the College of Medicine, King Faisal University. The two bacterial isolates were identified as Gram-positive MRSA-Methicillin-resistant *Staphylococcus aureus* and a *Streptococcus* sp. and supplied by Dr Munirah Aldayel, King Faisal University. The sensitivity of pathogenic strains to nanosilver was assessed by modified Kirby-Bauer Disk Diffusion Susceptibility method [21]. Sterile paper discs (6 mm in diameter) were saturated with 30 µL of nanosilver at the three concentrations examined. The discs were dried and placed on the surface of nutrient agar medium inoculated with a bacterial-suspension and kept for 24 h in an incubator at 37 °C. A positive control disc containing 30 µL antibiotic Chloramphenicol was also used. The diameter of the inhibition zones (mm) was measured in triplicates and the average and standard deviation were recorded [22]. As a negative control the supernatant without $AgNO_3$ was also used. The MIC/MBC was also determined to verify the bacteriostatic/bactericidal effect of the three types of the differently-sized nanoparticles.

3. Results

3.1. Description of Cyanobacterial Strain

The unicellular cyanobacterium tested appeared in pairs after division. No colonies form as there is no common mucilaginous envelope. Cells are coccoid and usually bright blue-green. All of these characteristics are typical of the genus *Cyanothece*. Molecular characterization revealed only 88% of similarity to *Cyanothece* sp. Due to the lack of sufficient diagnostic phenotypic characters of the coccoid cyanobacterium, it was designated *Cyanothece*-like cyanobacterium.

3.2. Preparations of Different Silver Nanoparticles Samples

The culture inoculum contained 4×10^4 cells/mL taken from one month old culture, The three different concentrations of silver nitrate applied to the cyanobacterial culture were; 1×10^{-4} M, 5×10^{-4} M and 1×10^{-3} M. The samples showed a gradual color change from faint brown to, light brown and then to dark brown, corresponding to the volumes of the precursor materials, i.e., 0.2, 1, and 2 mL, respectively. Each mixture was left in at room temperature, and the cell-free supernatant was taken from external solution of the cultures after incubation period of three days. The solution was microfiltered using Millipore filters of pore diameter size of 0.2 µm. The cell-free filtrates of the three samples were used in further analyses.

3.3. UV-Visible Spectroscopy

Nano-silver particles were successfully synthesized both intra and extra-cellularly. The extracellular formation of AgNPs was confirmed by UV-Vis absorbance spectra of AgNPs of the three samples with the highest peak belonging to highest concentration of nanosilver followed by the medium concentration and lowest concentration, respectively (Figure 1). All the three concentrations of biogenic nanosilver showed a strong specific peak for the synthesized AgNPs at 450 nm (Figure 1). There was congruence for the peaks detected for the three concentrations. This coincided with the plasmon resonance characteristic of nanosilver particles [23].

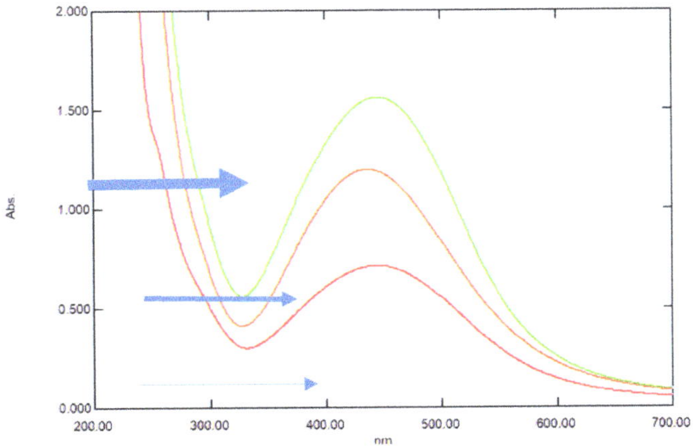

Figure 1. UV-Visible spectrum of the three samples of AgNPs with different heights and width according to the nanoparticle size. The lowest peak (red) denoted by the lower thin arrow belongs to the smallest-sized AgNP, the middle peak (red) denoted by the middle arrow belongs to the medium-sized particles and the highest peak (green) belongs to the largest sized silver nanoparticle. The X-axis denotes the wavelength whereas y-axis denotes the absorbance.

3.4. FTIR Spectroscopy

The FTIR spectra of the three samples of AgNPs had the same pattern exemplified in Figure 2 and clearly exhibited the characteristic signals of AgNPs at 3356–3350 cm^{-1} overlapping with OH signal. At 1636–1637 cm^{-1} there was a clear signal corresponding to C-H stretching [24]. No other functional groups were detected in all samples. (Figure 2 and Supplementary Materials Figures S1 and S2).

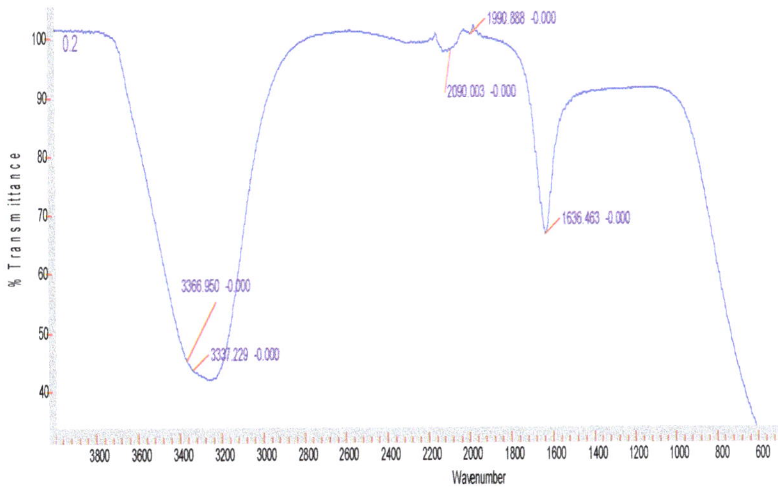

Figure 2. FTIR of the smallest−sized AgNPs. The x-axis denotes the wavenumber, The y-axis denotes % transmittance.

3.5. Dynamic Light Scattering

According to the DLS analysis, the size range for the smallest-sized silver nanoparticles (1×10^{-4} M) was 23–47 nm, with an average of 33.9 nm. The coefficient of var-iation (%) was 27.5, the polydispersity index (%) was 71.7, and rms was 0.36329. The size of the medium-sized silver nanoparticles (5×10^{-4} M) was in the range of 35–122 nm, with an average of 67 nm. The coefficient of variation (%) was 47, the polydisper-sity index (%) was 130.5, and rms was 0.02417. The size range for the highest-sized sil-ver nanoparticles was 78–108 nm, with some nanoparticles above 100 nm (Figure 3), with an average of 92.5 nm. The coefficient of variation (%) was 12.5, the polydispersi-ty index (%) was 32.3 (Figure 3)

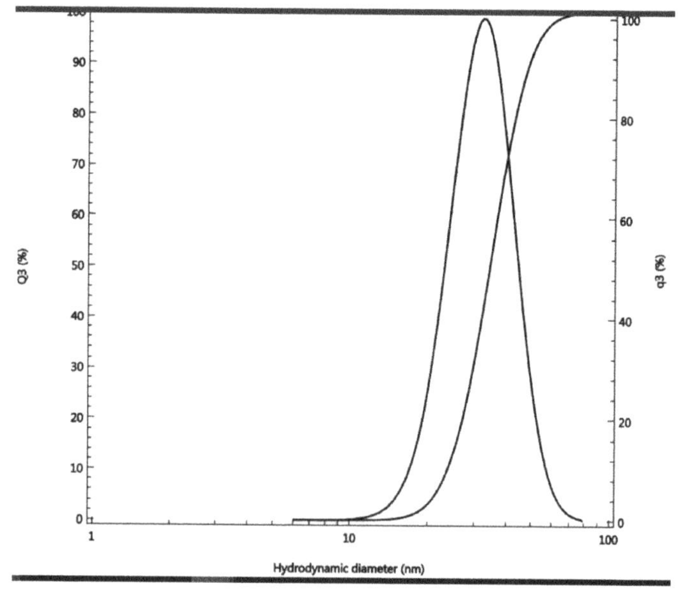

(a)

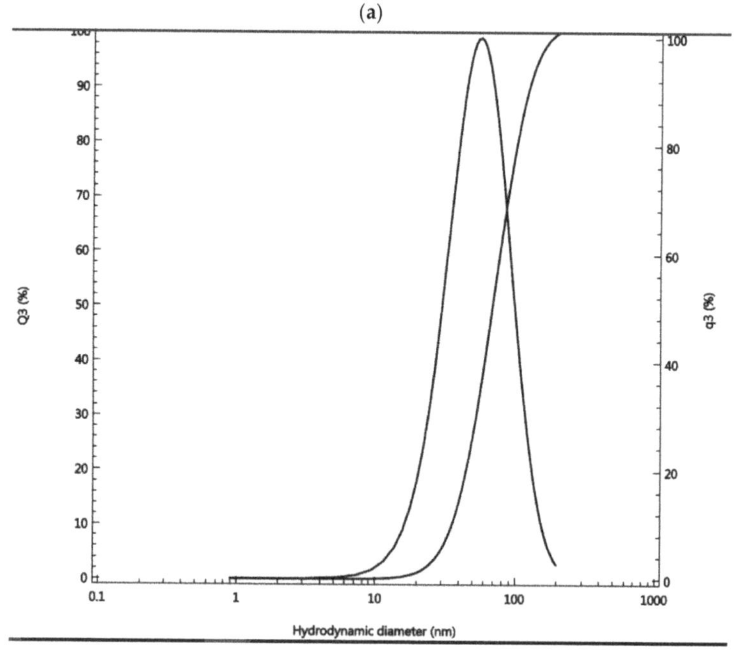

(b)

Figure 3. *Cont.*

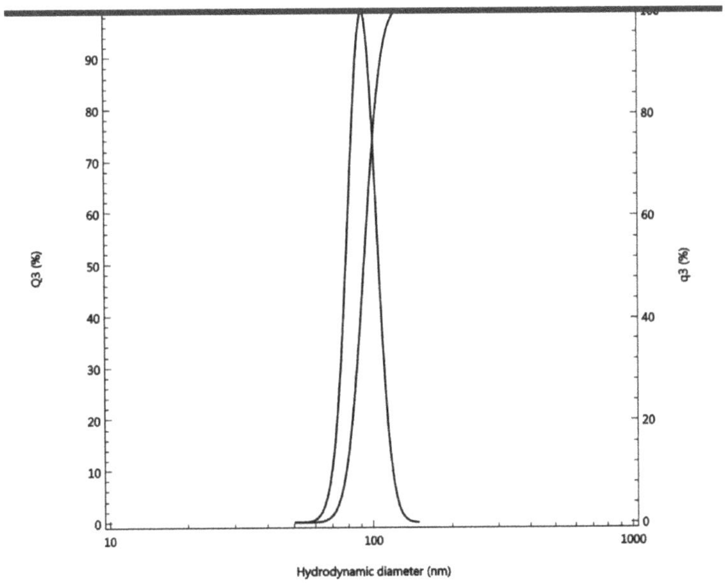

(c)

Figure 3. Hydrodynamic diameter of the (**a**) smallest-, (**b**) medium-, (**c**) largest-sized AgNPs.

3.6. Antimicrobial Bioassay

The three different nanosilver concentrations showed strong antimicrobial effects, with the smallest-sized nanoparticles being the most efficient in inhibiting bacterial growth, showing an inhibition zone diameter of 1.8 cm for the *Streptococcus* sp. and of 1.6 cm for *Staphylococcus aureus* (MRSA). The second potent antimicrobial concentration was that of medium-sized nanosilver particles (derived from 1 mL of $AgNO_3$ solution), with an inhibition zone of 1.4 cm for *Staphylococcus aureus* (MRSA) and of 1.3 cm for the Streptococcus sp. The least effective were the largest sized-nanosilver particles (derived from 2 mL of $AgNO_3$ solution), with an inhibition zone of 1.2 cm for the *Streptococcus* sp. and of 1.1 for *Staphylococcus aureus* (MRSA). (Table 1) The bactericidal and bacteriostatic effects for the three sized AgNPs were verified using an antimicrobial bioassay and calculating the MIC/MBC in µg/mL (Table 2).

Table 1. Antibacterial impact of differently sized nanosilver particles.

Strain	Inhibition Zone Diameter (mm) (Nanosilver Particles Derived from a High Precursor Concentration)	Inhibition Zone Diameter (mm) (Nanosilver Particles Derived from a Medium Precursor Concentration)	Inhibition Zone Diameter (mm) (Nanosilver Particles Derived from a Low Precursor Concentration)	Inhibition Zone Diameter (mm) of Control Disc (chloramphenicol)
Staphylococcus aureus (MRSA)	11 ± 2	13 ± 1	16 ± 1	28
Streptococcus sp.	12 ± 2 ± 2	14 ± 1	18 ± 1	25

Table 2. Antimicrobial Bioassay Determning MIC/MBC in μg/mL.

Bacteria	Smallest		Medium		Largest	
	MIC	MBC	MIC	MBC	MIC	MBC
Streptococcus sp.	1	1.5	2.5	2.5	4	4.5
(MRAS) *Staphylococcus aureus*	2	2.5	3.5	3.5	6	7

3.7. Antimicrobial Bioassay Determning MIC/MBC in μg/mL

The minimum inhibitory concentration (MIC) and minimum bactericidal concentration (MBC) for the silver nanoparticles of different particle size were investigated. The nanoparticles exhibited both bacteriostatic and bactericidal capabilities (Table 2). The lowest MIC and MBC values were for the smallest-sized silver nanoparticles, indicating their effectiveness at the lowest concentration.

4. Discussion

The ability of cyanobacteria to bind bulk ions from solutions followed by further reduction and nano-formation is mainly related to surface entities on cyanobacteria as well as to the polysaccharide sheath present on the cyanobacterial surface and, sometimes, in solution [3]. Indeed, our previous work showed the biosynthetic ability of cyanobacteria for gold nanoparticles [3], with the possibility of customized biosynthesis as well [4]. Cell-free media were found to be required for the synthesis of nanoparticles, as their content of enzymes, antioxidants, and phenolic and ions facilitates the bio-reduction of NPs. Thus, they are involved in the bio-fabrication of metallic NPs [6]. DLS is accurate for measuring the size of nanoparticles. According to [19], the size of a particle is related to the scattering time, as small-size molecules scatter faster than larger-sized molecules. The manipulation of the size of synthesized particles can be affected by the concentration of bio-reductants, which have an important influence on the shape and size of AgNPs [25]. It was also found that the extracellular formation of AgNPs depends on the dose of silver nitrate. This is in complete agreement with our results. Moreover, it was reported that the biosynthesis of AgNPs in cyanobacteria takes place both inside the cells (with particle size <10 nm) and in solution (with particle size of 1–200 nm), leading to spherical and octahedral particles over time [26]. This is again in total agreement with our results, as the range of the nanoparticles' size was within the one previously reported. As the volume decreased, the chance to produce freely dispersed nanoparticles became much higher. As the volume increased, more nanoparticles were produced in the external solution. The smallest the nanosilver particles, the higher their antibacterial activity. This is in accordance with previous results [20]. The preferred method to determine the size of nanoparticles is DLS. [19] compared the accuracy of the DLS method to TEM (transmission electron microscopy). They clearly showed that measurements by DLS were more accurate compared to TEM results. Dynamic Light Scattering (DLS) is based on the collision of dispersed particles with solvent molecules, leading to random movement (Brownian movement) and causing light scattering. The smaller the particles are, the faster their diffusion. UV–visible plasmon resonance showed excellent congruence within the peak range characteristic of silver nanoparticles, which indicated the purity of the three samples of silver nanoparticles; the peak heights of the samples varied, corresponding to their size difference. It is important to highlight the fact that the surface of metals is similar to plasma because of the presence of free electrons in the conduction band and positively charged nuclei [18]. Therefore, metallic nanoparticles have characteristic optical absorption spectra in the UV–visible region, as the vibration modes of electrons are limited by the size and shape of the particles [18]. Indeed, [18] were able to synthesize differently sized silver nanoparticles with the same absorption band but with varying band intensities and widths, due to the varying size of the AgNPs. This is in complete agreement with our results, which showed a similar trend. The poor detection of functional groups associated with silver nanoparticles by FTIR only indicates their absence in the cell-free filtrates. The small pore diameter of the Millipore filter used allowed the removal of any aggregates or bulky materials. Indeed, [6] reviewed reports showing that cell-free supernatants contain reducing ions/moieties that transformed silver ions into their reduced nanoform.

With regard to the antimicrobial bioassay, the pathogenic bacterium *Streptococcus* sp., which is a Gram-positive bacterium, has a wide range of hosts. It usually exhibits a high degree of resistance against antibiotics [27,28]. It was shown [29] that there is a noticeable rise in antibiotic-resistant Streptococci strains that are capable of infecting humans and animals, causing morbidity and fatalities. This rise was associated with several mechanisms including the activity of efflux pumps [30], the modifications of the antimicrobial targets by methylation of rRNA (*erm* genes) or target mutations, and enzymatic inactivation [31]. Another mechanism of multidrug resistance could be horizontal gene transfer or chromosomal point mutations caused by the excessive use of antimicrobials. Streptococcal strains also produce biofilms which are highly resistant to antibiotics. Similar, but more contagious, is the Methicillin-resistant *Staphylococcus aureus* (MRSA), which is also a Gram-positive bacterium. Through the years, this bacterium has become multi-drug resistant through both mutations and the gain of exogenous genes that have successfully converted *Staphylococcus aureus* into Methicillin-resistant *Staphylococcus aureus*, which is resistant to all β-lactam antibiotics. This, therefore decreases the efficacy of antibiotics and increases the mortality rates following infection by this [32]. Clearly, strains that are resistant to antibiotics need alternative biocontrol strategies. In that regard, biogenic silver nanoparticles represent the most favorable biocontrol alternative due to their minimal cost, possibility of massive production, lack of toxic by-products, rapid preparation, and broad-spectrum host range. Another good reason is that their biosynthesis can be simply manipulated and customized to suit the desired application, as demonstrated here. Nonetheless and most importantly, the ability of silver nanoparticles to combat multi-drug resistant pathogens, as shown also in our study, makes them plausible candidates for future antibacterial drugs. Indeed, current antibiotics can be potentiated by the addition of silver nanoparticles to increase their antibacterial potential [14]. Silver nanoparticles have both bactericidal and bacteriostatic effects against bacteria. We showed that the smallest-sized particles were the most effective at low concentration, in accordance with previous studies [33,34]. These previous works reviewed the bactericidal effect of silver nanoparticles and summarized their mechanism of action in the following steps: (1) adhesion of silver nanoparticles on surface layers, (2) AgNPs penetration in bacterial cells and damage to intracellular structures and macromolecules (protein, lipids, and DNA), (3) induction of oxidative stress by reactive oxygen species (ROS) and free radicals, and (4) modulation of signal transduction pathways. AgNPs may modulate the human immune system, promoting bacterial inhibition [35]. Silver nanoparticles have both bacteriostatic and bactericidal effect. It was shown [36] that the bacteriostatic action is more easily observed under aerobic conditions, whereas the bactericidal action is more intense under anaerobic conditions. It was suggested that AgNPs may act by decreasing the integrity of the cell membrane. Their bactericidal effect stems from their ability to penetrate the bacterial cell and disrupt its machinery and structure. The larger the nanosilver particles, the less successful they are in penetrating bacterial cells [20]. Small-size AgNPs in the range of 10 and 15 nm show more stability, biocompatibility, and enhanced antimicrobial activity [37]. A study conducted by [38] showed that silver nanoparticles synthesized by the microwave-assisted method had a size of 55 ± 10 nm, showing that they were effective against *Escherichia coli*, the Gram-negative bacterium which is hard to combat due to its outer lipid layer. The synthesis of differently shaped nanoparticles of different sizes was also reported. It should be noted that this synthetic method yielded silver nanoparticles whose size was within a range and not fixed and uniform. The manipulation of biological systems to produce nanoparticles of a specific size is difficult. Biogenic synthesis deals with a living organism, where the biosynthesis process is dependent on many factors including pH, temperature, nature of the synthesis process, precursor concentration, and bio-reductant [6]. The success in manipulating a biological system for the synthesis of nanoparticles of a certain size opens the door for many applications that need "tailored" nanoparticles. Unfortunately, studies on customized biosynthesis are scarce, especially those on species form rather pristine habitats whose microflora is underexplored and underexploited. It must be noted that it is highly needed to study unexplored organisms with potentially exceptional biosynthetic potentials and opportunities of exploitation. Our study provides novel data on the topic and describes a simple approach to reach that target. All the different independent analyses we performed confirmed the synthesis of biogenic silver nanoparticles of three different sizes and showed that those particles had different antimicrobial activity due to their size difference. Indeed, the smallest-sized biogenic silver nanoparticles were the most effective antibacterial agents. In agreement with this, a study showed that the antibacterial action of AgNPs is higher against *S. aureus* when nanoparticles of small size are used [39]. The inhibitory effect of nanosilver is due to the damage it causes to several cellular components, including cell wall and plasma membrane, through the generation of reactive oxygen species. This disrupts cellular respiration and permeability [40] as reactive oxygen species (ROS), such as superoxide or hydrogen peroxide, interact with lipids, proteins, or DNA, inducing cell lysis [41]. Nanosilver particles' adverse effects also include downregulating the enzymes responsible for the bacterial secretion system [42]. Silver nanoparticles can also inhibit proteins as well as DNA replication by accumulating at the membrane and interacting with sulfur and phosphorus needed for the synthesis of DNA [6,43]. Eventually, all those damages will result in microbial cell death. Hence, antimicrobial nanoparticles can serve as a substitute for antibiotics, at least in some cases. Indeed, Ref. [29] recommended the use of nanomedicine tools as an alternative to antibiotics to counteract multidrug-resistant pathogenic bacteria. It was also shown that current antibiotics can be potentiated with silver nanoparticles. The rationale behind this is that antibiotics and nanosilver particles use different mechanisms, and their combination would prevent the development of resistance [29]. For example, rifampicin associated with silver nanoparticles increased the antibiotic bioactivity against methicillin-resistant bacteria [44]. Ref. [45] used green biosynthesized silver nanoparticles in dental applications for better biocompatibility and antibacterial impact against *S. mutans*, responsible for caries. Green biosynthesized silver caused a reduction in lactic acid and polysaccharides in bacterial biofilms [46]. In our case, the absence of the biological cyanobacterial protein corona allowed an easy and smooth penetration of the silver nanoparticles into the vicinity of the bacterial cells. Nonetheless, there is always the possibility to further develop a commercial product that can be bionically coated to allow the commercialization and application of the product. Indeed, a study [47] used proteins (Retinin or Retinin-like proteins from insects) as a base for nanocoatings that allowed the particles to be more reactive with metallic ions. This facilitated reduction, coalescence, and nucleation of the metal nanoparticles. It was shown that these bionic nanocoatings, with metal nanoparticles, achieved higher antimicrobial activity compared to pure metallic coatings. The method provided in our research is very simple and versatile and opens the door for different ways of exploitation and manipulation.

In conclusion, antimicrobial silver nanoparticles can be produced using cyanobacteria as a green platform for their biosynthesis. Biogenic silver nanoparticles proved to be effective against multi-drug resistant pathogens. They are highly recommended to be used in the future either as a supplement to antibiotics or as an alternative treatment.

5. Conclusions

Nanosilver particles can be biosynthesized by cyanobacteria (green synthesis). Size manipulation is achievable by controlling the concentration of the precursor. Smallest-sized nanosilver particles are recommended as an antibacterial agent. Nanosilver particles are successful biocontrol agents against multi-drug resistant bacterial strains. Future antibacterial drugs can be based partially or entirely on nanosilver particles.

Supplementary Materials: The following are available online at https://www.mdpi.com/article/10.3390/antibiotics11081003/s1, Figure S1: FTIR from middle-sized AgNPs; Figure S2: FTIR from largest sized AgNPs

Author Contributions: N.A.E.S. was responsible for the conceptualization of the research, designing and conducting experiments. Isolation, culturing, and characterization of the Cyanobacterium, chemical analyses UV, FTIR, DLS, bioassays, and the writing of the manuscript were all per-formed by N.A.E.S. N.A.E.S. was also responsible for the analysis and interpretation of the data in addition to the writing of the manuscript. Moreover, N.A.E.S. obtained the funding from Al Bild Bank Scholarly chair for food security in Saudi Arabia, Deanship of Scientific Research, Vice presidency for Graduate studies and Scientific Research, King Faisal University, Al-Ahsa 31982, KSA Grant number CHAIR 59. E.M.B. was responsible for the preparation of the nanosilver stock solution and briefly advised on the nanosilver synthesis experiment, which was conducted by N.A.E.S. N.A.E.S. was also responsible for the re-writing of the manuscript several times. N.A.E.S. did all the modifications in the several rounds of revisions. Including reply to reviewers. All authors have read and agreed to the published version of the manuscript.

Funding: This research was funded by Al Bild Bank Scholarly chair for food security in Saudi Arabia, Deanship of Scientific Research, Vice presidency for Graduate studies and Scientific Research, King Faisal University, Al-Ahsa 31982, KSA Chair Grant number: 59 awarded to N.A.E.S.

Data Availability Statement: All data generated or analyzed during this study are included.

Acknowledgments: The authors express their deepest gratitude to the Bank Al Bilad scholarly chair for food security Deanship of Scientific Research, King Faisal University P.O. box 400, Al-Ahsa, post code: 31982, Kingdom of Saudi Arabia. Grant number-CHAIR 59 for financial and moral support. N.A.E.S. would like to express her deepest gratitude to Kawther El Amer for help with DLS and to Aziza Ibrahim Alibrahim and AlAnoud AlHomaidy for their excellent technical assistance regarding UV and FTIR analyses. N.A.E.S. is quite thankful to Munirah Aldayel for her help with the antimicrobial bioassay.

Conflicts of Interest: The authors declare no conflict of interest.

References

1. Ali, D.M.; Sasikala, M.; Gunasekaran, M.; Thajuddin, N. Biosynthesis and characterization of silver nanoparticles using marine cyanobacterium, Oscillatoria willei NTDM01. *Dig. J. Nanomater. Biostruct.* **2011**, *6*, 385–390.
2. Xu, C.; Van Zalinge, H.; Pearson, J.L.; Glidle, A.; Cooper, J.M.; Cumming, D.R.S.; Haiss, W.; Yao, J.; Schiffrin, D.J.; Proupín-Pérez, M.; et al. A combined top-down bottom-up approach for introducing nanoparticle networks into nanoelectrode gaps. *Nanotechnology* **2006**, *17*, 3333–3339. [CrossRef] [PubMed]
3. Bakir, E.M.; Younis, N.S.; Mohamed, M.E.; El Semary, N.A. Cyanobacteria as Nanogold Factories: Chemical and Anti-Myocardial Infarction Properties of Gold Nanoparticles Synthesized by *Lyngbya majuscula*. *Mar. Drugs* **2018**, *16*, 217. [CrossRef] [PubMed]
4. Younis, N.S.; Bakir, E.M.; Mohamed, M.E.; El Semary, N.A. Cyanobacteria as Nanogold Factories II: Chemical Reactivity and anti-Myocardial Infraction Properties of Cust. Gold Nanoparticles Biosynthesized by *Cyanothece* sp. *Mar. Drugs* **2019**, *17*, 402. [CrossRef] [PubMed]
5. Brayner, R.; Barberousse, H.; Hemadi, M.; Djedjat, C.; Yéprémian, C.; Coradin, T.; Livage, J.; Fiévet, F.; Couté, A. Cyanobacteria as Bioreactors for the Synthesis of Au, Ag, Pd, and Pt Nanoparticles via an Enzyme-Mediated Route. *J. Nanosci. Nanotechnol.* **2007**, *7*, 2696–2708. [CrossRef]
6. Hamida, R.S.; Ali, M.A.; Redhwan, A.M.O.; Bin-Meferij, M.M. Cyanobacteria—A Promising Platform in Green Nanotechnology: A Review on Nanoparticles Fabrication and Their Prospective Applications. *Int. J. Nanomed.* **2020**, *15*, 6033–6066. [CrossRef]
7. Dahoumane, S.A.; Djediat, C.; Yéprémian, C. Species selection for the design of gold nanobioreactor by photosynthetic organisms. *J. Nanopart. Res.* **2012**, *14*, 883. [CrossRef]
8. Rajeshkumar, S.; Malarkodi, C.; Paulkumar, K.; Vanaja, M.; Gnanajobitha, G.; Annadurai, G. Algae mediated green fabrication of silver nanoparticles and examination of its antifungal activity against clinical pathogens. *Int. J. Met.* **2014**, *2014*, 692643. [CrossRef]
9. Mata, Y.; Torres, E.; Blázquez, M.; Ballester, A.; González, F.; Muñoz, J.A. Gold(III) biosorption and bioreduction with the brown alga *Fucus vesiculosus*. *J. Hazard. Mater.* **2009**, *166*, 612–618. [CrossRef]
10. Vijayan, S.R.; Santhiyagu, P.; Singamuthu, M.; Ahila, N.K.; Jayaraman, R.; Ethiraj, K. Synthesis and Characterization of Silver and Gold Nanoparticles Using Aqueous Extract of Seaweed, *Turbinaria conoides*, and Their Antimicrofouling Activity. *Sci. World J.* **2014**, *2014*, 1–10. [CrossRef]
11. Singh, H.; Du, J.; Singh, P.; Yi, T.H. Extracellular synthesis of silver nanoparticles by Pseudomonas sp. THG-LS1.4 and their antimicrobial application. *J. Pharm. Anal.* **2018**, *8*, 258–264. [CrossRef] [PubMed]

12. Kumar, S.A.; Abyaneh, M.K.; Gosavi, S.W.; Kulkarni, S.K.; Pasricha, R.; Ahmad, A.; Khan, M.I. Nitrate reductase-mediated synthesis of silver nanoparticles from $AgNO_3$. *Biotechnol. Lett.* **2007**, *29*, 439–445. [CrossRef] [PubMed]
13. Abyaneh, M.K.; Kumar, S.A.; Gosavi, S.W.; Kulkarni, S.K.; Ahmad, A.; Khan, M.I. Sulfite reductase-mediated synthesis of gold nanoparticles capped with phytochelatin. *Biotechnol. Appl. Biochem.* **2007**, *47*, 191–195. [CrossRef]
14. Younis, N.S.; El Semary, N.A.; Mohamed, M.E. Silver nanoparticles green synthesis via cyanobacterium *Phormidium* sp.: Characterization, wound healing, antioxidant, antibacterial, and anti-inflammatory activities. *Eur. Rev. Med. Pharmacol. Sci.* **2021**, *25*, 3083–3096. [CrossRef] [PubMed]
15. El-Semary, N.A.; Mabrouk, M.; Faraag, A.H.; Kilany, M.; Omran, S.H.; Ghramh, H.A.; Alshehri, A.; Ibrahim, E.H.; Morsy, K.; El-Kott, A.F.; et al. Green Synthesis of Silver Nanoparticles via *Phormidium* sp. nov. (Cyanophyceae): Amelioration, Characterization and Assessment of the Antibacterial Potential Against Methicillin Resistant *Staphylococcus aureus*. *Sci. Adv. Mater.* **2021**, *13*, 209–216. [CrossRef]
16. Younis, N.S.; Mohamed, M.E.; El Semary, N.A. Green Synthesis of Silver Nanoparticles by the Cyanobacteria *Synechocystis* sp.: Characterization, Antimicrobial and Diabetic Wound-Healing Actions. *Mar. Drugs* **2022**, *20*, 56. [CrossRef]
17. Bandyopadhyay, A.; Elvitigala, T.; Liberton, M.; Pakrasi, H.B. Variations in the Rhythms of Respiration and Nitrogen Fixation in Members of the Unicellular Diazotrophic Cyanobacterial Genus *Cyanothece*. *Plant Physiol.* **2012**, *161*, 1334–1346. [CrossRef]
18. Desai, R.; Mankad, V.; Gupta, S.; Jha, P. Size Distribution of Silver Nanoparticles: UV-Visible Spectroscopic Assessment. *Nanosci. Nanotechnol. Lett.* **2012**, *4*, 30–34. [CrossRef]
19. Souza, T.G.F.; Ciminelli, V.S.T.; Mohallem, N.D.S. A comparison of TEM and DLS methods to characterize size distribution of ceramic nanoparticles. *J. Phys. Conf. Ser.* **2016**, *733*, 012039. [CrossRef]
20. Jeong, Y.; Lim, D.W.; Choi, J. Assessment of Size-Dependent Antimicrobial and Cytotoxic Properties of Silver Nanoparticles. *Adv. Mater. Sci. Eng.* **2014**, *2014*, 1–6. [CrossRef]
21. Bauer, A.W.; Kirby, W.M.; Sherris, J.C.; Turck, M. Antibiotic susceptibility testing by a standardized single disk method. *Am. J. Clin. Pathol.* **1966**, *45*, 493–496. [CrossRef] [PubMed]
22. El Semary, N. The antimicrobial profile of extracts of a *Phormidium*-like cyanobacterium changes with phosphate levels. *World J. Microbiol. Biotechnol.* **2011**, *28*, 585–593. [CrossRef] [PubMed]
23. Yasin, S.; Liu, Y. Biosynthesis of silver nanoparticles by bamboo leaves extract and their antimicrobial activity. *J. Fiber Bioeng. Inform.* **2013**, *6*, 77–84.
24. Aziz, S.B.; Hussein, G.; Brza, M.A.; Mohammed, S.J.; Abdulwahid, R.T.; Saeed, S.R.; Hassanzadeh, A. Fabrication of Interconnected Plasmonic Spherical Silver Nanoparticles with Enhanced Localized Surface Plasmon Resonance (LSPR) Peaks Using Quince Leaf Extract Solution. *Nanomaterials* **2019**, *9*, 1557. [CrossRef]
25. Hamouda, R.A.; Hussein, M.H.; Abo-Elmagd, R.A.; Bawazir, S.S. Synthesis and biological characterization of silver nanoparticles derived from the cyanobacterium *Oscillatoria limnetica*. *Sci. Rep.* **2019**, *9*, 1–17. [CrossRef]
26. Lengke, M.F.; Fleet, M.E.; Southam, G. Biosynthesis of Silver Nanoparticles by Filamentous Cyanobacteria from a Silver(I) Nitrate Complex. *Langmuir* **2007**, *23*, 2694–2699. [CrossRef]
27. Lin, J.; Nishino, K.; Roberts, M.C.; Tolmasky, M.; Aminov, R.; Zhang, L. Mechanisms of antibiotic resistance. *Front. Microbiol.* **2015**, *6*, 34. [CrossRef]
28. Saed, H.A.E.R.; Ibrahim, H.M.M. Antimicrobial profile of multidrug-resistant *Steptococcus* spp. Isolated from dairy cows with clinincal mastitis. *J. Adv. Vetrinary Anim. Res.* **2020**, *7*, 186–197. [CrossRef]
29. Alves-Barroco, C.; Rivas-García, L.; Fernandes, A.R.; Baptista, P.V. Tackling Multidrug Resistance in Streptococci—From Novel Biotherapeutic Strategies to Nanomedicines. *Front. Microbiol.* **2020**, *11*, 579916. [CrossRef]
30. Martinezgarriga, B.; Vinuesa, T.; Hernandez-Borrell, J.; Vinas, M. The contribution of efflux pumps to quinolone resistance in Streptococcus pneumoniae clinical isolates. *Int. J. Med. Microbiol.* **2007**, *297*, 187–195. [CrossRef]
31. Petinaki, E.; Papagiannitsis, C. Resistance of Staphylococci to Macrolides-Lincosamides—Streptogramins B (MLSB): Epidemiology and Mechanisms of Resistance. *Staphylococcus aureus* **2019**, *117*. [CrossRef]
32. Gnanamani, A.; Hariharan, P.; Paul-Satyaseela, M. *Staphylococcus aureus*: Overview of bacteriology, clinical dis-eases, epidemiology, antibiotic resistance and therapeutic approach. *Front. Staphylococcus aureus* **2017**, *4*, 28.
33. Dong, Y.; Zhu, H.; Shen, Y.; Zhang, W.; Zhang, L. Antibacterial activity of silver nanoparticles of different particle size against Vibrio Natriegens. *PLoS ONE* **2019**, *14*, e0222322. [CrossRef] [PubMed]
34. Dakal, T.C.; Kumar, A.; Majumdar, R.S.; Yadav, V. Mechanistic basis of antimicrobial actions of silver nanoparticles. *Front. Microbiol.* **2016**, *7*, 1831. [CrossRef] [PubMed]
35. Tian, J.; Wong, K.K.; Ho, C.M.; Lok, C.N.; Yu, W.Y.; Che, C.M.; Cliu, J.F.; Tam, P.K. Topical delivery of silver nanoparticles promotes wound healing. *ChemMedChem* **2007**, *2*, 129–136. [CrossRef]
36. Chen, Z.; Yang, P.; Yuan, Z.; Guo, J. Aerobic condition enhances bacteriostatic effects of silver nanoparticles in aquatic environment: An antimicrobial study on *Pseudomonas aeruginosa*. *Sci. Rep.* **2017**, *7*, 1–8. [CrossRef]
37. Yacamán, M.J.; Ascencio, J.A.; Liu, H.B.; Gardea-Torresdey, J. Structure shape and stability of nanometric sized particles. *J. Vac. Sci. Technol. B Microelectron. Nanometer Struct.* **2001**, *19*, 1091. [CrossRef]
38. Hong, X.; Wen, J.; Xiong, X.; Hu, Y. Silver nanowire-carbon fiber cloth nanocomposites synthesized by UV curing adhesive for electrochemical point-of-use water disinfection. *Chemosphere* **2016**, *154*, 537–545. [CrossRef]

39. Collins, T.L.; Markus, E.A.; Hassett, D.J.; Robinson, J.B. The Effect of a Cationic Porphyrin on Pseudomonas aeruginosa Biofilms. *Curr. Microbiol.* **2010**, *61*, 411–416. [CrossRef]
40. Hamida, R.S.; Ali, M.A.; Goda, D.A.; Khalil, M.I.; Al-Zaban, M.I. Novel Biogenic Silver Nanoparticle-Induced Reactive Oxygen Species Inhibit the Biofilm Formation and Virulence Activities of Methicillin-Resistant *Staphylococcus aureus* (MRSA) Strain. *Front. Bioeng. Biotechnol.* **2020**, *8*, 433. [CrossRef]
41. Quinteros, M.A.; Aristizábal, V.C.; Dalmasso, P.R.; Paraje, M.G.; Páez, P.L. Oxidative stress generation of silver nanoparticles in three bacterial genera and its relationship with the antimicrobial activity. *Toxicol. Vitr.* **2016**, *36*, 216–223. [CrossRef] [PubMed]
42. Hamida, R.S.; Ali, M.A.; Goda, D.A.; Khalil, M.I.; Redhwan, A. Cytotoxic effect of green silver nanoparticles against ampicillin-resistant *Klebsiella pneumoniae*. *RSC Adv.* **2020**, *10*, 21136–21146. [CrossRef] [PubMed]
43. Ramkumar, V.S.; Pugazhendhi, A.; Gopalakrishnan, K.; Sivagurunathan, P.; Saratale, G.D.; Dung, T.N.B.; Kannapiran, E. Biofabrication and characterization of silver nanoparticles using aqueous extract of seaweed *Enteromorpha compressa* and its biomedical properties. *Biotechnol. Rep.* **2017**, *14*, 1–7. [CrossRef] [PubMed]
44. Farooq, U.; Ahmad, T.; Khan, A.; Sarwar, R.; Shafiq, J.; Raza, Y.; Ahmed, A.; Ullah, S.; Rehman, N.U.; Al-Harrasi, A. Rifampicin conjugated silver nanoparticles: A new arena for development of antibiofilm potential against methicillin resistant *Staphylococcus aureus* and *Klebsiella pneumoniae*. *Int. J. Nanomed.* **2019**, *14*, 3983–3993. [CrossRef]
45. Choi, S.-H.; Jang, Y.-S.; Jang, J.-H.; Bae, T.-S.; Lee, S.-J.; Lee, M.-H. Enhanced antibacterial activity of titanium by surface modification with polydopamine and silver for dental implant application. *J. Appl. Biomater. Funct. Mater.* **2019**, *17*. [CrossRef]
46. Yin, I.X.; Yu, O.Y.; Zhao, I.S.; Mei, M.L.; Li, Q.-L.; Tang, J.; Chu, C.-H. Developing biocompatible silver nanoparticles using epigallocatechin gallate for dental use. *Arch. Oral Biol.* **2019**, *102*, 106–112. [CrossRef]
47. Babgi, B.A.; Alsayari, J.H.; Davaasuren, B.; Emwas, A.-H.; Jaremko, M.; Abdellattif, M.H.; Hussien, M.A. Synthesis, Structural Studies, and Anticancer Properties of [CuBr (PPh3) 2 (4, 6-Dimethyl-2-Thiopyrimidine-κ S]. *Crystals* **2021**, *11*, 688. [CrossRef]

antibiotics

MDPI

Article

Antimicrobial Activity of Biogenic Silver Nanoparticles from *Syzygium aromaticum* against the Five Most Common Microorganisms in the Oral Cavity

Erika Alejandra Jardón-Romero [1], Edith Lara-Carrillo [1], María G. González-Pedroza [2], Víctor Sánchez-Mendieta [3], Elías Nahum Salmerón-Valdés [4], Víctor Hugo Toral-Rizo [4], Oscar F. Olea-Mejía [3], Saraí López-González [1] and Raúl A. Morales-Luckie [4,*]

1 Center for Advanced Studies and Research on Dentistry, Autonomous University of the State of Mexico (UAEMex), Toluca 50200, Mexico; erika_abisma@hotmail.com (E.A.J.-R.); elarac@uaemex.mx (E.L.-C.); sarailogo@hotmail.com (S.L.-G.)
2 Faculty of Sciences, Department of Biotechnology, Autonomous University of the State of Mexico (UAEMex), Toluca 50200, Mexico; mggonzalezp@uaemex.mx
3 Joint Center for Research in Sustainable Chemistry (CCIQS), Department of Material Science, Autonomous University of the State of Mexico (UAEMex), Toluca 50200, Mexico; vsanchezm@uaemex.mx (V.S.-M.); ofoleam@uaemex.mx (O.F.O.-M.)
4 School of Dentistry, Autonomous University of the State of Mexico (UAEMex), Toluca 50130, Mexico; ensalmeronv@uaemex.mx (E.N.S.-V.); vhtoralr@uaemex.mx (V.H.T.-R.)
* Correspondence: ramluckie@gmail.com or rmoralesl@uaemex.mx

Citation: Jardón-Romero, E.A.; Lara-Carrillo, E.; González-Pedroza, M.G.; Sánchez-Mendieta, V.; Salmerón-Valdés, E.N.; Toral-Rizo, V.H.; Olea-Mejía, O.F.; López-González, S.; Morales-Luckie, R.A. Antimicrobial Activity of Biogenic Silver Nanoparticles from *Syzygium aromaticum* against the Five Most Common Microorganisms in the Oral Cavity. *Antibiotics* **2022**, *11*, 834. https://doi.org/10.3390/antibiotics11070834

Academic Editors: Christina N. Banti and Sotiris K. Hadjikakou

Received: 3 May 2022
Accepted: 15 June 2022
Published: 21 June 2022

Publisher's Note: MDPI stays neutral with regard to jurisdictional claims in published maps and institutional affiliations.

Copyright: © 2022 by the authors. Licensee MDPI, Basel, Switzerland. This article is an open access article distributed under the terms and conditions of the Creative Commons Attribution (CC BY) license (https://creativecommons.org/licenses/by/4.0/).

Abstract: *Syzygium aromaticum* (clove) has been used as a dental analgesic, an anesthetic, and a bioreducing and capping agent in the formation of metallic nanoparticles. The main objective of this study was to evaluate the antimicrobial effect in oral microorganisms of biogenic silver nanoparticles (AgNPs) formed with aqueous extract of clove through an ecofriendly method "green synthesis". The obtained AgNPs were characterized by UV-Vis (ultraviolet-visible spectroscopy), SEM-EDS (scanning electron microscopy–energy dispersive X-ray spectroscopy), TEM (transmission electron microscopy), and ζ potential, while its antimicrobial effect was corroborated against oral Gram-positive and Gram-negative microorganisms, as well as yeast that is commonly present in the oral cavity. The AgNPs showed absorption at 400–500 nm in the UV-Vis spectrum, had an average size of 4–16 nm as observed by the high-resolution transmission electron microscopy (HR-TEM), and were of a crystalline nature and quasi-spherical form. The antimicrobial susceptibility test showed inhibition zones of 2–4 mm in diameter. Our results suggest that AgNPs synthesized with clove can be used as effective growth inhibitors in several oral microorganisms.

Keywords: biosynthesis; silver nanoparticles; *Syzygium aromaticum*; oral microorganisms

1. Introduction

The ideal properties of an antibacterial coating include prolonged activity, high levels of bactericidal and bacteriostatic activity, ability to act against a wide spectrum of bacteria, biocompatibility, and low in vivo toxicity [1–5]. Historically, silver compounds and ions have been extensively used for hygienic and healing purposes. However, over time, their application as an anti-infection agent has dwindled due to the advent of antibiotics and other disinfectants [2,3,6,7]. Recently, there has been renewed interest in manufactured silver nanomaterials, thanks to their unusually strong physicochemical properties and biological activities compared to their bulk parent materials [1,4]. Today, it should be noted that nanomaterials have managed to enter different regulatory and safety fields (such as clinical trials and good manufacturing practices) [6], this to regulate and guarantee the quality of different areas of management and production [7]. In addition, silver nanoparticles (AgNPs) synthesized by green methods have many other applications in different biotechnological areas such as water filtration agents [8], disinfection and preservation of foods [9,10] and various materials [11], the production of cosmetics [10,12], nanoinsecticides and nanopesticides [13], nanocomposites [14], amongst others [15–20]. Depending on the technique used, their synthesis can be divided into the following: chemical methods involving the reduction or precipitation of metals in the presence of stabilizing agents; physical methods such as thermolysis, photochemical, and sonochemistry; finally, biological methods [16,21–25]. Biological methods are extremely important, since reducing agents of a chemical nature are not required. In biological methods, reducing agents are obtained from compounds present in natural extracts. This is the case of the synthesis being used in this research, which also has great advantages, because it is an easy and low-cost method [26]. Compared to chemical methods, biological methods represent less toxicity and are more respectful of the environment, since the most used reducing agents at the industrial level and within chemical methods are sodium borohydride, hydrazine and hypophosphite, which can increase environmental toxicity or biological hazards. In addition, capping agents, such as polyvinyl alcohol, must be used to prevent the AgNPs from aggregating. Another issue is that the high temperature may also increase the production cost [16,27]. Fortunately, the biological synthesis of nanoparticles (NPs), also known as "green synthesis," has allowed the formation of metallic nanostructures from the use of bacteria, fungi, plants, or their extracts, meaning that this approach to synthesis is a non-toxic and environmentally friendly alternative. Sometimes, the deployment of this synthesis equals or exceeds the expectations of NPs synthesized by physical and chemical methods, in terms of cost and characteristics, as previously described [27–30]. Green extracts contain molecules that carry hydroxyl moieties in their functional groups, mainly of the phenolic type, which can be used for the reduction of metal ions and formation of stable complexes with metallic NPs [1,31,32]. AgNPs show efficient antimicrobial properties compared to other metallic NPs, due to their large surface area, which provides better contact with microorganisms. Although AgNPs have been reported to be involved in a wide range of molecular processes within microorganisms, the mechanism of action is still being studied [33], It is important to note that there are not only bacteria that cause conditions in the oral cavity, but also some yeasts such as *Candida albicans* [34], Recently, it has been shown that AgNPs induce alterations in fungal cells and the formation of pores on the cell surface, in addition to changes in membrane fluidity, all of which may be related to changes in the lipid constitution of the plasma membrane and membrane depolarization [35]. Regarding the mechanism of action of AgNPs on bacteria, it has been studied that they have the ability to anchor and subsequently penetrate the bacterial cell wall, which causes structural changes and cell death [36]. The formation of free radicals by the AgNPs may be considered another mechanism by which cells die. Diverse studies, in which electron spin resonance spectroscopy was used, suggest that free radicals form when the NPs come into contact with the bacteria [37]; these radicals are able to damage the cell membrane, rendering it porous, which can ultimately lead to death [2]. It is also true that the interaction of the AgNPs with the sulfur and phosphorus major components of DNA can lead to problems in the DNA's replication of bacteria (such as epigenetic changes) [28]. Moreover, when it comes to microbial flora, the oral cavity is one of the most densely populated sites of the human body [38]. Over 700 bacterial species or phylotypes have been detected in this location, of which more than half have not been cultivated [39]. This means that microorganisms of the oral community should display extensive interactions when forming biofilm structures, carrying out physiological functions, and inducing pathogenesis [40,41]. The oral cavity is a complex ecosystem that is inhabited by more than 300 bacterial species. Some of these have been implicated in oral diseases such as caries and periodontitis, which are among the most common bacterial infections in humans [42]. The bacteria colonize the teeth in a reasonably predictable sequence. The first or primary colonizers tend to be aerobic (especially *streptococci*, which constitute 47–85% of the cultivable cells found during the first four hours after professional tooth cleaning); as plaque oxygen levels fall, the proportions of Gram-negative microorganisms tend to increase [43]. In the normal oral cavity, several species of the genera *Streptococcus, Lactobacillus, Lactococcus, Enterococcus, Staphylococcus, Corynebacterium, Veillonella* and Bacteroides stand out for being responsible for various oral conditions [44]. Silver compounds and NPs have already been used as dental restorative material, endodontic retrofill cements, dental implants, and caries inhibitory solutions. Despite the effectiveness shown by AgNPs in dental practice, controversy remains over their toxicity in biological and ecological systems, due to the cytotoxicity caused by high concentrations or the specific size of the nanoparticles [45,46]. *Syzygium aromaticum* (Clove) is employed in Indian ayurvedic medicine, Chinese medicine, and western herbalism, while in dentistry, its essential oil is used as an anodyne (painkiller) for emergencies. Clove oil is a pale-yellow liquid with a characteristic odor and taste. It consists of 81–95% phenols (eugenol with about 3% of acetyl eugenol), sesquiterpenes (a- and b-calyophyllenes), and small quantities of esters, alcohols, and ketones [24,27–30,36,41,45–52]. The high level of eugenol present in clove oil endows it with intense biological and antimicrobial activity [53], which is why it was explored here as a stabilizing and reducing agent in the synthesis of AgNPs. Therefore, it is important to develop dental materials with antibacterial activity and better mechanical properties, which could be manufactured and employed in future clinical applications.

Herein, AgNPs were synthesized using aqueous clove extract, as a reducing and stabilizing agent, taking advantage of the AgNPs antimicrobial properties in synergy with the clove dental applications as an analgesic and anesthetic. The green-synthesized AgNPs were characterized by ultraviolet-visible (UV-Vis) spectrometry, scanning electron microscopy–energy dispersive X-ray spectroscopy (SEM-EDS), transmission electron microscopy (TEM) and ζ potential. The antibacterial activity of the AgNPs against Gram-positive and Gram-negative microorganisms and yeast were tested by the disc-diffusion method.

2. Results

2.1. Synthesis of Silver Nanoparticles

The reduction of silver ions and formation of AgNPs occurred after silver nitrate solution was placed in contact with the clove aqueous extract, followed immediately by a change in color of the suspension, from colorless to yellow. This process was carried out under ambient conditions. It is relevant to mention that no stabilizing or capping agent was used.

2.2. Characterization of Synthesized Silver Nanoparticles

2.2.1. UV-Vis Spectroscopy

The spectra were recorded when both the color and absorption intensity of the colloidal samples remained constant, as shown in Figure 1. Each surface plasmon resonance (SPR) maximum was ubicated in the range of 431–447 nm. In addition, no surface plasmon resonance was observed at more than 500 nm, indicating that most of the AgNPs obtained were of small size and similar shape. As time increases, AgNPs stabilize and reach their maximum growth.

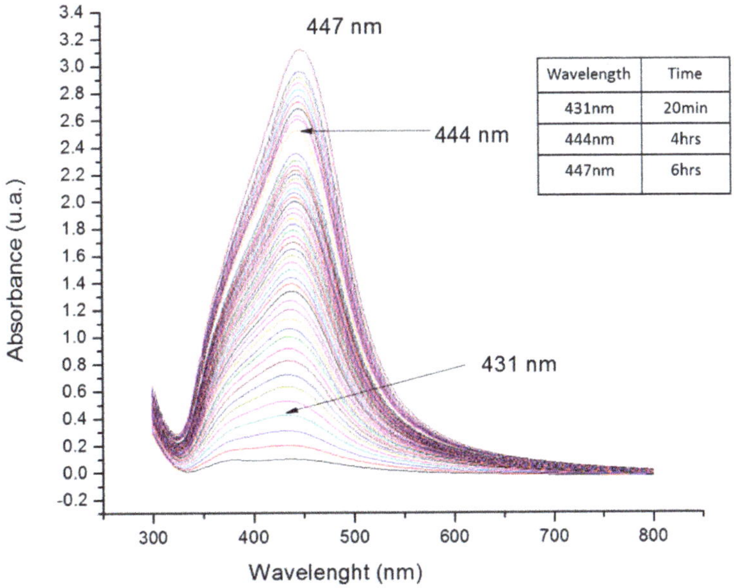

Figure 1. UV-Vis absorption spectrum of synthesized silver nanoparticles by *S. aromaticum* extract at different time intervals.

2.2.2. SEM-EDS Analysis

The characterization technique was consistent with the EDS analysis. AgNPs generally showed absorption peaks at approximately 3 keV (Figure 2).

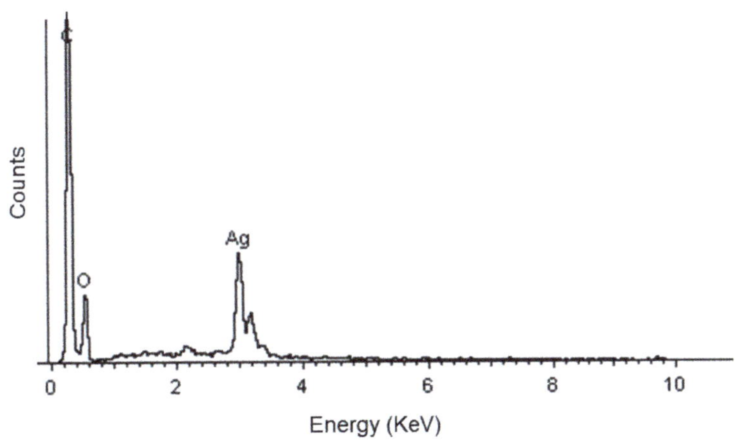

Figure 2. Scanning electron microscopy–energy dispersive X-ray spectrum of synthesized silver nanoparticles.

2.2.3. TEM

TEM was employed to characterize the size, shape, and morphology of the synthesized AgNPs, which resulted to be quasi-spherical, with sizes ranging 4–16 nm, with a mean size of 10 nm and a standard deviation of 4.27 nm (Figure 3).

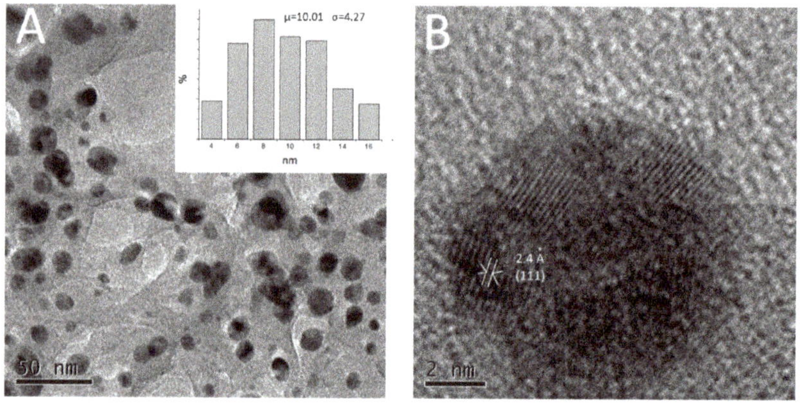

Figure 3. Characterization of synthesized biogenic silver nanoparticles from clove extract. (**A**) Micrograph of transmission electron microscopy, the insert shows size distribution and (**B**) high-resolution transmission electron microscopy micrograph.

2.2.4. ζ Potential

The electrostatic stabilization of the AgNPs was estimated by measuring their ζ potential values, which were found to be in the range of −15.7 to −16.2 mV.

2.3. Antimicrobial Activity

2.3.1. Disk Diffusion Test

The biogenic AgNPs were tested for antimicrobial activities against Gram-positive, Gram-negative bacteria and yeast, presenting similar inhibition zones in all cultures. The inhibition zones produced by the AgNPs displayed halos of 2–4 mm in diameter. The control group (clove extract) did not show any antibacterial effect only antifungal effect (Figure 4). This last datum provides information regarding the efficacy provided by the clove extract only against *Candida albicans*, which turns out to be really important. In Figure 4E, it is easy to perceive this important antifungal activity, and it is observed to be slightly enhanced by the AgNPs. Regarding the results obtained in bacteria, as previously described, it is possible to see the activity of the potentiated extract.

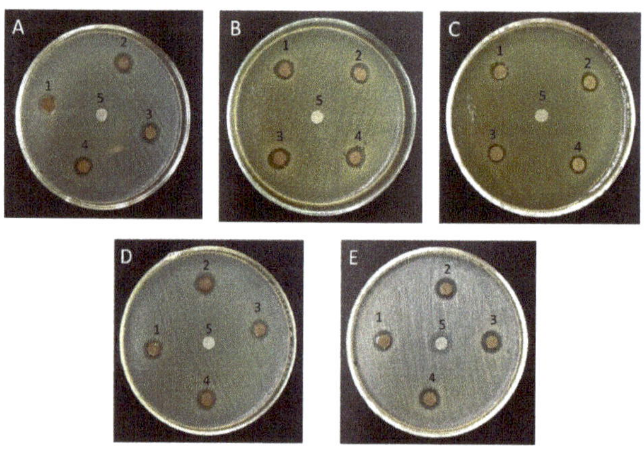

Figure 4. Antimicrobial activity against: (**A**) *Streptococcus mutans*, (**B**) *Staphylococcus aureus*, (**C**) *Escherichia coli*, (**D**) *Enterococcus faecalis*, (**E**) *Candida albicans*. Disk 1, 24 h after synthesis; disk 2, 7 days after synthesis; disk 3, 14 days after synthesis; disk 4, 21 days after synthesis; disk 5, control disk only with clove extract.

2.3.2. Microdilution in Broth Test

Through this study, we were able to determine both the maximum effective concentration (MEC), as well as the minimum inhibitory concentration (MIC), where we can observe the effectiveness of the NPs at different concentrations against the different microorganisms mentioned above (Figure 5a and Table 1). Specifically, in Figure 5a, it is possible to observe the complete inhibition at the highest concentration; likewise, it is reflected in Figure 5b how Microbial growth after exposure to determined concentrations of synthesized silver nanoparticles. This information is reflected in Table 1. A very important result is that of the activity it has against the bacteria *E. coli*, where the antibacterial activity is completely inhibited until the last treatment.

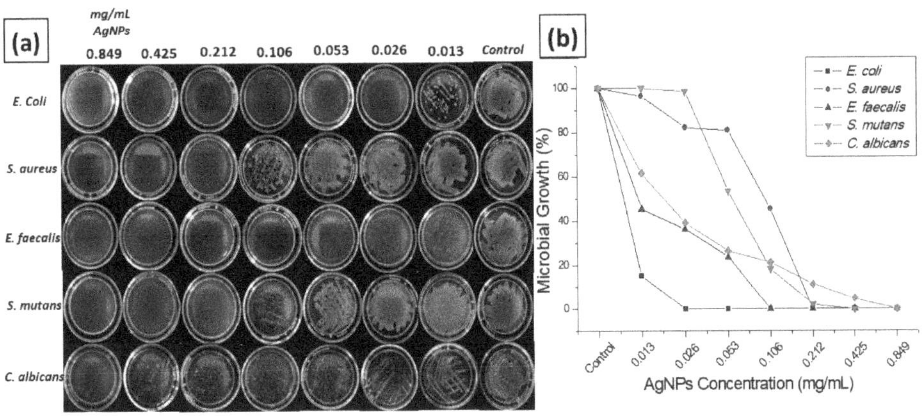

Figure 5. Microdilution in broth: (**a**) inhibition of microbial growth, (**b**) graph of antimicrobial growth against the applied concentration of nanoparticles.

Table 1. Minimum inhibitory concentration and maximum effective concentration of silver nanoparticles.

	AgNPs				
AgNPs (mg/mL)	*E. coli*	*S. aureus*	*E. faecalis*	*S. mutans*	*C. albicans*
0.849	-	-	-	-	-
0.425	-	-	-	-	+
0.212	-	-	-	+	+
0.106	-	+	-	+	+
0.053	-	+	+	+	+
0.026	-	+	+	+	+
0.013	+	+	+	+	+
Control	+	+	+	+	+

+ Microbial growth, - Without microbial growth MIC ● MEC ●.

Table 1 shows, for the five types of microorganisms, the minimum inhibitory concentration and the concentrations in milligrams/mL, as well as the maximum effective concentration of the AgNPs.

3. Discussion

Clove is one of the most valuable of all spices, and it has been used for centuries as a food preservative and for medicinal purposes, as an analgesic and antispasmodic, with eugenol being the main compound responsible for this activity. This plant represents one of the richest sources of phenolic compounds such as eugenol, eugenol acetate, and gallic acid (75–77%) [54–57]. Several authors [51,54,56] have studied the antimicrobial effects of clove extract. For example, Duhan used various plant extracts with high concentration of eugenol, including clove, and compared them with 3% sodium hypochlorite against *Enterococcus faecalis*. Meanwhile, Mansourian and Rana used clove extract against *C. albicans* and compared it to nystatin. In both cases, the results obtained with the clove extracts were more effective than the substances they were compared to. The most probable pathway for AgNPs biosynthesis is that the flavonoids, from the clove buds, act as reducing agents when they encounter Ag^{+} ions. The electrons are transferred and the reduction process occurs, leading to the formation of AgNPs. It is well known that flavonoids also prevent agglomeration and stabilize the AgNPs in aqueous solution [16,21–23,58]. This makes a strong case for the involvement of flavonoids in rapid biosynthesis and for the stability of metallic NPs in the aqueous medium [17,59]. Upon storage of the bio-functionalized AgNPs that were synthesized using macerated clove solution, it was observed that the colloidal solution maintained its stability and uniformity. The biosynthesis of AgNPs by clove extract is carried out rapidly, the solution changes color as a function of time, the kinetics of formation (Figure 1) indicates that the reaction stabilizes after 6 h. The process of biosynthesis in this study was carried out under ambient conditions. Qian Sun [60], Deshpande [49], and Bajpai [55] performed different studies to generate AgNPs using a similar green synthesis approach, obtaining antibacterial properties such as those in our study under the same conditions. In addition, Jeevika [48] produced silver nanowires via biosynthesis using clove oil. It is well known that AgNPs show a yellowish-brown color in aqueous solution; this color is a result of the excitation of surface plasmon vibrations in the metal NPs [48,49].

According to UV-Vis spectroscopy results, in the first 20 min of reaction, there was an incipient reduction in which the SPR could already be observed. This may be attributed to the high availability of reducing agents such as eugenol in the aqueous extract. During the next six hours of reaction, the SPR was clearly observed at 447 nm, indicating the formation of AgNPs. After six hours of reaction, SPR with a defined shape appears at 447 nm, indicating that the NPs now have a defined shape and size, compared to plasmons at 431 nm at short reaction times, indicating incipient nanoparticle formation. Measurements were made approximately once every five minutes to verify the AgNPs kinetics of formation.

EDS analysis of the AgNPs confirmed the presence of metallic Ag (see Figure 2), according to [61,62]. It is important to point out that these biogenic AgNPs exhibited certain low size polydispersity and similar shapes, making our method distinct from other bio-reductors; this was corroborated by the TEM micrographs, whereby the image in Figure 3A showed that AgNPs prepared using clove extract have a considerably low polydispersity and an almost spherical morphology. The particle size distribution histogram was determined using TEM images. AgNPs diameters range between 4 and 16 nm, with a standard deviation of 4.2 nm (Figure 3A).

The HR-TEM image in Figure 3B shows AgNPs of 8–10 nm in diameter. This image possesses atomic resolution; therefore, several crystalline planes are distinguishable and the interplanar distances can be measured. The interplanar distances shown in the micrograph of these NPs, wherein the corresponding crystalline planes were specified. The interplanar distances and their corresponding crystalline planes match those of metallic Ag (face-centered cubic structure). The measured interplanar distance was 2.4 Å, which corresponds to the plane [111]. Generally, a suspension that exhibits an absolute zeta potential of 0–100 mV is ubicated on the threshold of agglomeration, and the stability of the particles from the solution will be increased with higher values of approximately −100 mV [63,64]. In this study, the zeta potential of AgNPs was in the range of −15.7 to −16 mV. These values allow us to corroborate an acceptable stability of the AgNPs colloidal suspension.

The microorganisms selected in this study for microbiological tests are part of the normal and pathogenic microflora of the oral cavity. *Streptococcus mutans*, the main causal agent of tooth decay and periodontitis, is found in 70–90% of the population. Meanwhile, *Staphylococcus aureus* is found in patients with bacterial endocarditis and cellulite. *Enterococcus faecalis* is present in most failed root canal treatments. *Escherichia coli* has been isolated from salivary gland infections, while *Candida albicans* is found in one in every 1000 patients attending dental consultation and also in stomatitis associated with the use of dental prostheses. Taken together, these bacteria are responsible for various opportunistic infections in immunocompromised patients. In this study, the antibacterial activity of AgNPs against four types of bacteria and one yeast was investigated. The inhibition zones were similar in all cell cultures. Biological tests of AgNPs against test strains showed that they have a significant effect on the growth of Gram-positive and Gram-negative bacteria. The latter have a layer of lipopolysaccharides at the exterior, while underneath lies a thin layer of peptidoglycan. Thus, a potent antimicrobial effect can be determined in the microdilution in broth study that emphasizes the antibacterial and antifungal effectiveness of the AgNPs as shown in Figure 5b, taking into account the efficacy in *Escherichia coli* having microbial effect until the last concentration of NPs. On the other hand, we observed bacterial growth of the *Streptococcus mutans* strain from the third highest concentration, and also the following *Staphylococcus aureus* and *Enterococcus faecalis*; on the other hand, we see that the antifungal effectiveness of the AgNPs against *Candida albicans* is low but not null. Several studies indicate [1,30,65,66] that the antibacterial activity of AgNPs is based on the NPs attachment to the bacterial cell wall, or the formation of free radicals. In addition, the silver ions released from AgNPs may play a vital role in the antibacterial activity, due to their interaction with the thiol groups of enzymes. Although the lipopolysaccharides are composed of covalently linked lipids and polysaccharides, they have a lack of strength and rigidity. The negative charges on them are attracted toward the weak positive charge available on AgNPs [67,68]. This indicates that AgNPs have great potential to be used in biomedical applications.

4. Materials and Methods

4.1. Synthesis of AgNPs

In this experimental study, the chemical precursor used was silver nitrate ($AgNO_3$) (Sigma-Aldrich, St. Louis, MO, USA). The AgNPs were directly synthesized with the clove (Terana, Especias selectas, Ciudad de México, México) extract using a simple green synthesis procedure. Initially, various concentrations were tested until the optimum conditions were established.

A 10 mM solution of $AgNO_3$ was prepared using deionized water. To prepare the reducing agent, 1 g of chopped clove was added to 100 mL of boiling deionized water for five minutes. The mixture was allowed to cool before being filtered into a vacuum flask using a Buchner funnel and Whatman filter paper no. 5. The aqueous extract was mixed, at room temperature, with $AgNO_3$ in 3:1 ratio. The resultant mixture was kept undisturbed in a dark place. After a couple of hours, the color of the solution changed due to the formation of AgNPs. The process of biosynthesis was carried out under ambient environmental conditions (that is, at room temperature and under atmospheric pressure); the reaction was completed within a few minutes.

4.2. Characterization of Synthesized Silver Nanoparticles

4.2.1. Spectroscopy UV-Vis

The UV-Vis analysis was performed using a spectrophotometer (CARY 5000 Conc UV-Vis spectrophotometer, Varian, Inc., Palo Alto, CA, USA), which was operated at a resolution of 1 nm at room temperature. The spectral range was 300–800 nm. In this way, the kinetics of the reduction were followed until stable NPs were obtained.

4.2.2. SEM-EDS

The final product was sonicated for 30 min to break up larger nanoparticle agglomerates; then, the particles were dried in a vacuum at room temperature (20 °C) prior to analysis. The NPs were attached to aluminum stubs with conductive tape, coated with carbon, and observed under SEM (JEOL, JSM-6510LV at 20 kcV, Tokyo, Japan) with secondary electrons at ×100, ×500, and ×3000 magnification that was operating at 20 kV. EDS analysis was developed.

4.2.3. TEM

TEM was obtained via a JEOL JEM-2100 microscope (Tokyo, Japan). Samples for the TEM examination were prepared by placing a drop of the suspension on a copper grid (300 mesh) coated with carbon film, and allowing it to dry under ambient conditions.

4.2.4. Z Potential

Zeta potential measurements were determined using the Zetasizer 2000 (Malvem Instruments Ltd., Worcestershire, UK). The voltage applied to drive the electrodes of the zeta cell was 150 V capillary electrophoresis. The sample was prepared by injecting approximately 2 mL solution into the cell.

4.3. Antimicrobial Activity

The bacterial strains used in this study were obtained from the stock culture collection of the Biochemistry laboratory at the School of Dentistry within the National Autonomous University of Mexico (UNAM). The strains were originally collected from clinical samples of the oral cavity of patients from the previously mentioned institution. These strains are native to central Mexico, having all been characterized through a set of culture media assays and biochemical tests. They included Gram-positive (*Staphylococcus aureus*, *Streptococcus mutans*, *Enterococcus faecalis*) and Gram-negative microorganisms (*Escherichia coli*) and yeast (*Candida albicans*), which are commonly present in the oral cavity and are responsible for important conditions in oral health. The experiments into antimicrobial activity were carried out as proscribed by the Clinical and Laboratory Standards Institute [69].

4.3.1. Disk Diffusion Test

The NPs antibacterial properties were measured by the Kirby–Bauer disc diffusion method against the Gram-positive and Gram-negative bacteria and yeast. The inoculum was prepared by diluting the colonies with 0.9% of NaCl to 0.5 according to the McFarland scale, before they were applied to Muller–Hinton agar (Bioxon BD Mueller–Hinton II Agar) plates using sterile cotton swabs. The sterile paper discs were saturated with 10 μL of AgNPs (0.849 mg/mL) that had been prepared 24 h previously. In addition, three tubes with AgNPs that had been prepared 7, 14, and 21 days before the antimicrobial tests were used to saturate the paper discs and placed on agar plates. The disc impregnated with extract of clove was used as a positive control. After 24 and 48 h of incubation at 37 °C, the microbial susceptibility was determined through the measurement of any noticed inhibition zones. The assays were performed in triplicate [70].

The research protocol was reviewed and approved by the Ethics Committee of the Center for Research and Advanced Studies in Dentistry.

4.3.2. Microdilution in Broth

The antimicrobial capacity of AgNPs were determined following the dilution method of the broth [71]. Selective media were used to grow each strain and then cultured on non-selective media. Samples were initially incubated at 37 °C for 24 h for fresh bacterial cultures, which were used to prepare McFarland standards. Then, 100 μL of Mueller–Hinton broth medium and positive control and a negative control (Mueller–Hinton broth and NPs as sterility control) were used. Each well was aseptically inoculated with 5 μL of the bacterial suspension (final concentration approximately 5×10^5 CFU/mL) excluding controls. Subsequently, 100 μL of AgNPs (0.849 mg/mL) were placed at the beginning and seven serial microdilutions were made from the 100 μL of AgNPs. These assays were performed in triplicate in four wells for each concentration and strain. Inoculated microplates were incubated at 37 °C with continuous shaking at ~200 rpm for 24 h. The presence or absence of turbidity in each well was presented to the naked eye. The minimum concentration in the wells is taken as the MIC and the maximum effective concentration of MEC was identified by determining the lowest concentration of antimicrobial agent that reduces the viability of the initial bacterial inoculum by 99.9% or a log reduction in inoculum count [72].

5. Conclusions

AgNPs synthesized from aqueous extracts of *Sizygium aromaticum* show strong antibacterial activity. It is important to highlight that the nanoparticle generation method does not represent a strong environmental impact, it is very easy to carry out and, above all, very economical. In addition, this research has shown that AgNPs synthesized by this method are effective against the five most common microorganisms present in the oral cavity, these microorganisms are responsible for many oral health disorders, and we are sure that this research can contribute to the development of new antimicrobial agents for dental health or medical use, the routes of administration and cytotoxic assays will be exposed in future research because the expectations of this material are very high.

Author Contributions: Writing—preparation of the original draft and experimentation, E.A.J.-R. and E.N.S.-V.; review and edition-methodology-microbiology test, E.L.-C. and V.H.T.-R.; formal analysis in the synthesis of nanoparticles-broth microdilution test, M.G.G.-P. and O.F.O.-M.; statistical analysis, S.L.-G.; writing—preparation of the original draft and discussion, V.S.-M.; research-interpretation, discussion, design of the reduction agent and the nanoparticle synthesis, discussion proposal, R.A.M.-L. All authors have read and agreed to the published version of the manuscript.

Funding: This research received no external funding.

Conflicts of Interest: The authors declare no conflict of interest.

References

1. Chaloupka, K.; Malam, Y.; Seifalian, A.M. Nanosilver as a new generation of nanoproduct in biomedical applications. *Trends Biotechnol.* **2010**, *28*, 580–588. [CrossRef] [PubMed]
2. McShan, D.; Ray, P.C.; Yu, H. Molecular toxicity mechanism of nanosilver. *J. Food Drug Anal.* **2014**, *22*, 116–127. [CrossRef]
3. Zhang, J.; Shi, W.; Ma, Q.; Cui, H.; Zhang, L. Application of Nanotechnology in Immunity against Infection. *Coatings* **2021**, *11*, 430. [CrossRef]
4. Gomaa, E.Z. Silver nanoparticles as an antimicrobial agent: A case study on Staphylococcus aureus and Escherichia coli as models for Gram-positive and Gram-negative bacteria. *J. Gen. Appl. Microbiol.* **2017**, *63*, 36–43. [CrossRef] [PubMed]
5. Barabadi, H.; Mojab, F.; Vahidi, H.; Marashi, B.; Talank, N.; Hosseini, O.; Saravanan, M. Green synthesis, characterization, antibacterial and biofilm inhibitory activity of silver nanoparticles compared to commercial silver nanoparticles. *Inorg. Chem. Commun.* **2021**, *129*, 108647. [CrossRef]
6. Souto, E.B.; Silva, G.F.; Dias-Ferreira, J.; Zielinska, A.; Ventura, F.; Durazzo, A.; Lucarini, M.; Novellino, E.; Santini, A. Nanopharmaceutics: Part I—Clinical trials legislation and good manufacturing practices (GMP) of nanotherapeutics in the EU. *Pharmaceutics* **2020**, *12*, 146. [CrossRef]
7. Yeung, A.W.K.; Souto, E.B.; Durazzo, A.; Lucarini, M.; Novellino, E.; Tewari, D.; Wang, D.; Atanasov, A.G.; Santini, A. Big impact of nanoparticles: Analysis of the most cited nanopharmaceuticals and nanonutraceuticals research. *Curr. Res. Biotechnol.* **2020**, *2*, 53–63. [CrossRef]
8. Sanchooli, N.; Sanchooli, E.; Khandan Barani, H. Investigating the effect of water filter made using polyurethane foam containing silver nanoparticles on controlling Yersinia ruckeri in Oncorhynchus. *Iran. J. Fish. Sci.* **2022**, *21*, 500–515.
9. Liu, J.; Ma, Z.; Liu, Y.; Zheng, X.; Pei, Y.; Tang, K. Soluble soybean polysaccharide films containing in-situ generated silver nanoparticles for antibacterial food packaging applications. *Food Packag. Shelf Life* **2022**, *31*, 100800. [CrossRef]
10. Pardo, L.; Arias, J.; Molleda, P. Elaboración de nanopartículas de plata sintetizadas a partir de extracto de hojas de romero (*Rosmarinus officinalis* L.) y su uso como conservante. *LA GRANJA Rev. Cienc. Vida* **2022**, *35*, 45–58. [CrossRef]
11. Wangpraseurt, D.; You, S.; Sun, Y.; Chen, S. Biomimetic 3D living materials powered by microorganisms. *Trends Biotechnol.* **2022**. [CrossRef] [PubMed]
12. Ong, W.T.J.; Nyam, K.L. Evaluation of silver nanoparticles in cosmeceutical and potential biosafety complications. *Saudi J. Biol. Sci.* **2022**. [CrossRef] [PubMed]
13. Awad, M.A.; Eid, A.M.; Elsheikh, T.M.Y.; Al-Faifi, Z.E.; Saad, N.; Sultan, M.H.; Selim, S.; Al-Khalaf, A.A.; Fouda, A. Mycosynthesis, Characterization, and Mosquitocidal Activity of Silver Nanoparticles Fabricated by Aspergillus niger Strain. *J. Fungi* **2022**, *8*, 396. [CrossRef] [PubMed]

14. Demchenko, V.; Kobylinskyi, S.; Iurzhenko, M.; Riabov, S.; Vashchuk, A.; Rybalchenko, N.; Zahorodnia, S.; Naumenko, K.; Demchenko, O.; Adamus, G. Nanocomposites based on polylactide and silver nanoparticles and their antimicrobial and antiviral applications. *React. Funct. Polym.* **2022**, *170*, 105096. [CrossRef]
15. Iravani, S. Green synthesis of metal nanoparticles using plants. *Green Chem.* **2011**, *13*, 2638–2650. [CrossRef]
16. Amooaghaie, R.; Saeri, M.R.; Azizi, M. Synthesis, characterization and biocompatibility of silver nanoparticles synthesized from Nigella sativa leaf extract in comparison with chemical silver nanoparticles. *Ecotoxicol. Environ. Saf.* **2015**, *120*, 400–408. [CrossRef]
17. Pani, A.; Lee, J.H.; Yun, S.-I. Autoclave mediated one-pot-one-minute synthesis of AgNPs and Au–Ag nanocomposite from Melia azedarach bark extract with antimicrobial activity against food pathogens. *Chem. Cent. J.* **2016**, *10*, 15. [CrossRef]
18. Jagtap, U.B.; Bapat, V.A. Green synthesis of silver nanoparticles using Artocarpus heterophyllus Lam. seed extract and its antibacterial activity. *Ind. Crops Prod.* **2013**, *46*, 132–137. [CrossRef]
19. Huq, M.A.; Ashrafudoulla, M.; Rahman, M.M.; Balusamy, S.R.; Akter, S. Green synthesis and potential antibacterial applications of bioactive silver nanoparticles: A review. *Polymers* **2022**, *14*, 742. [CrossRef]
20. Salem, S.S.; Fouda, A. Green synthesis of metallic nanoparticles and their prospective biotechnological applications: An overview. *Biol. Trace Elem. Res.* **2021**, *199*, 344–370. [CrossRef]
21. Nagajyothi, P.; Sreekanth, T.; Lee, J.-i.; Lee, K.D. Mycosynthesis: Antibacterial, antioxidant and antiproliferative activities of silver nanoparticles synthesized from Inonotus obliquus (Chaga mushroom) extract. *J. Photochem. Photobiol. B Biol.* **2014**, *130*, 299–304. [CrossRef] [PubMed]
22. Kumar, D.A.; Palanichamy, V.; Roopan, S.M. Green synthesis of silver nanoparticles using Alternanthera dentata leaf extract at room temperature and their antimicrobial activity. *Spectrochim. Acta Part A Mol. Biomol. Spectrosc.* **2014**, *127*, 168–171. [CrossRef] [PubMed]
23. Morales-Luckie, R.A.; Sanchez-Mendieta, V.; Arenas-Alatorre, J.A.; López-Castañares, R.; Perez-Mazariego, J.L.; Marquina-Fabrega, V.; Gómez, R.W. One-step aqueous synthesis of stoichiometric Fe–Cu nanoalloy. *Mater. Lett.* **2008**, *62*, 4195–4197. [CrossRef]
24. Tolaymat, T.M.; El Badawy, A.M.; Genaidy, A.; Scheckel, K.G.; Luxton, T.P.; Suidan, M. An evidence-based environmental perspective of manufactured silver nanoparticle in syntheses and applications: A systematic review and critical appraisal of peer-reviewed scientific papers. *Sci. Total Environ.* **2010**, *408*, 999–1006. [CrossRef] [PubMed]
25. Saravanan, A.; Kumar, P.S.; Karishma, S.; Vo, D.-V.N.; Jeevanantham, S.; Yaashikaa, P.R.; George, C.S. A review on biosynthesis of metal nanoparticles and its environmental applications. *Chemosphere* **2021**, *264*, 128580. [CrossRef] [PubMed]
26. González-Pedroza, M.G.; Argueta-Figueroa, L.; García-Contreras, R.; Jiménez-Martínez, Y.; Martínez-Martínez, E.; Navarro-Marchal, S.A.; Marchal, J.A.; Morales-Luckie, R.A.; Boulaiz, H. Silver nanoparticles from Annona muricata peel and leaf extracts as a potential potent, biocompatible and low cost antitumor tool. *Nanomaterials* **2021**, *11*, 1273. [CrossRef]
27. Mystrioti, C.; Xanthopoulou, T.; Tsakiridis, P.; Papassiopi, N.; Xenidis, A. Comparative evaluation of five plant extracts and juices for nanoiron synthesis and application for hexavalent chromium reduction. *Sci. Total Environ.* **2016**, *539*, 105–113. [CrossRef]
28. Mandal, D.; Dash, S.K.; Das, B.; Chattopadhyay, S.; Ghosh, T.; Das, D.; Roy, S. Bio-fabricated silver nanoparticles preferentially targets Gram positive depending on cell surface charge. *Biomed. Pharmacother.* **2016**, *83*, 548–558. [CrossRef]
29. Kokila, T.; Ramesh, P.; Geetha, D. Biosynthesis of AgNPs using Carica Papaya peel extract and evaluation of its antioxidant and antimicrobial activities. *Ecotoxicol. Environ. Saf.* **2016**, *134*, 467–473. [CrossRef]
30. Yee, M.S.-L.; Khiew, P.-S.; Chiu, W.S.; Tan, Y.F.; Kok, Y.-Y.; Leong, C.-O. Green synthesis of graphene-silver nanocomposites and its application as a potent marine antifouling agent. *Colloids Surf. B Biointerfaces* **2016**, *148*, 392–401. [CrossRef]
31. Dhand, V.; Soumya, L.; Bharadwaj, S.; Chakra, S.; Bhatt, D.; Sreedhar, B. Green synthesis of silver nanoparticles using Coffea arabica seed extract and its antibacterial activity. *Mater. Sci. Eng. C* **2016**, *58*, 36–43. [CrossRef] [PubMed]
32. Gama-Lara, S.A.; Morales-Luckie, R.; Argueta-Figueroa, L.; Hinestroza, J.P.; García-Orozco, I.; Natividad, R. Synthesis, characterization, and catalytic activity of platinum nanoparticles on bovine-bone powder: A novel support. *J. Nanomater.* **2018**, *2018*, 6482186. [CrossRef]
33. Ferreyra Maillard, A.P.V. Estudio del Efecto Antibacteriano de bio (Nano) Materiales Contra Microorganismos Patógenos Productores de Mastitis Bovina. 2019. Available online: https://ri.conicet.gov.ar/handle/11336/145722 (accessed on 2 May 2022).
34. Fonseca, M.S.; Rodrigues, D.M.; Sokolonski, A.R.; Stanisic, D.; Tomé, L.M.; Góes-Neto, A.; Azevedo, V.; Meyer, R.; Araújo, D.B.; Tasic, L. Activity of Fusarium oxysporum-Based Silver Nanoparticles on *Candida* spp. Oral Isolates. *Nanomaterials* **2022**, *12*, 501. [CrossRef] [PubMed]
35. Radhakrishnan, V.S.; Mudiam, M.K.R.; Kumar, M.; Dwivedi, S.P.; Singh, S.P.; Prasad, T. Silver nanoparticles induced alterations in multiple cellular targets, which are critical for drug susceptibilities and pathogenicity in fungal pathogen (*Candida albicans*). *Int. J. Nanomed.* **2018**, *13*, 2647. [CrossRef] [PubMed]
36. Revati, S.; Bipin, C.; Chitra, P.B.; Minakshi, B. In vitro antibacterial activity of seven Indian spices against high level gentamicin resistant strains of enterococci. *Arch. Med. Sci. AMS* **2015**, *11*, 863. [CrossRef]
37. He, W.; Liu, Y.; Wamer, W.G.; Yin, J.-J. Electron spin resonance spectroscopy for the study of nanomaterial-mediated generation of reactive oxygen species. *J. Food Drug Anal.* **2014**, *22*, 49–63. [CrossRef]
38. Gao, L.; Xu, T.; Huang, G.; Jiang, S.; Gu, Y.; Chen, F. Oral microbiomes: More and more importance in oral cavity and whole body. *Protein Cell* **2018**, *9*, 488–500. [CrossRef]

39. Parahitiyawa, N.B.; Scully, C.; Leung, W.K.; Yam, W.C.; Jin, L.J.; Samaranayake, L.P. Exploring the oral bacterial flora: Current status and future directions. *Oral Dis.* **2010**, *16*, 136–145. [CrossRef]
40. Aas, J.A.; Paster, B.J.; Stokes, L.N.; Olsen, I.; Dewhirst, F.E. Defining the normal bacterial flora of the oral cavity. *J. Clin. Microbiol.* **2005**, *43*, 5721–5732. [CrossRef]
41. Kuramitsu, H.K.; He, X.; Lux, R.; Anderson, M.H.; Shi, W. Interspecies interactions within oral microbial communities. *Microbiol. Mol. Biol. Rev.* **2007**, *71*, 653–670. [CrossRef]
42. AlSheikh, R.; Albagieh, H.N.; Abdouh, I.; Zaki, H.; Alzahrani, A.M.; Halawany, H.S.; Al-Khalifa, K.S. In vitro activity of caffeic acid phenethyl ester against different oral microorganisms. *Appl. Sci.* **2022**, *12*, 3959. [CrossRef]
43. Harris, N.O.; Garcia-Godoy, F. *Primary Preventive Dentistry*; Pearson Education: Upper Saddle River, NJ, USA, 2004.
44. Alghamdi, S. Isolation and identification of the oral bacteria and their characterization for bacteriocin production in the oral cavity. *Saudi J. Biol. Sci.* **2022**, *29*, 318–323. [CrossRef] [PubMed]
45. García-Contreras, R.; Argueta-Figueroa, L.; Mejía-Rubalcava, C.; Jiménez-Martínez, R.; Cuevas-Guajardo, S.; Sánchez-Reyna, P.A.; Mendieta-Zeron, H. Perspectives for the use of silver nanoparticles in dental practice. *Int. Dent. J.* **2011**, *61*, 297–301. [CrossRef] [PubMed]
46. Takahashi, N.; Nyvad, B. The role of bacteria in the caries process: Ecological perspectives. *J. Dent. Res.* **2011**, *90*, 294–303. [CrossRef]
47. Uju, D.E.; Obioma, N.P. Anticariogenic potentials of clove, tobacco and bitter kola. *Asian Pac. J. Trop. Med.* **2011**, *4*, 814–818. [CrossRef]
48. Jeevika, A.; Shankaran, D.R. Seed-free synthesis of 1D silver nanowires ink using clove oil (*Syzygium aromaticum*) at room temperature. *J. Colloid Interface Sci.* **2015**, *458*, 155–159. [CrossRef]
49. Raghunandan, D.; Bedre, M.D.; Basavaraja, S.; Sawle, B.; Manjunath, S.; Venkataraman, A. Rapid biosynthesis of irregular shaped gold nanoparticles from macerated aqueous extracellular dried clove buds (*Syzygium aromaticum*) solution. *Colloids Surf. B Biointerfaces* **2010**, *79*, 235–240. [CrossRef]
50. Fayaz, A.M.; Balaji, K.; Girilal, M.; Yadav, R.; Kalaichelvan, P.T.; Venketesan, R. Biogenic synthesis of silver nanoparticles and their synergistic effect with antibiotics: A study against gram-positive and gram-negative bacteria. *Nanomed. Nanotechnol. Biol. Med.* **2010**, *6*, 103–109. [CrossRef]
51. Al-Mariri, A.; Safi, M. In vitro antibacterial activity of several plant extracts and oils against some gram-negative bacteria. *Iran. J. Med. Sci.* **2014**, *39*, 36.
52. Rana, I.S.; Rana, A.S.; Rajak, R.C. Evaluation of antifungal activity in essential oil of the *Syzygium aromaticum* (L.) by extraction, purification and analysis of its main component eugenol. *Braz. J. Microbiol.* **2011**, *42*, 1269–1277. [CrossRef]
53. Miyoshi, J.H.; Castro, J.C.; Fenelon, V.C.; Garcia, F.P.; Nakamura, C.V.; Nogueira, A.C.; Ueda-Nakamura, T.; de Souza, H.M.; Mangolim, C.S.; Moura-Costa, G.F. Essential oil characterization of *Ocimum basilicum* and *Syzygium aromaticum* free and complexed with β-cyclodextrin. Determination of its antioxidant, antimicrobial, and antitumoral activities. *J. Incl. Phenom. Macrocycl. Chem.* **2022**, *102*, 117–132. [CrossRef]
54. Cortés-Rojas, D.F.; de Souza, C.R.F.; Oliveira, W.P. Clove (*Syzygium aromaticum*): A precious spice. *Asian Pac. J. Trop. Biomed.* **2014**, *4*, 90–96. [CrossRef]
55. Bajpai, S.; Kumari, M. A green approach to prepare silver nanoparticles loaded gum acacia/poly (acrylate) hydrogels. *Int. J. Biol. Macromol.* **2015**, *80*, 177–188. [CrossRef] [PubMed]
56. Mandal, S.; DebMandal, M.; Saha, K.; Pal, N.K. In vitro antibacterial activity of three Indian spices against methicillin-resistant Staphylococcus aureus. *Oman Med. J.* **2011**, *26*, 319. [CrossRef]
57. Gupta, A.; Duhan, J.; Tewari, S.; Sangwan, P.; Yadav, A.; Singh, G.; Juneja, R.; Saini, H. Comparative evaluation of antimicrobial efficacy of *Syzygium aromaticum*, *Ocimum sanctum* and *Cinnamomum zeylanicum* plant extracts against *Enterococcus faecalis*: A preliminary study. *Int. Endod. J.* **2013**, *46*, 775–783. [CrossRef]
58. Chaudhari, A.N.; Ingale, A.G. *Syzygium aromaticum* extract mediated, rapid and facile biogenic synthesis of shape-controlled (3D) silver nanocubes. *Bioprocess Biosyst. Eng.* **2016**, *39*, 883–891. [CrossRef]
59. Morales-Luckie, R.A.; Lopezfuentes-Ruiz, A.A.; Olea-Mejía, O.F.; Liliana, A.-F.; Sanchez-Mendieta, V.; Brostow, W.; Hinestroza, J.P. Synthesis of silver nanoparticles using aqueous extracts of Heterotheca inuloides as reducing agent and natural fibers as templates: Agave lechuguilla and silk. *Mater. Sci. Eng. C* **2016**, *69*, 429–436. [CrossRef]
60. Sun, Q.; Cai, X.; Li, J.; Zheng, M.; Chen, Z.; Yu, C.-P. Green synthesis of silver nanoparticles using tea leaf extract and evaluation of their stability and antibacterial activity. *Colloids Surf. A Physicochem. Eng. Asp.* **2014**, *444*, 226–231. [CrossRef]
61. Patra, J.K.; Baek, K.-H. Biosynthesis of silver nanoparticles using aqueous extract of silky hairs of corn and investigation of its antibacterial and anticandidal synergistic activity and antioxidant potential. *IET Nanobiotechnol.* **2016**, *10*, 326–333. [CrossRef]
62. Hema, J.A.; Malaka, R.; Muthukumarasamy, N.P.; Sambandam, A.; Subramanian, S.; Sevanan, M. Green synthesis of silver nanoparticles using Zea mays and exploration of its biological applications. *IET Nanobiotechnol.* **2016**, *10*, 288–294. [CrossRef]
63. Badawy, A.M.E.; Luxton, T.P.; Silva, R.G.; Scheckel, K.G.; Suidan, M.T.; Tolaymat, T.M. Impact of environmental conditions (pH, ionic strength, and electrolyte type) on the surface charge and aggregation of silver nanoparticles suspensions. *Environ. Sci. Technol.* **2010**, *44*, 1260–1266. [CrossRef] [PubMed]
64. El Badawy, A.M.; Scheckel, K.G.; Suidan, M.; Tolaymat, T. The impact of stabilization mechanism on the aggregation kinetics of silver nanoparticles. *Sci. Total Environ.* **2012**, *429*, 325–331. [CrossRef] [PubMed]

65. Seil, J.T.; Webster, T.J. Antimicrobial applications of nanotechnology: Methods and literature. *Int. J. Nanomed.* **2012**, *7*, 2767.
66. Rai, M.; Yadav, A.; Gade, A. Silver nanoparticles as a new generation of antimicrobials. *Biotechnol. Adv.* **2009**, *27*, 76–83. [CrossRef]
67. Salleh, A.; Naomi, R.; Utami, N.D.; Mohammad, A.W.; Mahmoudi, E.; Mustafa, N.; Fauzi, M.B. The potential of silver nanoparticles for antiviral and antibacterial applications: A mechanism of action. *Nanomaterials* **2020**, *10*, 1566. [CrossRef]
68. Hemlata, P.R.M.; Singh, A.P.; Tejavath, K.K. Biosynthesis of silver nanoparticles using cucumis prophetarum aqueous leaf extract and their antibacterial and antiproliferative activity against cancer cell lines. *ACS Omega* **2020**, *5*, 5520. [CrossRef]
69. Daniel, W.; Tholen, M. *Protocols for Determination of Limits of Detection and Limits of Quantitation*; Clinical and Laboratory Standards Institute (CLSI): Wayne, PA, USA, 2004.
70. Clinical and Laboratory Standards Institute. *Performance Standards for Antimicrobial Disk Susceptibility Tests for Bacteria Isolated from Animals: CLSI Supplement VET01S*; Replaces VET01-S2; Clinical and Laboratory Standards Institute: Wayne, PA, USA, 2015.
71. Wayne, P. *Reference Method for Broth Dilution Antifungal Susceptibility Testing of Yeasts, Approved Standard*; CLSI Document M27-A2; Clinical and Laboratory Standards Institute: Wayne, PA, USA, 2002.
72. Kim, K.S.; Anthony, B.F. Importance of bacterial growth phase in determining minimal bactericidal concentrations of penicillin and methicillin. *Antimicrob. Agents Chemother.* **1981**, *19*, 1075–1077. [CrossRef]

antibiotics

MDPI

Article

Optimized Synthesis of Small and Stable Silver Nanoparticles Using Intracellular and Extracellular Components of Fungi: An Alternative for Bacterial Inhibition

Elvira Ivonne Murillo-Rábago [1], Alfredo R. Vilchis-Nestor [2], Karla Juarez-Moreno [3], Luis E. Garcia-Marin [1], Katrin Quester [4] and Ernestina Castro-Longoria [1,*]

1 Department of Microbiology, Center for Scientific Research and Higher Education of Ensenada (CICESE), Carr. Tijuana-Ensenada 3918, Zona Playitas, Ensenada 22860, Mexico; emurillo@cicese.edu.mx (E.I.M.-R.); luisgm@cicese.edu.mx (L.E.G.-M.)

2 Sustainable Chemistry Research Joint Center UAEM—UNAM (CCIQS) Carr. Toluca-Atlacomulco Km 14.5, San Cayetano, Toluca 50200, Mexico; arvilchisn@uaemex.mx

3 Center for Applied Physics and Advanced Technology, UNAM, Blvd. Juriquilla 3001, Juriquilla La Mesa, Juriquilla, Queretaro 76230, Mexico; kjuarez@fata.unam.mx

4 Center for Nanoscience and Nanotechnology, UNAM, Carr. Tijuana-Ensenada Km 107, Ensenada 22860, Mexico; quester@ens.cnyn.unam.mx

* Correspondence: ecastro@cicese.mx

Citation: Murillo-Rábago, E.I.; Vilchis-Nestor, A.R.; Juarez-Moreno, K.; Garcia-Marin, L.E.; Quester, K.; Castro-Longoria, E. Optimized Synthesis of Small and Stable Silver Nanoparticles Using Intracellular and Extracellular Components of Fungi: An Alternative for Bacterial Inhibition. *Antibiotics* **2022**, *11*, 800. https://doi.org/10.3390/antibiotics11060800

Academic Editors: Sotiris K Hadjikakou and Christina N. Banti

Received: 4 May 2022
Accepted: 4 June 2022
Published: 14 June 2022

Publisher's Note: MDPI stays neutral with regard to jurisdictional claims in published maps and institutional affiliations.

Copyright: © 2022 by the authors. Licensee MDPI, Basel, Switzerland. This article is an open access article distributed under the terms and conditions of the Creative Commons Attribution (CC BY) license (https://creativecommons.org/licenses/by/4.0/).

Abstract: Silver nanoparticles (AgNPs) represent an excellent option to solve microbial resistance problems to traditionally used antibiotics. In this work, we report optimized protocols for the production of AgNPs using extracts and supernatants of *Trichoderma harzianum* and *Ganoderma sessile*. AgNPs were characterized using UV-Vis spectroscopy and transmission electron microscopy, and the hydrodynamic diameter and Z potential were also determined. The obtained AgNPs were slightly larger using the fungal extract, and in all cases, a quasi-spherical shape was obtained. The mean sizes of AgNPs were 9.6 and 19.1 nm for *T. harzianum* and 5.4 and 8.9 nm for *G. sessile* using supernatant and extract, respectively. The AgNPs were evaluated to determine their in vitro antibacterial effect against *Escherichia coli, Pseudomonas aeruginosa* and *Staphylococcus aureus*. The minimum inhibitory concentration (MIC) was determined, and in all cases the AgNPs showed an antimicrobial effect, with a MIC varying from 1.26–5.0 µg/mL, depending on the bacterial strain and type of nanoparticle used. Cytotoxicity analyses of AgNPs were carried out using macrophages and fibroblast cell lines. It was determined that the cell viability of fibroblasts exposed for 24 h to different concentrations of AgNPs was more than 50%, even at concentrations of up to 20 µg/mL of silver. However, macrophages were more susceptible to exposure at higher concentrations of AgNPs as their viability decreased at concentrations of 10 µg/mL. The results presented here demonstrate that small AgNPs are obtained using either supernatants or extracts of both fungal strains. A remarkable result is that very low concentrations of AgNPs were necessary for bacterial inhibition. Furthermore, AgNPs were stable for more than a year, preserving their antibacterial properties. Therefore, the reported optimized protocol using fungal supernatants or extracts may be used as a fast method for synthesizing small AgNPs with high potential to use in the clinic.

Keywords: silver nanoparticles; green synthesis; *Trichoderma harzianum*; *Ganoderma sessile*; bacterial inhibition; cytotoxicity

1. Introduction

The production of nanoparticles (NPs), with a scale of 0.1 to 100 nm in at least one of its three dimensions [1], has allowed the manufacture of a variety of nanomaterials for their use in optical, magnetic, electrical, as well as in the food, cosmetic, pharmaceutical and diagnostic industries, among others [2,3]. Silver NPs (AgNPs) in particular have important biomedical applications; they have been evaluated as antimicrobial agents in dentistry and surgical materials, as well as in cancer diagnosis and therapy [2,4–7]. AgNPs are regularly produced by chemical or physical methods, which are associated with high environmental impact due to the use of toxic chemicals, high-energy consumption and the use of high temperatures during their manufacture [3]. The most common chemical method is the chemical reduction, using three main components: metal precursors, reducing agents and stabilizing/capping agents. The advantage of chemical methods is the high yield, compared to physical methods; however, the use of toxic and hazardous materials is a clear disadvantage. The most important physical methods are evaporation-condensation and laser ablation. Here, the clear advantage is that no hazardous chemicals are involved, but these methods have low yield and a high energy consumption [3].

In the past decades, the synthesis of NPs using biological materials has been extensively studied because it has been recognized as a simple and rapid method, which allows obtaining non-toxic, eco-friendly organic molecules to produce metallic NPs under optimized conditions [3,8,9]. Among the biological materials that have been reported for the synthesis of NPs, fungi are considered excellent candidates because they secrete a large number of extracellular enzymes and secondary metabolites, which serve as bio-reducing and stabilizing agents for the production of metallic NPs.

Biogenic AgNPs are potentially useful in the clinic; they have been proposed as antimicrobial agents to offer an alternative treatment to solve problems of resistance to antibiotics traditionally used against pathogenic strains [10,11]. However, for the application of biogenic NPs in the medical area, it is necessary that the biological material used for the synthesis does not come from pathogenic microorganisms, containing potentially toxic molecules for humans [10].

The biosynthesis of NPs using living microorganisms such as fungi can occur intracellular or extracellularly [11,12]. Although the exact mechanism for the synthesis of nanoparticles has not been fully described, some well-described mechanisms have been proposed to intervene in the process. Some of the suggested mechanisms are an electron transfer dependent on enzymatic processes, electron charges, reducing agents and oxidative stress, among others, where the metal ions are converted into a less toxic form in the fungal mycelium [11,13]. On the other hand, the use of extracts and/or supernatants excreted by the mycelium has also been described for the biogenic synthesis of NPs [8]. Here, the biosynthesis process is dependent on the presence of bioactive metabolites, which intervene in the synthesis and stabilization of NPs [14–16]. According to the reports, each of these compounds can influence the particular characteristics of the obtained NPs, their properties and their potential applications [15,16].

Most of the studies for the biogenic synthesis of metallic NPs using fungi report the use of extracellular components [8]. The extracellular or supernatant fraction of fungal liquid cultures contain extracellular proteins and other secreted metabolites, while the intracellular components contain the released proteins, enzymes, primary and secondary metabolites, DNA and RNA fragments, etc., after mechanically disrupting the fungal washed biomass [17]. The use of both intracellular and extracellular components has only been reported for several thermophilic fungi [17] and various basidiomycetes [18]. Differences in size and polydispersity were observed when using the intracellular and extracellular fractions of fungal liquid cultures. Bigger and polydisperse NPs are produced when using the intracellular components, compared with smaller polydisperse NPs when using the extracellular fraction [17]. In addition, shape, size and aggregation of synthesized nanoparticles depend on the species used and the type of reduction agent, i.e., extracellular and intracellular fractions [17,18].

However, one of the main challenges when using biogenic precursors is to achieve the reproducibility of the synthesis method [19], which is essential to be able to reproduce the specific characteristics and properties of the synthesized nanoparticles.

Therefore, the objective of this study is to obtain and define a reproducible protocol for the biosynthesis of small silver NPs (1–20 nm), using extracts and supernatants of fungi and compare their antibacterial and biocompatibility properties. For this, *T. harzianum* and

G. sessile were used since they are considered not harmful for humans. *Trichoderma* spp. are commonly used to stimulate plant growth and for the bio-control of phytopathogenic fungi in economically important crops [20]. *Ganoderma* spp. are wood degrading fungi, and their crude extracts are traditionally used and consumed worldwide because of their antioxidant and immune-stimulant properties [21]. Therefore, toxic or harmful effects due to the biological material used for the synthesis were prevented. After defining a reproducible protocol, the synthesized AgNPs were evaluated in in vitro assays to determine their antimicrobial effect against pathogenic bacteria. In addition, their biocompatibility in murine cell models of fibroblasts and macrophages was determined.

2. Materials and Methods

2.1. Strains and Culture Conditions

Strains of *Ganoderma sessile* and *Trichoderma harzianum* were obtained from the stock of the Microbiology Department of the Ensenada Center for Scientific Research and Higher Education (CICESE). Strains were cultured in potato dextrose agar (PDA) in Petri plates at 26 °C and stocks conserved at 4 °C.

To obtain fungal biomass in liquid cultures, the strains were cultivated in potato-dextrose broth (PDB) and incubated at 26 °C under gentle agitation in an orbital shaker (Orbit Environ Shaker) at 125 rpm for 7 days. To inoculate the liquid cultures, plugs from fresh Petri plate cultures were placed in 500 mL flasks containing 300 mL of medium. After incubation, biomass was collected and weighted.

2.2. Fungal Extracts

To obtain the intracellular components of fungal strains, the biomass of each fungus was separated from the culture broth, washed with distilled water and triturated in an agate mortar with deionized water in a ratio of 1:1 (g/mL) until obtaining a homogeneous mixture [22]. Afterward, the resulting mixture was centrifuged at 10,000 rpm for 15 min, the pellet was discarded, and the aqueous extract was filtered with a 0.22 μm nitrocellulose membrane to eliminate any biomass residues.

2.3. Fungal Supernatants

To obtain the extracellular components of fungal strains, the biomass of each fungus was washed with deionized water, weighed and placed in a flask with 100 mL of deionized water. Cultures were maintained at 26 °C for 3 days in a shaker at 125 rpm. Afterward, biomass was removed by filtration and the resulting supernatant was centrifuged and subsequently filtered with a 0.22 μm nitrocellulose membrane.

2.4. Biosynthesis of Silver Nanoparticles (AgNPs)

The optimized protocol for the synthesis of AgNPs was carried out with the supernatants and extracts of *T. harzianum* and *G. sessile*. Final conditions, according to the results previously obtained (Supplementary Materials), were as follows: Supernatant was obtained using 10 g of previously washed biomass that was incubated in 100 mL of deionized water for 3 days. Fungal extracts were obtained using the protocol reported by Quester et al. [22]. The formation of NPs was best achieved using a ratio of 1:3 v/v of (extract or supernatant)/1 mM $AgNO_3$ and incubation for 3 days at 60 °C.

2.5. UV-Vis Spectroscopy

Each synthesis was evaluated by UV-Vis spectroscopy at 300 to 700 nm after 24 h; at this time the color of the reaction changed from clear to pale brown, which is characteristic of the reaction kinetics in the synthesis of AgNPs. Curves of absorbance for the optimized protocol are reported after 72 h of incubation; at this time, no further change in absorbance was detected.

2.6. Transmission Electron Microscopy

To determine the size and shape of synthesized nanoparticles, 5 μL of the samples were placed on formvar/carbon-coated copper grids and allowed to dry. Samples were analyzed under transmission electron microscopy (TEM) (Hitachi H7500, Hitachi Ltd. Tokyo, Japan) at 100 kV. Samples from the optimized protocol were placed on Lacey carbon copper grids and analyzed under high-resolution transmission electron microscopy (HRTEM) (JEM0-2100 from JEOL, JEOL Ltd. Tokyo, Japan) at 200 kV to obtain the crystalline structure of the particles.

2.7. Dynamic Light Scattering (DLS) Characterization

For optimized synthesis, the zeta potential and the hydrodynamic diameter of the synthesized AgNPs were measured with a Zetasizer Nano ZS instrument (Malvern Panalytical Inc., Westborough, MA, USA).

2.8. FTIR Analysis of Synthesized AgNPs

To determine the functional groups from the supernatants/extracts involved in the reduction and stabilization of silver nanoparticles, the samples were placed on a glass slide and allowed to dry at room conditions and finally the Fourier transform infrared spectroscopy (FTIR) spectra were collected using a Bruker Tensor 27 FT-IR spectrometer (SpectraLab Scientific Inc., Markham, ON, Canada) with resolution of 0.5 cm^{-1} in transmission mode. For each spectrum, an average of 5 scans was recorded over the 450–4000 cm^{-1} wavelength range.

2.9. Evaluation of Antibacterial Effect

Minimum inhibitory concentration (MIC) and minimum bactericidal concentration (MBC) was evaluated with AgNPs obtained with the optimized protocol. Amount of silver was calculated for serial dilutions and approximate concentrations were used (0.31, 0.63, 1.2, 2.5, 5.1, 10.1, 20.2 and 40.4 μg/mL). The bacteria strains used in this study were *Staphylococcus aureus* (ATCC 25923), *Pseudomonas aeruginosa* (ATCC 27853) and *Escherichia coli* (ATCC 25922). To prepare the inoculum, fresh Petri plate cultures were used. Each microorganism was suspended separately in sterile saline solution (0.9% NaCl) and adjusted by UV-Vis spectroscopy to an optical density (OD) corresponding to a 0.5 McFarland standard (1.5×10^8 CFU / mL).

For the inhibition assays, the Mueller Hinton liquid culture medium was used. The bacteria were cultured in 96-well plates at 37 °C for 24 h. To evaluate qualitatively and quantitatively the degree of inhibition in bacterial growth, bacteria were exposed to serial dilutions of the produced AgNPs according to concentrations mentioned above. In each well, 50 μL of the culture medium of the corresponding strain was placed and 50 μL of the corresponding AgNP dilution was added. Samples were incubated for 24 h, and then 2.5 μL of the sample was taken from each culture and spread on Mueller Hinton agar medium and incubated at 37 °C for 24 h. As negative controls, the supernatants and extracts of each AgNP synthesis were used. These studies were performed in triplicate at different times for each strain. For qualitative evaluation, changes in coloration and turbidity in the culture media were taken into account, as well as the presence of colony growth when the cultures were spread on agar plates.

2.9.1. Cell Viability Assay

Bacteria were seeded in a 96-well plate for 24 h at 37°C in the presence of different silver concentrations (0.31, 1.26, and 10 μg/mL) or with the supernatant from *T. harzianum* (TS) or *G. sessile* (GS), in a final volume of 200 μL per well. After this, cells were washed thrice with PBS 1x and then 10 μL of MTT (3-(4,5-dimethyl-2-thiazolyl)-2,5-diphenyl-2H-tetrazolium bromide, 0.5 μg/μL) was added to each well in 90 μL LB media and incubated in darkness for 4 h at 37°C. The reduction of MTT was used to assess cell viability, as reported by [23]. Briefly, the absorbance of the samples was read in an ELISA plate reader (Thermo Fisher

Scientific, Waltham, MA, USA). The background absorbance of the cell viability test was measured at 690 nm and subtracted from the absorbance values at 570 nm. A control from cell viability was taken from cells cultivated in LB media without treatment, while positive control was carried out by exposing the cells to Ciprofloxacin at a concentration of 50 μg/mL. The absorbance of negative control cells was used to calculate cell viability from the obtained data from three experiments.

2.9.2. Reactive Oxygen Species (ROS) Quantification by Fluorimetry

Cells were seeded in a 96-well plate at a concentration of 1×10^5 cells per well and incubated for 24 h at 37 °C, with different silver concentrations (0.31, 1.26, and 10 μg/mL) and with the supernatant from *T. harzianum* (TS) or *G. sessile* (GS). After this, cells were washed thrice with 200 μL of PBS 1× and incubated in darkness with DCFDA (2′,7′-dichlorofluorescein diacetate, 45 μM) for 60 min at 37°C. Then the fluorescence was recorded with a Cary Eclipse fluorescence spectrophotometer (Agilent Technologies CA, USA) using a 485 nm excitation laser and 530 nm emission laser. Bacteria without silver concentrations were used as a negative control, while cells incubated with 1mM of H_2O_2 were considered positive for generating higher ROS levels. The results were plotted by comparing the mean ± standard deviation fluorescence of three experiments with the positive control of ROS generation (H_2O_2).

2.10. Biocompatibility Evaluation

To evaluate the biocompatibility of AgNPs obtained with optimized protocol, cell viability tests were assessed in the cell lines from L929 fibroblasts and RAW 264.7 macrophages. Cytotoxicity assays were performed using the standard MTT metabolic reduction method, based on the reduction of MTT. The cell lines were obtained from ATCC and cultured in Dulbecco's Modified Eagle's Medium (DMEM) supplemented with 10% fetal bovine serum, 1% L-glutamine, 1% antibiotic/antifungal and 1.5 g of sodium bicarbonate. Cells grown in DMEM medium without AgNPs were used as a positive control, and as negative viability control, cells were incubated with 1% TritonX-100 in PBS to induce cell death (negative viability control). For cytotoxicity assays, dilution series of biogenic AgNPs were carried out, according to the results obtained in the bacterial inhibition assays. The cytotoxicity of the precursors was also measured ($AgNO_3$ and reducing agents). For the assays, 10,000 cells were seeded per well in a 96-well plate and incubated for 24 h at 37 °C and at atmosphere of 5% CO_2. Subsequently, the different concentrations of AgNPs were added in a final volume of 100 μL per well and incubated for 24 h at 37 °C and in an atmosphere of 5% CO_2. Afterward, the medium was removed and the cells were washed three times with 200 μL of PBS. Thereafter, 10 μL of MTT (0.25 mg/mL) and 90 μL of DMEM medium were added; the plates were incubated for 4 h at 37 °C and an atmosphere of 5% CO_2. Finally, 100 μL of isopropanol were added to dissolve the formazan, a product of MTT metabolism. The absorbance of the culture plates was read at 570 and 690 nm. The resulting values were taken as an indirect measure of cell viability, taking the value of absorbance of cells grown in DMEM media as 100% of live cells. Three independent assays were conducted.

2.11. Statistical Analysis

Mean and standard deviation of nanoparticle size were calculated after measuring 1000 nanoparticles of each synthesis. Inhibition and biocompatibility experiments were carried out in triplicate and were expressed as the mean and standard deviation.

3. Results

3.1. Optimized Synthesis of AgNPs Using Fungal Extract and Supernatant of T. harzianum and G. sessile

Synthesis of AgNPs in all reactions was observed after 24 h of incubation; however, it was after three days that no further change in the reaction was detected, i.e., no further change in color or agglomeration was observed and the peaks of absorbance by UV-Vis analysis were similar (values were 438 and 442 nm for *G. sessile* and 453 and 460 nm for *T. harzianum*, for AgNPs synthesized with the supernatant and extract, respectively). In all cases, the reaction changed gradually from clear/pale yellow to dark brown, and the characteristic curve of absorbance for AgNPs was obtained. Analysis by UV-Vis spectroscopy indicated the SPR (surface plasmon resonance) absorption maximum in the range of 400–500 nm (Figures 1–4). Analysis by TEM revealed polydispersed quasi-spherical nanoparticles in all synthesis reactions (Figures 1–4). AgNPs obtained with the supernatant of *T. harzianum* (AgNPs-TS) presented an average size of 9.6 ± 4.6 nm and size range of 1 to 33 nm (Figure 1C), with a hydrodynamic diameter of 22 nm and a zeta potential of −18.5 mV (Figure 1D). AgNPs obtained with the extract of *T. harzianum* (AgNPs-TE) were slightly larger, presenting an average size of 19.1± 12.6 nm and a size range of 3 to 59 nm (Figure 2C) with a hydrodynamic diameter of 33 nm and a zeta potential of −11.8 mV (Figure 2D). Average size for AgNPs obtained with the supernatant of *G. sessile* (AgNPs-GS) was 5.4 ± 3.0 nm with size range of 1–50 nm (Figure 3C); with a hydrodynamic diameter of 24 nm and zeta potential of −23.3 mV (Figure 3D). Nanoparticles synthesized using the extract of *G. sessile* (AgNPs-GE) presented an average size of 8.9 ± 7.7 nm and size range of 1–38 nm (Figure 4C), with a hydrodynamic diameter of 47 nm and a zeta potential of −33.2 mV (Figure 4D). Characterization of AgNPs obtained with the optimized protocol, using the supernatant and extract of both fungi is summarized in (Table 1).

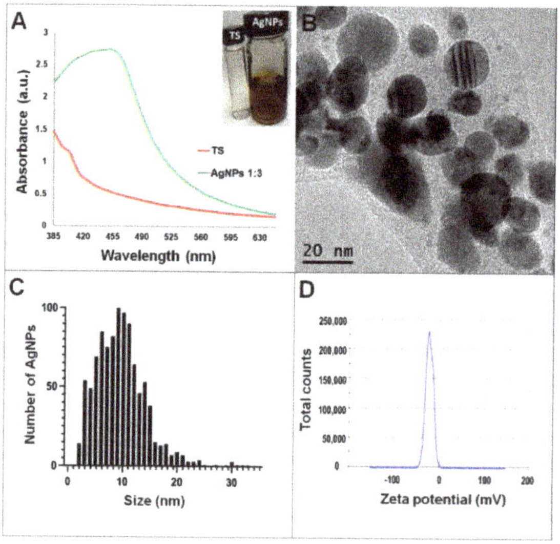

Figure 1. Silver nanoparticles obtained with the optimized protocol using the supernatant of *T. harzianum* (**A**) UV-Vis analysis of AgNPs-TS after 72 h of reaction, (**B**) TEM micrograph showing the morphology of NPs, (**C**) size distribution histogram of AgNPs, (**D**) Zeta potential. Inset in (**A**) shows the fungal supernatant and synthesized NPs. TS = *Trichoderma* supernatant.

After one year of synthesis, NPs synthesized using the optimized protocol were analyzed by UV-Vis and TEM, and in none of the cases aggregates or precipitation of the nanoparticles were observed. Furthermore, NPs were analyzed under high-resolution transmission electron microscopy confirming quasi-spherical shape. In addition, the lattice fringe spacing of AgNPs was measured confirming the planes (111) (Figure 5A,C). Additionally, the corresponding diffraction patterns of pure Ag particles with a crystalline structure were obtained (Figure 5B,D). HRTEM images (Figure 5E) and (Figure 5F) show the lattice spacing was about 0.23 nm between the (111) planes, consistent with FCC structure of Ag. Furthermore, the SAED patterns in Figure 5G,H indicated that the samples have a polycrystalline nature, with a d-spacing of 0.23 nm, which could be indexed as (111) reflection corresponding to FCC silver structure.

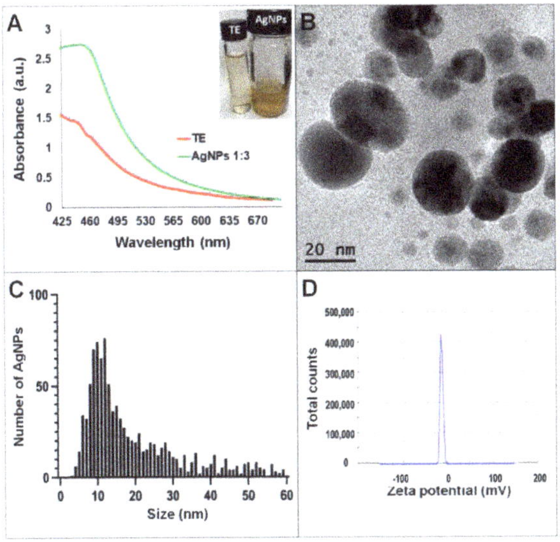

Figure 2. Silver nanoparticles obtained with the optimized protocol using the extract of *T. harzianum* (**A**) UV-Vis analysis of AgNPs-TE after 72 h of reaction, (**B**) TEM micrograph showing the morphology of NPs, (**C**) size distribution histogram of AgNPs, (**D**) Zeta potential. Inset in (**A**) shows the fungal extract and synthesized NPs. TE = *Trichoderma* extract.

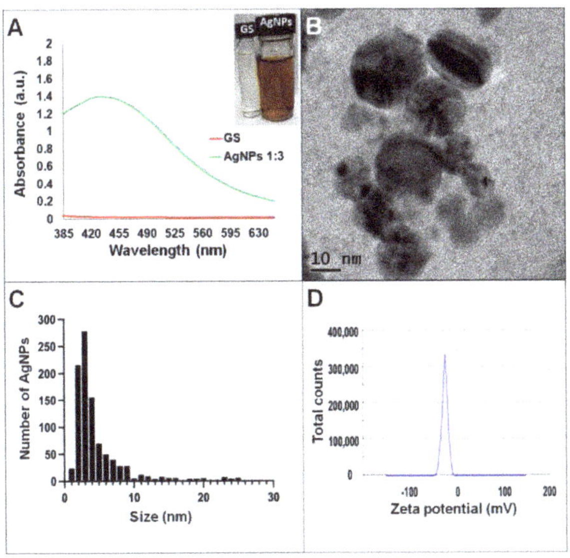

Figure 3. Silver nanoparticles obtained with the optimized protocol using the supernatant of *G. sessile*. (**A**) UV-Vis analysis of AgNPs-GS after 72 h of reaction, (**B**) TEM micrograph showing morphology of NPs, (**C**) size distribution histogram of AgNPs, (**D**) Zeta potential. Inset in (**A**) shows the fungal supernatant and synthesized NPs. GS = *Ganoderma* supernatant.

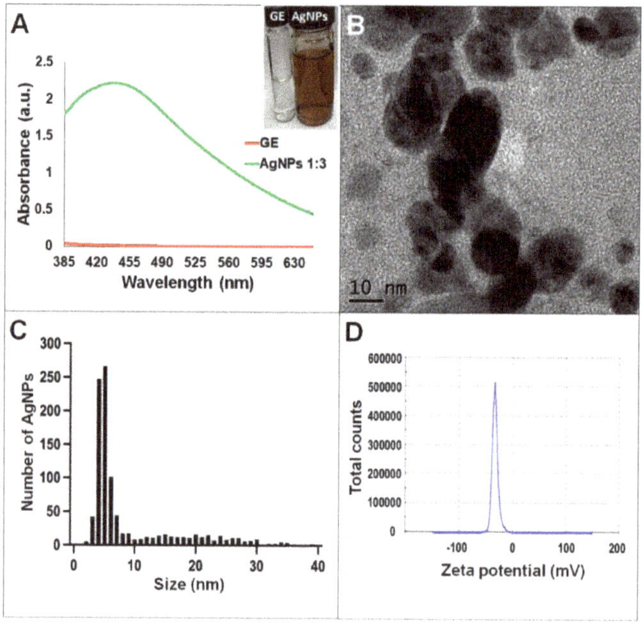

Figure 4. Silver nanoparticles obtained with the optimized protocol using the extract of *G. sessile*. (**A**) UV-Vis analysis of AgNPs-GE after 72 h of reaction, (**B**) TEM micrograph showing morphology of NPs, (**C**) size distribution histogram of AgNPs, (**D**) Zeta potential. Inset in (**A**) shows the fungal extract and synthesized NPs. GE = *Ganoderma* extract.

Table 1. Characterization of silver nanoparticles obtained with the optimized protocol, using a ratio of 1:3 *v/v* of reduction agent/$AgNO_3$ 1mM and incubation time of three days at 60 °C.

Fungus	Reduction Agent	UV-Vis Peak (nm)	Average Size (nm)	Size Range (nm)	Z Potential (mV)	Hydrodynamic Diameter (nm)
Trichoderma harzianum	Supernatant	450	9.6	1–33	−18.5	22
	Extract	451	19.1	3–59	−11.8	33
Ganoderma sessile	Supernatant	435	5.4	1–25	−23.3	24
	Extract	437	8.9	1–38	−33.2	47

3.2. FTIR Analysis of Synthesized AgNPs

Fourier transform infrared spectroscopy (FTIR) was used to elucidate the possible functional groups of the supernatants and extracts of *T. harzianum* and *G. sessile* that are involved in the reduction and stabilization of AgNPs. Samples of the produced AgNPs were analyzed with a FTIR spectrometer in the region of 4000–500 cm^{-1}. The spectrum of AgNPs using both the supernatant and extract of *T. harzianum* revealed a similar profile (Figure 6A,B). The bands around 3257 and 3244 cm^{-1} correspond to the N–H stretching vibrations, the bands at 2916 and 2914 cm^{-1} correspond to the C–H stretching vibrations. The absorption peaks at 1618 and 1620 cm^{-1} show the C=C stretching and N-H bending; the peaks at 1370 and 1386 cm^{-1} correspond to the –C–N stretching vibrations; and finally, the peaks at 1037 and 1015 cm^{-1} correspond to the C–O bending and C-N stretching.

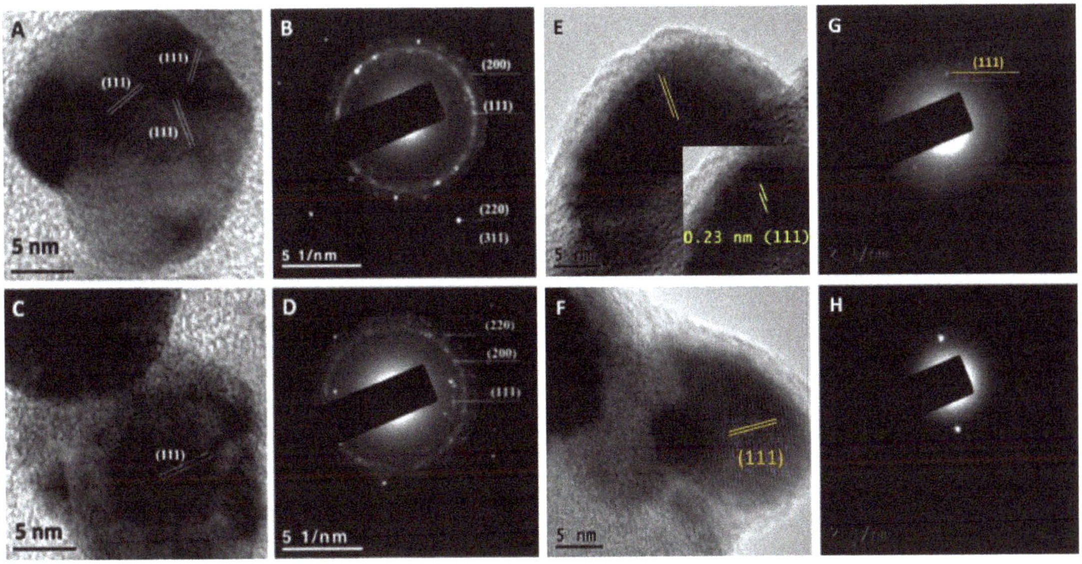

Figure 5. High-resolution transmission electron microscopy of AgNPs synthesized using the supernatant of *T. harzianum* (**A**) and *G. sessile* (**C**) with planes (111) and the corresponding diffraction pattern (**B**,**D**) indicating the crystalline structure. HRTEM image (**E**) and SAED pattern (**G**) of the Ag nanoparticles obtained with *T. harzianum* supernatant and HRTEM micrograph (**F**) and SAED pattern (**H**) of Ag synthetized with *G. sessile* supernatant, respectively.

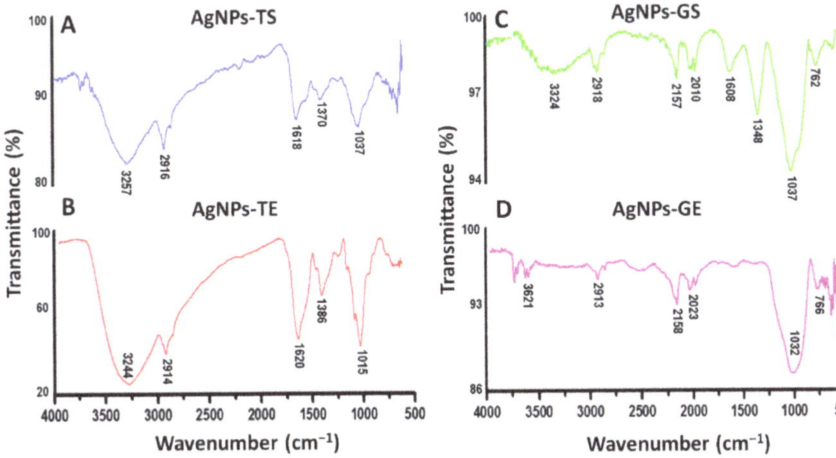

Figure 6. FTIR spectra of synthesized AgNPs. (**A**,**B**) AgNPs synthesized using the supernatant (TS) and extract (TE) of *T. harzianum*, respectively, (**C**,**D**) AgNPs synthesized using the supernatant (GS) and extract (GE) of *G. sessile*, respectively.

The spectra of AgNPs using the supernatant and extract of *G. sessile* were similar; however, in AgNPs-GE the peaks at 1608 and 1348 cm^{-1} were hardly noticeable (Figure 6D). The absorption peaks at 3324 and 3621 cm^{-1}, correspond to the O–H stretching vibrations. The absorption peaks at 2918 and 2913 cm^{-1} correspond to stretching vibrations of C–H bonds. The absorption peaks at 2157 and 2158 cm^{-1} correspond to C=C conjugated and C≡ stretch for alkynes. Bands at 2010 and 2023 cm^{-1} correspond to C≡C–H stretch,

1608 cm^{-1} indicates the presence C=C stretching and N-H bending vibrations. The peak at 1348 cm^{-1} indicates the presence of –C–N stretching, the strong absorption peaks at 1037 and 1032 cm^{-1} correspond to the C–O bending and C–N stretching, and the absorption peaks at 762 and 766 cm^{-1} correspond to the C–H (rocking) and N–H rocking vibrations.

The spectrum of the AgNPs synthesized with the extracts showed more intense peaks; this could be due to the presence of a higher concentration of biomolecules. In general, FTIR spectra of AgNPs synthesized using supernatant and extract of the same fungal species were similar, while some band assignments were not detected in both species (Table 2).

Table 2. FTIR spectrum absorption bands and the corresponding functional groups involved in the synthesis of AgNPs.

AgNPs-TS	AgNPs-TE	AgNPs-GS	AgNPs-GE	Band Assignment
3257	3244	3324	3621	N–H and O–H stretching
2916	2914	2918	2913	C–H stretching
-	-	2157	2158	C=C conjugated and C≡ stretch
-	-	2010	2023	C≡ C–H stretch
1618	1620	1608	-	C=C stretching and N–H bending
1370	1386	1348	-	–C–N stretching (Aromatic amines)
1037	1015	1037	1032	C–O bending and C–N stretching
-	-	762	766	C–H (rocking) and N–H rocking

TS = *T. harzianum* supernatant, TE = *T. harzianum* extract, GS = *G. sessile* supernatant, GE = *G. sessile* extract.

3.3. Evaluation of the Bactericidal Properties of AgNPs

All bacterial strains were inhibited at low silver concentrations using the obtained AgNPs (Table 3). Inhibition of *S. aureus* was observed from concentrations of approximately 2.5 µg/mL with AgNPs-GS, the MBC was obtained at a concentration of 10.0 µg/mL with AgNPs of both fungi. In the case of *P. aeruginosa*, inhibition was obtained with AgNPs at concentrations from 1.26 µg/mL; the MBC was 2.5 µg/mL for all AgNPs. For *E. coli*, inhibition was observed also from a concentration of 1.26 µg/mL and the MBC was 2.5–5.0 µg/mL depending on the type of particle. After 12 months of storage, inhibition of all bacteria was observed using the disk diffusion method (Figure 7A), and the CMI was determined with these NPs (Table 3). Fresh supernatants were used in the disk diffusion method, but no evident inhibition was observed (Figure 7A). Inhibition with AgNPs-TS was better against *E. coli* and *P. aeruginosa*, although inhibition by Ciprofloxacin was superior to those of AgNPs, However for *S. aureus* better inhibition was obtained with AgNPs-GS, even compared with Ciprofloxacin (Figure 8).

Table 3. Minimum inhibitory concentration (MIC) and minimum bactericidal concentration (MBC) of AgNPs synthesized with the supernatants and extracts of *T. harzianum* and *G. sessile*.

Bacterial Strains	MIC (µg/mL)				MBC (µg/mL)			
	AgNPs-TS	AgNPs-TE	AgNPs-GS	AgNPs-GE	AgNPs-TS	AgNPs-TE	AgNPs-GS	AgNPs-GE
E. coli	1.26	1.26	1.26	2.5	2.5	2.5	2.5	5.0
P. aeruginosa	1.26	1.26	1.26	1.26	2.5	2.5	2.5	2.5
S. aureus	5.0	5.0	2.5	5.0	10.0	10.0	10.0	10.0

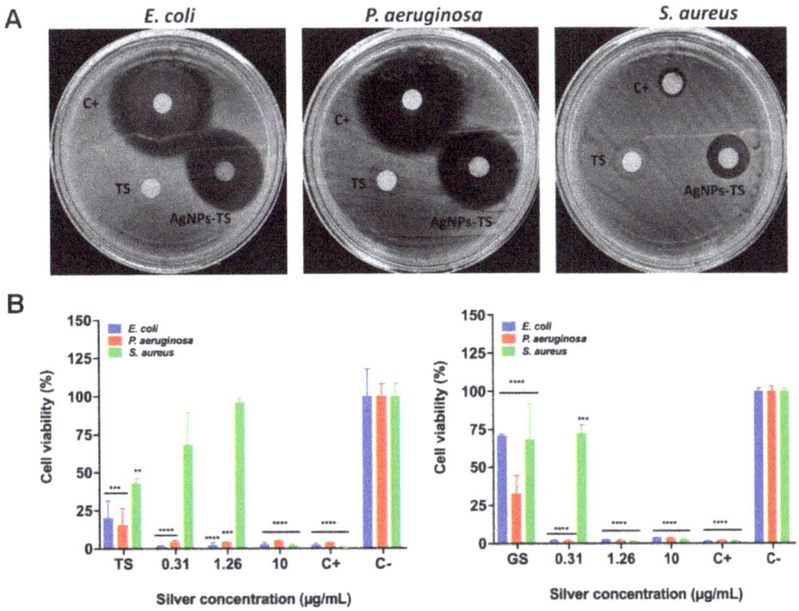

Figure 7. Representative images of bacterial inhibition strains with biosynthesized AgNPs (**A**) and cell viability of bacteria (**B**) after exposed to AgNPs for 24 h. Results are expressed as the mean ± SD (n = 3) ** $p < 0.01$; *** $p < 0.001$; **** $p > 0.0001$ using two-way ANOVA with a Dunnett's multiple comparisons tests.

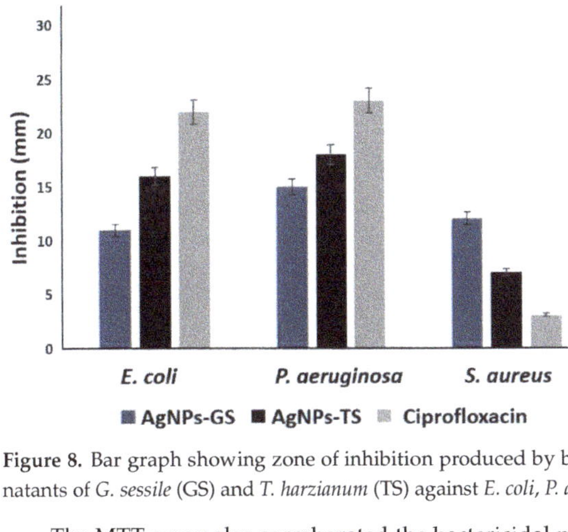

Figure 8. Bar graph showing zone of inhibition produced by biosynthesized AgNPs using supernatants of *G. sessile* (GS) and *T. harzianum* (TS) against *E. coli*, *P. aeruginosa* and *S. aureus*.

The MTT assay also corroborated the bactericidal properties of AgNPs (Figure 7B). The growth of *E. coli* and *P. aeruginosa* was inhibited with all the silver concentrations used. However, the inhibition of *S. aureus* was only decreased in the presence of 10.0 µg/mL of silver from AgNPs obtained from the extract of *T. harzianum*. In addition, the supernatant from *T. harzianum* (TS) significantly diminished the growth of the three bacteria. In contrast, all the silver concentrations from AgNPs obtained by the extract from *G. sessile* exhibited a more prominent toxic effect for *E. coli* and *P. aeruginosa*. Indeed, the growth of *S. aureus* decreased after 1.26 µg/mL of silver. However, the supernatant from *G. sessile* (GS) induced an apparent cytotoxic effect for *P. aeruginosa* rather than for *E. coli* and *S. aureus*.

The ROS production was scored to investigate whether the bactericidal effect elicited by AgNPs was due to oxidative stress. As observed in Figure 9, it is clear that overproduction of ROS correlated with the bactericidal effect. In the case of AgNPs produced with the extract from T. harzianum, all the concentrations of silver induced an abrupt overproduction of ROS in both E. coli and P. aeruginosa. However, the fact that ROS overproduction in S. aureus is less could be attributed to a growth inhibitory effect after 0.31 μg/mL of silver. In addition, the supernatant from T. harzianum (TS) did not induce any overproduction of ROS. Interestingly, all the silver concentrations used herein, from the AgNPs produced by the extract of G. sessile, also induced ROS overproduction in all the tested bacteria, especially for S. aureus; the diminishment of growth was more significant than for E. coli and P. aeruginosa. Worth mentioning is that the supernatant from G. sessile (GS) also induced overproduction of ROS in all the tested bacteria.

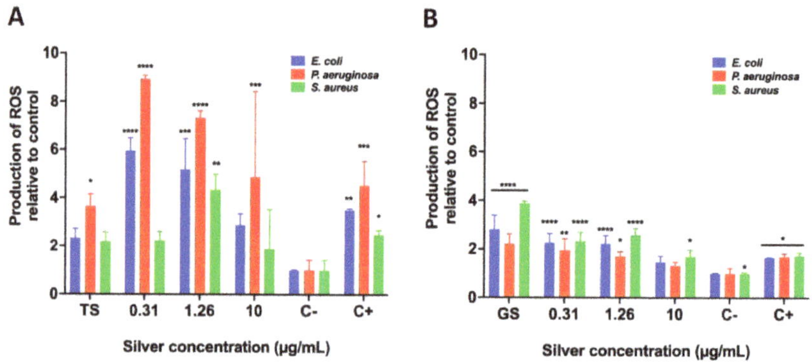

Figure 9. ROS production by biosynthesized AgNPs using supernatants of *T. harzianum* (**A**) and *G. sessile* (**B**) against *E. coli*, *P. aeruginosa* and *S. aureus*. TS = *T. harzianum* supernatant, GS = *G. sessile* supernatant. Results are expressed as the mean ± SD (n = 3) * $p < 0.05$; ** $p < 0.01$; *** $p < 0.001$, **** $p < 0.0001$ using two-way ANOVA with a Dunnett's multiple comparisons tests.

3.4. Biocompatibility Evaluation of AgNPs in Mammalian Cell Lines

For biocompatibility analysis, cell lines were exposed to AgNPs obtained from the optimized protocol using the extract and supernatant of both fungi. The concentration range evaluated was determined according to the concentration at which bacteria were inhibited, including a higher concentration of silver. Serial dilutions of AgNPs were used in all cases at the following approximate concentrations of silver: 0.0, 0.63, 1.26, 2.53, 5.06, 10.11 and 20.23 μg/mL.

In the cytotoxicity assays on fibroblasts, it was observed that cell viability was higher at the lowest concentration (0.63 μg/mL) of silver in cells exposed to AgNPs synthetized using both fungi. Results obtained at all other concentrations and types of NPs showed a degree of cytotoxic effect; however, cell viability was greater than 55% (Figure 10A).

In the biocompatibility tests carried out on macrophages, it was observed that in general, cell viability decreased in a dependent manner with the increase in the concentration of silver. Although macrophages were more sensitive at higher concentrations of silver, cell viability was greater than 80% in concentrations of up to 20.23 μg/mL with AgNPs synthesized with the supernatant of *T. harzianum*. Good viability also was observed at concentrations up to 5.06 μg/mL for AgNPs-TS and AgNPs-TE. However, with AgNPs-GS and AgNPs-GE cell viability decreased drastically at concentrations of 5 μg/mL and higher (Figure 10B).

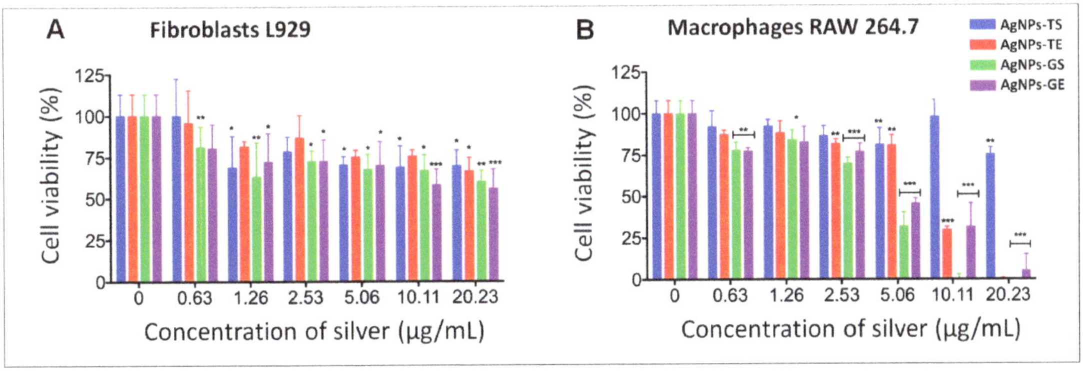

Figure 10. Cell viability assay in fibroblasts (**A**) and macrophages (**B**) exposed to AgNPs synthesized using the supernatant and extract of *T. harzianum* and *G. sessile*. The bars represent the mean and standard deviations of experiments performed in triplicate. * $p \leq 0.05$, ** $p \leq 0.01$, *** $p \leq 0.001$.

4. Discussion

It is well-documented that silver nanoparticles (AgNPs) can be produced using extracts of numerous biological materials, searching for eco-friendly protocols. Furthermore, by using extracts of biological material to synthesize AgNPs, the potential use in medical treatments is increased since the risk of adverse reactions to harmful chemicals is avoided. Silver nanoparticles possess excellent antimicrobial properties; hence, their use in difficult-to-treat cutaneous infections could be a feasible option if the reducing agent used for the synthesis does not provoke unfavorable effects. Among the biological materials used to synthesize nanoparticles, fungi are reported as excellent reducing agents for the production of metallic nanoparticles [8]. However, few studies have reported the stability of the produced nanoparticles and the biocompatibility with animal cells. Therefore, in this work we report an optimized and highly reproducible protocol for the production of small stable AgNPs using *Trichoderma harzianum* and *Ganoderma sessile*.

The use of *T. harzianum* for the synthesis of AgNPs has been reported previously [24–31]; however, in all cases, only the extracellular supernatant has been used as a reducing agent. Using extracellular supernatants of fungi is widely applied in the synthesis of metallic nanoparticles due to the straightforward protocol; thus, it is the most common fungal reducing agent [8]. To our best knowledge, there are only two reports using the extracellular and intracellular fractions for the synthesis of metallic NPs. For instance, Molnár et al. [17] used the extracellular and intracellular components and the autolyzates of thermophilic filamentous fungi to synthesize gold nanoparticles. Although the protocols used by [17] have significant differences compared with the ones used in this work and the synthesized NPs were gold, not silver, in general terms the results are comparable; they found smaller, less polydispersed NPs using the supernatants, compared with the intracellular fractions. In this work, AgNPs synthesized using the intracellular extracts of *T. harzianum* were bigger than those obtained with the supernatants (average size of 19 and 9 nm, respectively). In addition, in most reports the supernatants used for the synthesis of nanoparticles contain traces of the culture media [17]. They demonstrated that those remnants contribute to the formation of NPs. In this work, the biomass was thoroughly washed before incubation in deionized water to obtain the supernatant, and the same was done before obtaining the extracts; therefore, only fungal compounds were responsible for the formation and stabilization of NPs. In the study of [18], the use of intracellular and extracellular components of *Lentinus edodes*, *Ganoderma lucidum*, *Pleurotus ostreatus* and *Grifola frondosa* was reported for the synthesis of silver and gold NPs. They found large and aggregated AgNPs using the extracellular fraction of all basidiomycetes; however, the culture conditions and synthesis protocol differ largely from the one used in this work.

Synthesis of metallic NPs using *G. sessile* has not been previously reported; therefore, the optimized protocol used for *T. harzianum* was applied. A similar protocol was used to synthesize NPs using *G. lucidum* by [32], who reported AgNPs of 10.72 nm using the supernatant of fruiting bodies. The authors described as optimum conditions the use of 1 mM $AgNO_3$ and a reaction temperature of 85 °C. However, in the case of *G. sessile* AgNPs were approximately 50% smaller, compared to those obtained with *T. harzianum*. Differences in size can be due to the nature and/or number of excreted metabolites. In fact, differences in the production of AgNPs were detected even when using fungi of the same genera, as reported by Devi et al. [33], who explored the synthesis of AgNPs using the same protocol for different species of *Trichoderma*. Although the size range for each species was not reported, they found differences in the synthesis reaction; the highest UV-Vis absorption band was found for *T. virens* followed by *T. longibranchiatum*, *T. asperellum*, *T. pseudokoningii* and *T. harzianum*. The authors explored the role of reductases in the biosynthesis of AgNPs by the nitrate reductase assay. However, they did not find a direct relationship with the nitrate reductase activity and the production of AgNPs between species.

The analysis by DLS revealed a bigger size of nanoparticles in all cases, compared with the size obtained by TEM. This was expected since DLS measures the hydrodynamic diameter, which includes the diameter of the nanoparticle plus the layer of molecule(s) attached or absorbed on their surface, whereas the analysis by TEM provides the projected surface area of dry particles [34]. Thus, the size of particles obtained by DLS is substantially increased by the capping agents, which are bound to biological synthesized NPs [30,35]. Zeta potential (ZP) measurements were negative for all obtained AgNPs; values were similar to previously reported [30,36,37]. Considering the commonly used guidelines [34], NPs are relatively stable and moderately stable for *T. harzianum* and *G. sessile*, respectively. Values of ± 30 mV are considered as highly stable; however, there are stable colloids with low ZP and vice versa, since stability depends not only on electrostatic repulsive forces but also on the van der Waals attractive forces [34]. ZP values for *G. sessile* were higher compared with those of *T. harzianum;* this could be due to the nature of the reducing agent, since ZP is influenced by several parameters such as pH, ionic strength and concentration [34]. Nevertheless, NPs did not display agglomeration even after one year of storage at room temperature.

FTIR spectroscopy provides information about the organic functional groups attached to AgNPs, which can be responsible for their synthesis and/or stabilization. The band assignments of the functional groups involved in AgNPs synthesis were resolved according to [38,39]. The spectrum of AgNPs-TS (Figure 6A) was found to be comparable to the spectrum reported for *T. Harzianum* [25,30]. The presence of alcohol, phenols, carbonyl, amines (both aromatic and aliphatic) and amide functional groups is observed in both spectra AgNPs-TS (Figure 6A,B) and the report of Ahluwalia et al. [30].

In the case of FTIR analysis for AgNPs synthesized using *G. sessile*, no previous studies report the production of metallic NPs or the analysis of the metabolites produced by this fungus. Recently, *G. sessile* was reported as a fast polysaccharide producer, achieving the highest biomass and polysaccharide yields, both in liquid culture and solid-state fermentation, compared with *Ganoderma lingzhi* and *Ganoderma oregonense* [40]. Thus, this fungus has high potential for the manufacture of dietary supplements [40] and for its use in the production of metallic NPs. Previous FTIR analyses on metallic NPs synthesized with other *Ganoderma* spp. confirm the presence of capping biomolecules that stabilize and affect the morphology of the nanostructures [36,37,40–47]. Strong bands in the region of 3650–1000 cm^{-1} were reported for AgNPs synthesized using *G. lucidum* [45], with the –OH and C–OH stretching vibrations of polysaccharides represented by the appearance of a strong absorption band at 1043 cm^{-1}. The metabolites responsible for the reduction of silver ions were polyphenols, polysaccharides, triterpenoids and sterols [45]. For AgNPs synthesized with *G. sessile*, the functional groups detected could be associated with proteins, aromatic amines and polysaccharides, which could be acting as reducing or/and capping agents.

As mentioned earlier, silver nanoparticles possess antimicrobial activity, and those synthesized using supernatants of *Trichoderma* spp. have been reported as possessing excellent antibacterial properties [26,30,48]. In this work, AgNPs synthesized using intracellular and extracellular components had similar antibacterial properties; in general, bacterial growth was significantly reduced in a dose-dependent manner. Usually, very low concentrations were necessary for bacterial inhibition; depending on the bacterial strain and the type of AgNPs, the necessary concentration for inhibition (MIC) was 1.26–5.0 μg/mL. Similarly, it was reported that AgNPs synthesized with the supernatant of *T. harzianum* were effective against *Staphylococcus aureus* and *Klebsiella aeruginosa*. In addition, it was found that bacterial growth was reduced in a dose-dependent manner using concentrations of 3–15 μg/mL [30]. AgNPs synthesized with intracellular components of *T. harzianum* have not been evaluated against human pathogens; however, they were evaluated against *Clavibacter michiganensis* subsp. *michiganensis*, which is the causative pathogen of tomato canker disease. The synthesized AgNPs were used for antimicrobial assays using the disk diffusion method, and the inhibitory effect improved with increased concentration of silver [26].

Studies reporting AgNPs synthesized using *Ganoderma* spp. have documented inhibitory properties against pathogenic bacteria. However, most studies for the synthesis of AgNPs using *Ganoderma* spp. have been carried out with *G. lucidum* [32,42,43,45–47]. Strong bactericidal activity against *S. aureus*, *E. coli*, *Staphylococcus* mutants, *Klebsilla pneumoniae* and *P. aeruginosa* was reported using AgNPs (5–30 nm) synthesized with the supernatant of *G. lucidum* [42]. Recently, other *Ganoderma* spp. have started to be investigated for their use in AgNPs synthesis and their potential applications as antimicrobial agents. For instance, AgNPs synthesized using *G. applanatum* (20–25 nm) exhibited high antioxidant capacity and in vitro antibacterial activity against *S. aureus* and *E. coli* [41]. AgNPs obtained using *G. sessiliforme* extract were effective against common food-borne bacteria, such as *E. coli*, *Bacillus subtilis*, *Streptococcus faecalis*, *Listeria innocua* and *Micrococcus luteus*, and it was concluded that the synthesized AgNPs could be used to control the growth of food-borne pathogens and thus have potential application in the food packaging industry [44]. Most studies with AgNPs synthesized using *Ganoderma* spp. have been carried out using the pulverized fruiting bodies to obtain the extract; however, *G. lucidum* mycelia has also been used for the synthesis of AgNPs, and strong bactericidal activity against *S. aureus* and *E. coli* was reported for AgNPs in the range of 10–70 nm [43]. Furthermore, the best inhibition of *E. coli*, *S. aureus* and *P. aeruginosa* was found with AgNPs (11.38 ± 5.51 nm) biosynthesized from *G. lucidum* extracts, compared with AgNPs synthesized by chemical methods [45]. In this work, AgNPs synthesized using *G. sessile* displayed strong bacterial inhibition at very low silver concentrations, and inhibition increased with NPs synthesized with the supernatant fraction. Furthermore, inhibition of *S. aureus* was higher with *G. sessile* AgNPs, and therefore we can assume that AgNPs are more effective to kill Gram-positive bacteria.

Although the exact mechanism of action of AgNPs on bacteria is not completely described, there is significant information on the interaction of AgNPs with bacteria and their effects. In general, the antibacterial activity of AgNPs apparently is closely related with the continuous release of Ag+; small AgNPs attach to the bacterial cell surface and release high concentrations of Ag+, causing physical modifications in the membrane [48]. AgNPs together with the continuous release of Ag+ inactivate several cell functions, causing: (a) Disruption of the cell wall and membrane, (b) Denaturation of ribosomes, inhibiting protein synthesis, (c) Interruption of ATP production, (d) Membrane disruption by ROS, (e) DNA damage, (f) Denaturation of membrane, and (g) Perforation of membrane [49]. In fact, disruption of the cell membrane, resulting in big gaps or holes in the cell surface and the consequent leakage of cellular materials, has been demonstrated [50,51]. ROS production by metallic NPs correlates with particle size, shape, surface area and chemistry [52]. High concentrations of ROS (including superoxide anion radicals, hydroxyl radicals and hydrogen peroxide) in bacterial cells can result in oxidative stress. It is reported that bacterial cells exposed to AgNPs causes overproduction of ROS attacking membrane lipids, leading to a breakdown of the membrane function [52]. ROS formation has been previously documented in bacteria when exposed to fungal synthesized AgNPs [53–55]. ROS production completely inhibited bacterial growth provoking membrane disruption, causing leakage of cell components and finally cell death [54,55]. In this work we also detected overproduction of ROS correlated with the bactericidal effect in bacteria exposed to different AgNPs concentrations, corroborating previous reports [53–55]. Cell viability assays also corroborated that very low concentrations of the biosynthesized AgNPs are necessary to inhibit bacteria and that AgNPs-GS have great potential for killing both, Gram-negative and Gram-positive bacteria.

Regarding human cell viability, we found that the pure extracts and supernatants of *T. harzianum* and *G. sessile* did not affect the viability of fibroblasts or macrophages (Figure 10). Therefore, cytotoxic effects on cell lines were due only to AgNPs, which were found to be dose-dependent. In the case of AgNPs obtained from the extract of *T. harzianum*, fibroblasts were more sensitive, presenting 80% of viability at a concentration of 2.53 μg/mL, and at higher concentrations, cell viability was up to 70 % (Figure 10A). AgNPs obtained from *T. harzianum* supernatant induced good cell viability at the higher concentrations tested (Figure 10B). Similarly, it was reported that AgNPs using the supernatant of *T. harzianum* presented cytotoxic and genotoxic effects depending on the cell line used and the exposure concentration [56]. In addition, cell lines respond differently when exposed to silver nanoparticles, some being more sensitive than others [57].

AgNPs synthesized using *G. sessile* had more cytotoxic effects in both cell lines than those synthesized using *T. harzianum*. In this case, macrophages were more susceptible than fibroblasts at the higher concentrations assayed (Figure 10B). The results obtained could be due to the smaller sizes obtained with the extracts and supernatants of *G. sessile*. In general, smaller AgNPs have been reported with more cytotoxic effects, as documented by Yen et al. [58], who reported that AgNPs of 3 nm showed more significant cytotoxicity compared to those of 25 nm at a concentration of 10.0 μg/mL in macrophage cell lines (J774 A1). AgNPs synthesized with the supernatant of *G. sessiliforme* were evaluated in fibroblast cells (L-929), and 80% of cell viability was found at high concentrations of silver (100 μg/mL); however, AgNPs had an average size of 45 nm [44].

An important characteristic of the obtained AgNPs is their stability; however, few studies document this parameter. In the case of *T. harzianum*, AgNPs were reported to be stable for 3 months [30]. In this study, all synthesized AgNPs were stable for more than 1 year, which is advantageous for increasing their shelf life.

5. Conclusions

A high reproducible and optimized protocol for the synthesis of small AgNPs using extracellular (supernatants) and intracellular (extracts) fungal components is reported. The results presented here suggest a harmless use of the extracts and supernatants of *T. harzianum* and *G. sessile* to synthesize nanoparticles, since no cytotoxic effects were detected in fibroblasts and macrophages. The resulting AgNPs were smaller using the supernatants of both fungi, and in general, NPs synthesized with *G. sessile* were smaller than those synthesized with *T. harzianum* (sizes were 9.6 and 19.1 nm for *T. harzianum* and 5.4 and 8.9 nm for *G. sessile* using the supernatant and extract, respectively). An important result is the low concentration of silver that was necessary for bacterial inhibition, with an MIC varying from 1.26–5.0 μg/mL and MBC from 2.5–10.0 μg/mL. Results indicate that AgNPs synthesized with *G. sessile* supernatant (AgNPs-GS) are effective against Gram-positive bacteria, killing *S. aureus* at a concentration of 10.0 μg/mL. In addition, the obtained AgNPs were stable for more than 1 year at ambient temperature, increasing their shelf life. Therefore, the potential application of AgNPs as topical antimicrobials can be further evaluated since the synthesized AgNPs preserve their antibacterial activity after a long time of storage.

Supplementary Materials: The following supporting information can be downloaded at: https://www.mdpi.com/article/10.3390/antibiotics11060800/s1.

Author Contributions: E.I.M.-R. contributed to the study design. E.C.-L. contributed with conceptualization and study design. A.R.V.-N., K.J.-M., L.E.G.-M. and K.Q. contributed to the implementation and supervision of experimental protocols. All authors have read and agreed to the published version of the manuscript.

Funding: This research received no external funding.

Acknowledgments: We thank CONACyT for a scholarship for E.I. Murillo-Rábago.

Conflicts of Interest: The authors declare no conflict of interest.

References

1. Whatmore, R.W. Nanotechnology—what is it? Should we be worried? *Occup. Med.* **2006**, *56*, 295–299. [CrossRef]
2. Wei, L.; Lu, J.; Xu, H.; Patel, A.; Chen, Z.S.; Chen, G. Silver nanoparticles: Synthesis, properties, and therapeutic applications. *Drug Discov. Today* **2015**, *20*, 595–601. [CrossRef]
3. Zhang, X.-F.; Liu, Z.-G.; Shen, W.; Gurunathan, S. Silver Nanoparticles: Synthesis, Characterization, Properties, Applications, and Therapeutic Approaches. *Int. J. Mol. Sci.* **2016**, *17*, 1534. [CrossRef]
4. Martinez-Andrade, J.M.; Avalos-Borja, M.; Vilchis-Nestor, A.R.; Sanchez-Vargas, L.O.; Castro-Longoria, E. Dual function of EDTA with silver nanoparticles for root canal treatment–A novel modification. *PLoS ONE* **2018**, *13*, 1–19. [CrossRef]
5. Qayyum, S.; Oves, M.; Khan, A.U. Obliteration of bacterial growth and biofilm through ROS generation by facilely synthesized green silver nanoparticles. *PLoS ONE* **2017**, *12*, 1–18. [CrossRef]
6. Zhang, X.-F.; Shen, W.; Gurunathan, S. Silver Nanoparticle-Mediated Cellular Responses in Various Cell Lines: An in Vitro Model. *Int. J. Mol. Sci.* **2016**, *17*, 1603. [CrossRef]
7. Huy, T.Q.; Huyen, P.; Le, A.T.; Tonezzer, M. Recent advances of silver nanoparticles in cancer diagnosis and treatment. *Anti-Cancer Agents Med. Chem. (Former. Curr. Med. Chem.-Anti-Cancer Agents)* **2020**, *20*, 1276–1287. [CrossRef]
8. Castro-Longoria, E. Fungal Biosynthesis of Nanoparticles, a Cleaner Alternative. In *Fungal Applications in Sustainable Environmental Biotechnology*; Springer: Berlin/Heidelberg, Germany, 2016; pp. 323–351. [CrossRef]
9. Vilchis-Nestor, A.R.; Sánchez-Mendieta, V.; Camacho-López, M.A.; Gómez-Espinosa, R.M.; Camacho-López, M.A.; Arenas-Alatorre, J.A. Solventless synthesis and optical properties of Au and Ag nanoparticles using *Camellia sinensis* extract. *Mater. Lett.* **2008**, *62*, 3103–3105. [CrossRef]
10. Castro-Longoria, E.; Garibo-Ruiz, D.; Martínez-Castro, S. Myconanotechnology to Treat Infectious Diseases: A Perspective. In *Fungal Nanotechnology*; Springer: Berlin/Heidelberg, Germany, 2017; pp. 235–261.
11. Khandel, P.; Kumar, S. Mycogenic nanoparticles and their bio-prospective applications: Current status and future challenges. *J. Nanostructure Chem.* **2018**, *8*, 369–391. [CrossRef]
12. Castro-Longoria, E.; Vilchis-Nestor, A.R.; Avalos-Borja, M. Biosynthesis of silver, gold and bimetallic nanoparticles using the filamentous fungus *Neurospora crassa*. *Colloids Surf. B Biointerfaces* **2011**, *83*, 42–48. [CrossRef]
13. Bahrulolum, H.; Nooraei, S.; Javanshir, N.; Tarrahimofrad, H.; Mirbagheri, V.S. Green synthesis of metal nanoparticles using microorganisms and their application in the agrifood sector. *J. Nanobiotechnol.* **2021**, *19*, 86. [CrossRef]
14. Ovais, M.; Khalil, A.T.; Ayaz, M.; Ahmad, I.; Nethi, S.K.; Mukherjee, S. Biosynthesis of Metal Nanoparticles via Microbial Enzymes: A Mechanistic Approach. *Int. J. Mol. Sci.* **2018**, *19*, 1–20. [CrossRef]
15. Shah, M.; Fawcett, D.; Sharma, S.; Tripathy, S.K.; Poinern, G.E.J. Green Synthesis of Metallic Nanoparticles via Biological Entities. *Materials* **2015**, *8*, 7278–7308. [CrossRef]
16. Vetchinkina, E.P.; Loshchinina, E.A.; Vodolazov, I.R.; Kursky, V.F.; Dykman, L.A.; Nikitina, V.E. Biosynthesis of nanoparticles of metals and metalloids by basidiomycetes. Preparation of gold nanoparticles by using purified fungal phenol oxidases. *Appl. Microbiol. Biotechnol.* **2017**, *101*, 1047–1062. [CrossRef]
17. Molnár, Z.; Bódai, V.; Szakacs, G.; Erdélyi, B.; Fogarassy, Z.; Sáfrán, G.; Varga, T.; Kónya, Z.; Tóth-Szeles, E.; Szucs, R.; et al. Green synthesis of gold nanoparticles by thermophilic filamentous fungi. *Sci. Rep.* **2018**, *8*, 1–12. [CrossRef]
18. Vetchinkina, E.; Loshchinina, E.; Kupryashina, M.; Burov, A.; Pylaev, T.; Nikitina, V. Green synthesis of nanoparticles with extracellular and intracellular extracts of basidiomycetes. *PeerJ* **2018**, *6*, e5237. [CrossRef]
19. Loshchinina, E.A.; Vetchinkina, E.P.; Kupryashina, M.A.; Kursky, V.F.; Nikitina, V.E. Nanoparticles synthesis by *Agaricus* soil basidiomycetes. *J. Biosci. Bioeng.* **2018**, *126*, 44–52. [CrossRef]
20. Saba, H.; Vibhash, D.; Manisha, M.; Prashant, K.S.; Farhan, H.; Tauseef, A. *Trichoderma*–a promising plant growth stimulator and biocontrol agent. *Mycosphere* **2012**, *3*, 524–531. [CrossRef]
21. Bishop, K.S.; Kao, C.H.; Xu, Y.; Glucina, M.P.; Paterson, R.R.M.; Ferguson, L.R. From 2000 years of *Ganoderma lucidum* to recent developments in nutraceuticals. *Phytochemistry* **2015**, *114*, 56–65. [CrossRef]
22. Quester, K.; Avalos-Borja, M.; Castro-Longoria, E. Controllable Biosynthesis of Small Silver Nanoparticles Using Fungal Extract. *J. Biomater. Nanobiotechnol.* **2016**, *07*, 118–125. [CrossRef]

23. Mosmann, T. Rapid colorimetric assay for cellular growth and survival: Application to proliferation and cytotoxicity assays. *J. Immunol. Methods* **1983**, *65*, 55–63. [CrossRef]
24. El-Waseif, A.A.-M.; Ibrahim, N.A.; Ahmed, S.A.E.-A.; Abdallah, N.A. Bioactivity Enhancement with Microbial Silver Nanoparticles Produced and Characterized from *Streptomyces* and *Trichoderma*. *Egypt. Soc. Exp. Biol.* **2017**, *13*, 439–445. [CrossRef]
25. El-Moslamy, S.H.; Elkady, M.F.; Rezk, A.H.; Abdel-Fattah, Y.R. Applying Taguchi design and large-scale strategy for mycosynthesis of nano-silver from endophytic *Trichoderma harzianum* SYA. F4 and its application against phytopathogens. *Sci. Rep.* **2017**, *7*, 1–22. [CrossRef]
26. Noshad, A.; Iqbal, M.; Folkers, L.; Hetherington, C.; Khan, A.; Numan, M.; Ullah, S. Antibacterial Effect of Silver Nanoparticles (AgNPs) Synthesized from *Trichoderma Harzianum* against *Clavibacter Michiganensis*. *J. Nano Res.* **2019**, *58*, 10–19. [CrossRef]
27. Consolo, V.F.; Torres-Nicolini, A.; Alvarez, V.A. Mycosinthesized Ag, CuO and ZnO nanoparticles from a promising *Trichoderma harzianum* strain and their antifungal potential against important phytopathogens. *Sci. Rep.* **2020**, *10*, 1–9. [CrossRef]
28. Singh, P.; Raja, R.B. Biological synthesis and characterization of silver nanoparticles using the fungus *Trichoderma harzianum*. *Asian J. Exp. Biol. Sci.* **2011**, *2*, 600–605.
29. Gherbawy, Y.A.; Shalaby, I.M.; El-sadek, M.S.A.; Elhariry, H.M.; Banaja, A.A. The anti-fasciolasis properties of silver nanoparticles produced by *Trichoderma harzianum* and their improvement of the anti-fasciolasis drug triclabendazole. *Int. J. Mol. Sci.* **2013**, *14*, 21887–21898. [CrossRef]
30. Ahluwalia, V.; Kumar, J.; Sisodia, R.; Shakil, N.A.; Walia, S. Green synthesis of silver nanoparticles by *Trichoderma harzianum* and their bio-efficacy evaluation against *Staphylococcus aureus* and *Klebsiella pneumonia*. *Ind. Crops Prod.* **2014**, *55*, 202–206. [CrossRef]
31. Konappa, N.; Udayashankar, A.C.; Dhamodaran, N.; Krishnamurthy, S.; Jagannath, S.; Uzma, F.; Jogaiah, S. Ameliorated antibacterial and antioxidant properties by *Trichoderma harzianum* mediated green synthesis of silver nanoparticles. *Biomolecules* **2021**, *11*, 535. [CrossRef]
32. Nguyen, V.P.; Le Trung, H.; Nguyen, T.H.; Hoang, D.; Tran, T.H. Synthesis of biogenic silver nanoparticles with eco-friendly processes using *Ganoderma lucidum* extract and evaluation of their theranostic applications. *J. Nanomater.* **2021**, *2021*, 6135920. [CrossRef]
33. Devi, T.P.; Kulanthaivel, S.; Kamil, D.; Borah, J.L.; Prabhakaran, N.; Srinivasa, N. Biosynthesis of silver nanoparticles from *Trichoderma* species. *Indian J. Exp. Biol.* **2013**, *51*, 543–547. [PubMed]
34. Bhattacharjee, S. DLS and zeta potential–what they are and what they are not? *J. Control. Release* **2016**, *235*, 337–351. [CrossRef] [PubMed]
35. Mukherjee, P.; Roy, M.; Mandal, B.P.; Dey, G.K.; Mukherjee, P.K.; Ghatak, J.; Tyagi, A.K.; Kale, S.P. Green synthesis of highly stabilized nanocrystalline silverparticles by a non-pathogenic and agriculturally important fungus *T. asperellum*. *Nanotechnology* **2008**, *19*, 75103. [CrossRef] [PubMed]
36. Dandapat, S.; Kumar, M.; Ranjan, R.; Sinha, M.P. Acute and sub-acute toxicity of *Ganoderma applanatum* (pres.) pat. extract mediated silver nanoparticles on rat. *Not. Sci. Biol.* **2019**, *11*, 351–363. [CrossRef]
37. Mohanta, Y.K.; Singdevsachan, S.K.; Parida, U.K.; Panda, S.K.; Mohanta, T.K.; Bae, H. Green synthesis and antimicrobial activity of silver nanoparticles using wild medicinal mushroom *Ganoderma applanatum* (Pers.) Pat. from Similipal Biosphere Reserve, Odisha, India. *IET Nanobiotechnol.* **2016**, *10*, 184–189. [CrossRef]
38. Socrates, G. *Infrared and Raman Characteristic Group Frequencies: Tables and Charts*; John Wiley & Sons: Hoboken, NJ, USA, 2004.
39. Sangeetha, B.; Krishnamoorthy, A.S.; Amirtham, D.; Sharmila, D.J.S.; Renukadevi, P.; Malathi, V.G. FT-IR Spectroscopic Characteristics of *Ganoderma lucidum* Secondary Metabolites. *J. Appl. Sci. Technol.* **2019**, *38*, 1–8. [CrossRef]
40. Viceconte, F.R.; Diaz, M.L.; Soresi, D.S.; Lencinas, I.B.; Carrera, A.; Prat, M.I.; Gurovic, M.S.V. *Ganoderma sessile* is a fast polysaccharide producer among *Ganoderma* species. *Mycologia* **2021**, *113*, 513–524. [CrossRef]
41. Jogaiah, S.; Kurjogi, M.; Abdelrahman, M.; Hanumanthappa, N.; Tran, L.S.P. *Ganoderma applanatum*-mediated green synthesis of silver nanoparticles: Structural characterization, and in vitro and in vivo biomedical and agrochemical properties. *Arab. J. Chem.* **2019**, *12*, 1108–1120. [CrossRef]
42. Kannan, M.; Muthusamy, P.; Venkatachalam, U.; Rajarajeswaran, J. Mycosynthesis, characterization and antibacterial activity of silver nanoparticles (Ag-NPs) from fungus *Ganoderma lucidum*. *Malaya J. Biosci.* **2014**, *1*, 134–142.
43. Karwa, A.S.; Gaikwad, S.; Rai, M.K. Mycosynthesis of silver nanoparticles using Lingzhi or Reishi medicinal mushroom, *Ganoderma lucidum* (W. Curt.: Fr.) P. Karst. and their role as antimicrobials and antibiotic activity enhancers. *Int. J. Med. Mushrooms* **2011**, *13*, 483–491. [CrossRef]
44. Mohanta, Y.K.; Nayak, D.; Biswas, K.; Singdevsachan, S.K.; Abd Allah, E.F.; Hashem, A.; Mohanta, T.K. Silver nanoparticles synthesized using wild mushroom show potential antimicrobial activities against food borne pathogens. *Molecules* **2018**, *23*, 655. [CrossRef] [PubMed]
45. Do Dat, T.; Viet, N.D.; Dat, N.M.; My, P.L.T.; Thinh, D.B.; Thy, L.T.M.; Hieu, N.H. Characterization and bioactivities of silver nanoparticles green synthesized from Vietnamese *Ganoderma lucidum*. *Surf. Interfaces* **2021**, *27*, 101453. [CrossRef]
46. Aygün, A.; Özdemir, S.; Gülcan, M.; Cellat, K.; Şen, F. Synthesis and characterization of Reishi mushroom-mediated green synthesis of silver nanoparticles for the biochemical applications. *J. Pharm. Biomed. Anal.* **2020**, *178*, 112970. [CrossRef] [PubMed]
47. Al-Ansari, M.M.; Dhasarathan, P.; Ranjitsingh, A.J.A.; Al-Humaid, L.A. *Ganoderma lucidum* inspired silver nanoparticles and its biomedical applications with special reference to drug resistant *Escherichia coli* isolates from CAUTI. *Saudi J. Biol. Sci.* **2020**, *27*, 2993–3002. [CrossRef]

48. Khalandi, B.; Asadi, N.; Milani, M.; Davaran, S.; Abadi, A.J.; Abasi, E.; Akbarzadeh, A. A review on potential role of silver nanoparticles and possible mechanisms of their actions on bacteria. *Drug Res.* **2017**, *11*, 70–76. [CrossRef]
49. Yin, I.X.; Zhang, J.; Zhao, I.S.; Mei, M.L.; Li, Q.; Chu, C.H. The Antibacterial Mechanism of Silver Nanoparticles and Its Application in Dentistry. *Int. J. Nanomed.* **2020**, *15*, 2555–2562. [CrossRef]
50. Li, W.R.; Xie, X.B.; Shi, Q.S.; Zeng, H.Y.; Ou-Yang, Y.S.; Chen, Y.B. Antibacterial activity and mechanism of silver nanoparticles on *Escherichia coli*. *Appl. Microbiol. Biotechnol.* **2010**, *85*, 1115–1122. [CrossRef]
51. Li, W.R.; Sun, T.L.; Zhou, S.L.; Ma, Y.K.; Shi, Q.S.; Xie, X.B.; Huang, X.M. A comparative analysis of antibacterial activity, dynamics, and effects of silver ions and silver nanoparticles against four bacterial strains. *Int. Biodeterior. Biodegrad.* **2017**, *123*, 304–310. [CrossRef]
52. Abdal Dayem, A.; Hossain, M.K.; Lee, S.B.; Kim, K.; Saha, S.K.; Yang, G.M.; Cho, S.G. The role of reactive oxygen species (ROS) in the biological activities of metallic nanoparticles. *Int. J. Mol. Sci.* **2017**, *18*, 120. [CrossRef]
53. Quinteros, M.A.; Aristizábal, V.C.; Dalmasso, P.R.; Paraje, M.G.; Páez, P.L. Oxidative stress generation of silver nanoparticles in three bacterial genera and its relationship with the antimicrobial activity. *Toxicol. Vitr.* **2016**, *36*, 216–223. [CrossRef]
54. Ninganagouda, S.; Rathod, V.; Singh, D.; Hiremath, J.; Singh, A.K.; Mathew, J. Growth kinetics and mechanistic action of reactive oxygen species released by silver nanoparticles from *Aspergillus niger* on *Escherichia coli*. *BioMed Res. Int.* **2014**, *2014*, 753419. [CrossRef] [PubMed]
55. Ramalingam, B.; Parandhaman, T.; Das, S.K. Antibacterial effects of biosynthesized silver nanoparticles on surface ultrastructure and nanomechanical properties of gram-negative bacteria viz. *Escherichia coli and Pseudomonas aeruginosa*. *ACS Appl. Mater. Interfaces* **2016**, *8*, 4963–4976. [CrossRef] [PubMed]
56. Guilger, M.; Pasquoto-Stigliani, T.; Bilesky-Jose, N.; Grillo, R.; Abhilash, P.C.; Fraceto, L.F.; Lima, R.D. Biogenic silver nanoparticles based on *Trichoderma harzianum*: Synthesis, characterization, toxicity evaluation and biological activity. *Sci. Rep.* **2017**, *7*, 1–13. [CrossRef] [PubMed]
57. Mukherjee, S.G.; O'claonadh, N.; Casey, A.; Chambers, G. Comparative in vitro cytotoxicity study of silver nanoparticle on two mammalian cell lines. *Toxicol. Vitr.* **2012**, *26*, 238–251. [CrossRef]
58. Yen, H.; Hsu, S.; Tsai, C. Cytotoxicity and Immunological Response of Gold and Silver Nanoparticles of Different Sizes. *Small* **2009**, *5*, 1553–1561. [CrossRef]

antibiotics

MDPI

Article

Comparison and Advanced Antimicrobial Strategies of Silver and Copper Nanodrug-Loaded Glass Ionomer Cement against Dental Caries Microbes

Amal Adnan Ashour [1], Mohammed Fareed Felemban [2], Nayef H. Felemban [3], Enas T. Enan [4], Sakeenabi Basha [5], Mohamed M. Hassan [6] and Sanaa M. F. Gad El-Rab [7,*]

1 Department of Oral and Maxillofacial Surgery and Diagnostic Sciences, Oral Pathology Division, Faculty of Dentistry, Taif University, Taif 26571, Saudi Arabia; a.a.ashour@tu.edu.sa
2 Department of Oral and Maxillofacial Surgery and Diagnostic Sciences, Periodontics Division, Faculty of Dentistry, Taif University, Taif 26571, Saudi Arabia; mfelemban@tudent.edu.sa
3 Preventive Dentistry Department, Faculty of Dentistry, Taif University, Taif 26571, Saudi Arabia; nfelemban@tudent.edu.sa
4 Department of Dental Biomaterials, Faculty of Dentistry, Mansoura University, Mansoura 35511, Egypt; enasenan275@mans.edu.eg
5 Department of Preventive and Community Dentistry, Faculty of Dentistry, Taif University, Taif 26571, Saudi Arabia; sakeena@tudent.edu.sa
6 Department of Biology, College of Science, Taif University, P.O. Box 11099, Taif 21944, Saudi Arabia; m.khyate@tu.edu.sa
7 Department of Botany and Microbiology, Faculty of Science, Assiut University, Assiut 71516, Egypt
* Correspondence: sanaafahmy@aun.edu.eg; Tel.: +20-102-547-5454

Citation: Ashour, A.A.; Felemban, M.F.; Felemban, N.H.; Enan, E.T.; Basha, S.; Hassan, M.M.; Gad El-Rab, S.M.F. Comparison and Advanced Antimicrobial Strategies of Silver and Copper Nanodrug-Loaded Glass Ionomer Cement against Dental Caries Microbes. *Antibiotics* **2022**, *11*, 756. https://doi.org/10.3390/antibiotics11060756

Academic Editors: Nicholas Dixon, Anthony William Coleman, Christina N. Banti and Sotiris K Hadjikakou

Received: 8 April 2022
Accepted: 30 May 2022
Published: 2 June 2022

Publisher's Note: MDPI stays neutral with regard to jurisdictional claims in published maps and institutional affiliations.

Copyright: © 2022 by the authors. Licensee MDPI, Basel, Switzerland. This article is an open access article distributed under the terms and conditions of the Creative Commons Attribution (CC BY) license (https://creativecommons.org/licenses/by/4.0/).

Abstract: Caries lesions during cement repairs are a severe issue, and developing a unique antimicrobial restorative biomaterial can help to reduce necrotic lesion recurrence. As a result, *Thymus vulgaris* extract was used to biosynthesize copper nanoparticles (TVE-CuNPs) exhibiting different characteristics (TVE). Along with TVE-CuNPs, commercial silver nanoparticles (AgNPs) and metronidazole were combined with glass ionomer cement (GIC) to test its antibacterial efficacy and compressive strength. FTIR, XRD, UV-Vis spectrophotometry, and TEM were applied to characterize the TVE-CuNPs. Additionally, AgNPs and TVE-CuNPs were also combined with metronidazole and GIC. The modified GIC samples were divided into six groups, where groups 1 and 2 included conventional GIC and GIC with 1.5% metromidazole, respectively; group 3 had GIC with 0.5% TVE-CuNPs, while group 4 had 0.5% TVE-CuNPs with metronidazole in 1.5%; group 5 had GIC with 0.5% AgNPs, and group 6 had 0.5% AgNPs with metronidazole at 1.5%. An antimicrobial test was performed against *Staphylococcus aureus* (*S. aureus*) and *Streptococcus mutans* (*S. mutans*) by the disc diffusion method and the modified direct contact test (MDCT). GIC groups 4 and 6 demonstrated a greater antimicrobial efficiency against the two tested strains than the other groups. In GIC groups 4 and 6, the combination of GIC with two antimicrobial agents, 1.5% metronidazole and 0.5% TVE-CuNPs or AgNPs, enhanced the antimicrobial efficiency when compared to that of the other groups with or without a single agent. GIC group specimens combined with nanosilver and nanocopper had similar mean compressive strengths when compared to the other GIC groups. Finally, the better antimicrobial efficacy of GIC boosted by metronidazole and the tested nanoparticles against the tested strains may be relevant for the future creation of more efficient and modified restorations to reduce dental caries lesions.

Keywords: antimicrobial efficiency; AgNPs; caries lesions; GIC; metronidazole; TVE-CuNPs

1. Introduction

Dental caries, one of the famous common chronic oral diseases in humans worldwide [1,2], is related to a wide range of Gram-positive and Gram-negative microorganisms. According to the contemporary caries etiology hypothesis, an imbalance of oral flora can cause acid accumulation and tooth demineralization, resulting in caries formation. According to the World Health Organization, dental caries affects roughly three-quarters of the world's population [3]. Caries treatment is still dependent on filling repair at the moment. Therefore, a hotspot in the field of caries suppression and treatment is the evolution of novel anti-caries materials. Metals such as silver, gold, copper, zinc, titanium, and others have been used as antimicrobial agents for centuries to solve this problem [4]. Each metal has unique properties and activity spectra. Because of their delivery capabilities, biocidal, and anti-adhesive properties, silver and copper nanoparticles have received the most attention in recent years [5]. In comparison to silver nanoparticles, few studies have reported copper's antimicrobial properties against oral microbes, and copper nanoparticles bind to SH groups and disrupt bacterial nucleic acids and key enzymes. The copper nanoparticles have a high thermal and electrical conductivity and are less expensive than silver nanoparticles [6]. The metal nanoparticles can be synthesized by biological methods using plants. Typically, *Thymus vulgaris* leaf oil or extract is used to treat sore throats, tonsillitis, gum disease, rheumatism, and arthritis [7–9] and can also be used in the biosynthesis of nanoparticles. AgNPs kill bacteria by a variety of mechanisms, including the rupture of the bacterial cell membrane via AgNP adherence and the penetration of AgNPs into the cell and nucleus, resulting in binding interactions with proteins and DNA and, ultimately, cell death [10]. Combinations of AgNPs with antibiotics have been reported to have synergistic antibacterial effects toward both nonresistant and resistant strains [11]. According to Xu et al. [12], nanoparticles such as those of silver or copper have good mechanical properties, allowing them to be used in dental restorative cements. Glass ionic cement (GIC) is a popular restorative material in dentistry due to its biocompatibility, antimicrobial action, adhesion to dental structures, and fluoride release [13]. The GIC surface erodes because of biofilm formation [14]. To solve this problem, restorative materials have been mixed with antibiotics in previous studies because they have antimicrobial properties [15]. Nonetheless, due to the emergence of multidrug resistant pathogens, traditional antimicrobials are becoming ineffective in treating oral infections. Metronidazole is one of these antibiotics that has antimicrobial activity and inhibits DNA synthesis in anaerobic microbes [16]. Metronidazole resistance has been described in dental abscesses and anaerobic streptococcus illness [17].

To the best of the authors' knowledge, the effect of antibiotic (metronidazole), AgNPs, and biosynthesized CuNPs in novel GIC biomaterial formulations against oral microbes has never been studied. This study uses *Thymus vulgaris*, which is used in dental applications, to produce TVE-CuNPs. The principal aim of this research is to investigate the antimicrobial efficacy of GIC after being treated with metronidatole, commercial AgNPs, and TVE-CuNPs to bio-fabricate a novel anti-microbial dental biomaterial. The researchers tested modified GIC as a novel anti-microbial dental filling biomaterial against caries-causing oral microbes. This research is significant because improving the antibacterial efficiency and compressive strength of GIC dental cements will result in modified GIC cements with superior properties that can increase the longevity of the restorations and reduce recurrent caries.

2. Materials and Methods

2.1. Experimental Design

This study used GIC. GIC was mixed with three antimicrobial agents: metronidazole, AgNPs, and TVE-CuNPs. Commercially available AgNPs powder of size 20 to 50 nm (Alibaba Company, Shanghai Xinglu Chemical Technology Co., LTD, Shanghai, China) was purchased. The antimicrobial efficiency of the modified dental cement on two common microbial strains, *S. mutans* and *S. aureus*, was determined using Kirby–Bauer agar diffusion and modified direct contact tests (Scheme 1).

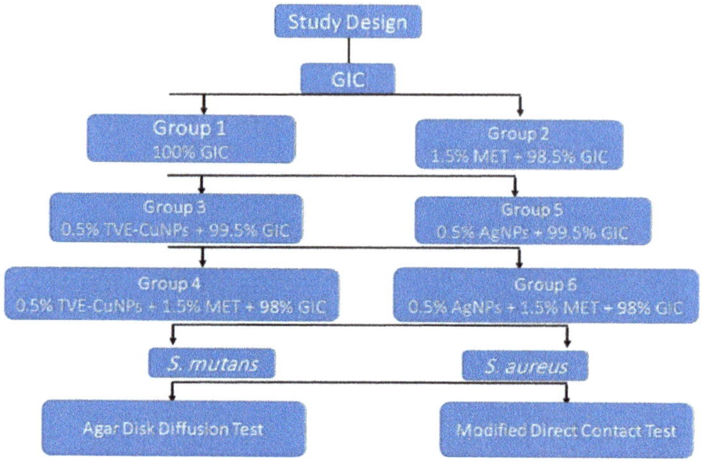

Scheme 1. Outline of the study design of conventional GIC and modified GIC against *S. mutans* and *S. aureus*. MET = Metronidazole.

2.2. Materials and Equipment

The GIC (GC Fuji IX, Tokyo, Japan) was used in this study, as well as three antimicrobial agents: commercial AgNPs (Sigma-Aldrich, Saint Louis, MO, USA), TVE-CuNPs, and metronidazole (VETRANAL®, analytical standard, Sigma-Aldrich, Saint Louis, MO, USA). Thymus vulgaris L extract was used to biosynthesize TVE-CuNPs. Before manipulation, the AgNPs, TVE-CuNPs, and metronidazole were weighed and mixed with GIC powder.

2.3. Biosynthesis of Copper Nanoparticles

Thymus vulgaris (25 g) was added to 100 mL of distilled water. The solution was heated at 80 °C for one hour to prepare an aqueous extract, as shown in our previous study by Gad El-Rab et al. [18]. In brief, 40 mL of copper chloride (2 mM) was mixed with 10 mL of Thymus vulgaris extract (TVE) and stirred for more than 2 h at room temperature with constant magnetic stirring (200 rpm). The solution mixture's color changed from deep brown to yellowish brown during the reaction. It is suggested that TVE-CuNPs were formed.

2.3.1. TVE-CuNP Characterization

The characterization of TVE-CuNPs was performed using the following methods.

UV-Visible Spectrum

This analysis was conducted using a UV–Vis spectrometer (Shimadzu UV-1650, Tokyo, Japan) in the range of 300–800 nm to confirm the manufacture of TVE-CuNPs [15]. After that, the final mixture was centrifuged at 10,000× *g* to separate the TVE-CuNPs. Finally, brown-black TVE-CuNPs were recovered.

Transmission Electron Microscopy (TEM)

TEM analysis was carried out on TEM JEOL at 100 kV, Tokyo, Japan (Assiut Electron Microscope Unit). Each sample for TEM analysis was prepared by placing a drop of the suspension on carbon-coated copper grids and allowing it to dry on the grid for 4 min. The shape and size of TVE-CuNPs were determined by a TEM micrograph [16].

X-ray Diffraction

A thin layer of well-grinded dry TVE-CuNPs was distributed on the glass slide that was introduced into the XRD chamber to study the XRD pattern. The phase of TVE-CuNPs was characterized using Shimadzu XRD (3A, Tokyo, Japan), and the spectra were recorded by CuKα radiation with a wavelength of 1.5406 Å in the 2θ (from the range of 20–80°) [15].

Fourier Transform Infrared Analysis (FT-IR)

To determine the functional groups that are responsible for TVE-CuNP formation and stabilization [19], as KBr pellets, the FTIR spectra of TVE and TVE-CuNPs were recorded on a Shimadzu IR-470 Spectrometer, Tokyo, Japan, in the range of 4000–500 cm^{-1}.

2.4. Preparation and Characterization of AgNPs

The AgNPs with diameters of 20–50 nm were obtained from Alibaba Company, Shanghai Xinglu Chemical Technology Co., LTD, Shanghai, China. The samples were re-suspended in deionized water at the concentration of 1 mg/mL. The UV–Vis spectra of the AgNPs were recorded using a UV–Vis spectrometer (Shimadzu UV-1650, Tokyo, Japan). The dimensions and sizes of the AgNPs were confirmed using a transmission electron microscope TEM JEOL at 100 kV (Assiut Electron Microscope Unit).

2.5. Preparation of Specimens

A split Teflon mold (3 mm in height and 6 mm in diameter) was used to create disc-shaped specimens of GIC cement (180 disc) [18]. All GIC specimens were prepared at room temperature. The GIC powder and liquid were apportioned and combined using a plastic spatula for 30 s to avoid dehydration of the GICs, as indicated by the manufacturer, before being inserted into the molds. The molds were covered on both sides with polyester strips and thick glass plates and let to set for 20 min before being taken from the mold and sanitized for 30 min with UV light. The modified GIC samples (n = 180) were divided into six groups of 30 specimens each, as follows:

1. (Group 1) GIC.
2. (Group 2) GIC and 1.5% metronidazole.
3. (Group 3) 99.5% GIC with 0.5% TVE-CuNPs.
4. (Group 4) 98% GIC with 0.5% TVE-CuNPs and 1.5% metronidazole.
5. (Group 5) 99.5% GIC with 0.5% AgNPs.
6. (Group 6) 98% GIC with 0.5% AgNPs and 1.5% metronidazole.

All groups were tested for antibacterial efficacy against *S. aureus* and *S. mutans* (at 1 day, 1 week, and 1 month).

2.6. Drug Release Determination

To determine metronidazole release from modified GIC, the samples were submerged in 5 mL of phosphate-buffered saline (PBS) at pH 7.4 and incubated at 37 °C for up to 30 days. A UV-Vis spectrophotometer (Shimadzu UV-1650 pc spectrophotometer, Tokyo, Japan) was used to quantify the concentration of metronidazole in the release medium at 340 nm using a calibration curve. The metronidazole release (%) was calculated using the equation: $C (\%) = B/A \times 100$, where C is the metronidazole release (%), B is the total quantity of metronidazole released in the solution, and A is the amount of loaded metronidazole in modified GIC samples.

2.7. Bacteria and Growth Conditions

Strains of *S. aureus* and *S. mutans* were obtained from our previous study by Enan et al. [18]. Before performing the antimicrobial test, fresh inoculums of *S. aureus* and *S. mutans* were inoculated in Mannitol Salt Agar medium and Brain Heart Infusion (BHI) medium (Oxoid Ltd., London, UK), respectively, and adjusted to the famous standard (i.e., 1.5×10^8 CFU/mL) with the help of BHI and Mannitol Salt medium, respectively, and the culture was incubated in growing conditions (18 h at 37 ± 2 °C).

2.7.1. Amicrobial Efficacy Using Agar Disc Diffusion Assay

The disc diffusion assay was followed to determine the inhibition zones of the test modified GIC [19]. Mueller–Hinton agar was prepared and added to the Petri plates upon autoclaving and kept for setting or solidifying for 10–15 min. The targeted test for *S. aureus* and *S. mutans* of 0.1 mL was added to media plates and spread with sterile cotton swabs evenly throughout the plates. The test modified GIC discs were placed on the plates, which were kept in an incubator at 37 °C for growth. The inhibition zones were recorded with the zone interpretation scale. All experiments were run in triplets. The antibacterial efficiency of modified GIC discs was evaluated at three time intervals: 24 h, 1 week, and 1 month.

2.7.2. Modified Direct Contact Test

The modified direct contact test (MDCT) [20] is based on determining the colony-forming unit (CFU) of bacterial growth in 96-well microtiter plates. A microtiter plate was held horizontally, and the floor of the wells was evenly coated with a thin layer of GIC. The side walls of 96-well microtiter plates were coated with a thin layer of GIC. In accordance with the manufacturer's recommendation, the modified luting cement was allowed to set. A total of 50 µL of *S. aureus* or *S. mutans* (approximately 10^6) were placed on the GIC with 100 µL of Mueller–Hinton broth or brain–heart infusion broth, respectively, for *S. aureus* or *S. mutans*. A positive control, as prepared by placing 50 µL of bacterial suspension along with 100 µL of Mueller–Hinton broth or brain–heart infusion broth in a separate well without the GIC cement, was considered. After 24 h, the *S. aureus* or *S. mutans* suspension was introduced in Mueller–Hinton agar or brain–heart infusion agar plates for *S. aureus* or *S. mutans*, respectively, and the CFUs of the suspension were compared. The data were recorded approximately 24 h after incubation. Additional tests were performed on a set of modified GIC that had been aged for one month.

2.8. *Compressive Strength Measurement*

GIC samples (4 mm in diameter and 6 mm in height) were produced for the compressive strength test (CS) using a Teflon mold. For 20 min, the GIC samples were left at room temperature. After 24 h of mixing, the CS of GIC cement was measured using a Material Test System (810 MTS Co., Minneapolis, MN, USA) at a crosshead speed of $0.5\ \text{mm/min}^{-1}$. Six samples were tested for each GIC sample group. The highest recorded force was measured at the fracture, and CS (N/mm^2) was calculated using the equation shown below [18].

$$CS = 4P/\pi d^2 \tag{1}$$

where P denotes the failure load and d denotes the sample's diameter.

2.9. *Statistical Analysis*

The mean difference was calculated using One-way Statistical Analysis of Variance (ANOVA) and Tukey's Post Hoc test in the Statistical Package for Social Sciences version 17 software (SPSS Inc., Chicago, IL, USA). All statistical tests were two-sided, with a significance level of $p < 0.05$.

3. Results

3.1. *Biosynthesis of TVE-CuNPs*

TVE-CuNPs were biosynthesized in the current study by reducing copper chloride to TVE-CuNPs using TVE. A change in color from blue to brown indicates the green-mediated synthesis of TVE-CuNPs in the reaction mixture. The formation of brown color in an aqueous solution of TVE-CuNPs was caused by the excitation of surface plasmon resonance (SPR) [21].

3.2. Characterization of TVE-CuNPs

3.2.1. UV-Vis Spectroscopy of TVE-CuNPs

After the formation of TVE-CuNPs, UV–Vis absorption was measured using a UV–visible spectrophotometer between 485 and 800 nm. Peaks at 573 nm were observed for pure TVE-CuNPs (Figure 1). Cu-NPs, in particular, generated noticeable absorption in the visible area in the range of 573–600 nm as a result of SPR.

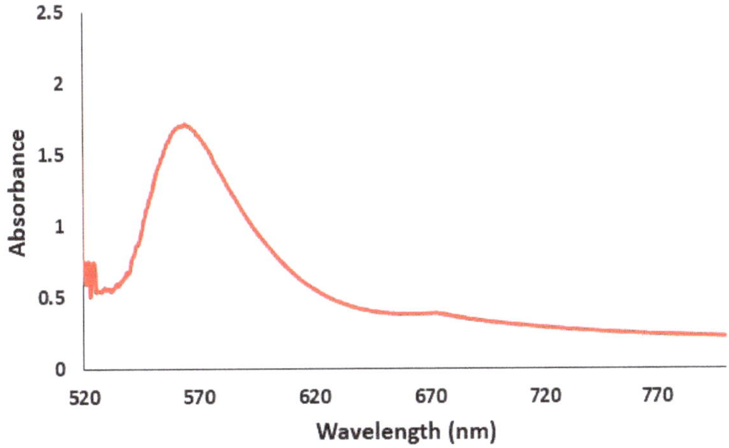

Figure 1. Biosynthesis of TVE-CuNPs using *Thymus vulgaris* extract.

3.2.2. Size and Shape of TVE-CuNPs

TEM was used to investigate the TVE-CuNPs (Figure 2). TEM analysis determined the size of the synthesized TVE-CuNPs to be 10–25 nm. The particle size distribution of the TVE-CuNP picture further revealed that the TVE-AgNP size distribution ranged from 10 to 25 nm. TVE-CuNPs were spherical and monodispersed (Figure 2).

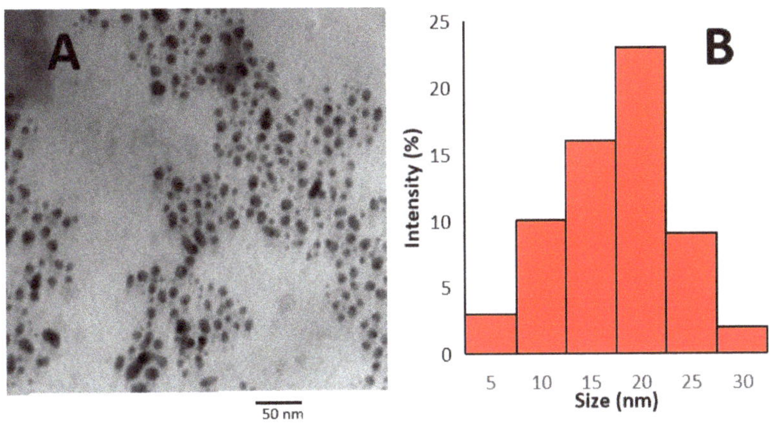

Figure 2. TEM images of the synthesized TVE-CuNPs (**A**) and percentage of the particle size distribution of TVE-CuNP (**B**).

3.2.3. X-ray Diffraction (XRD)

The XRD pattern explained the sharp and distinct peaks of (2θ) 46.3°, 53.5°, and 72.3° that correspond to the planes (111), (200), and (220) of the pure copper's face-centered cubic crystalline (FCC) structure (Figure 3). The purity of the TVE-CuNPs was reflected in the XRD results of this study.

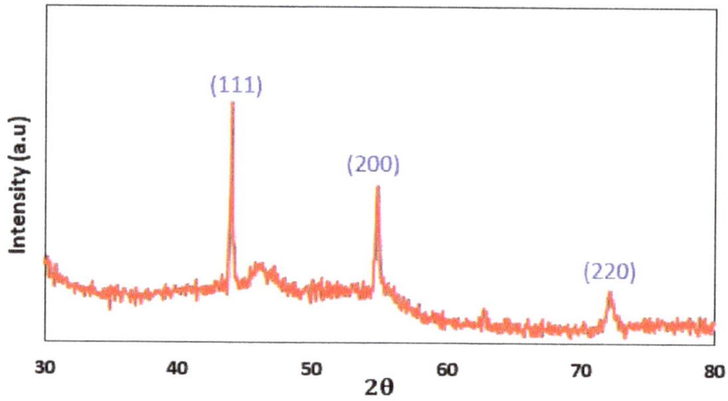

Figure 3. XRD patterns of the synthesized TVE-CuNPs.

3.2.4. Comparison of FTIR Spectra of *Thymus vulgaris* Extract and TVE-CuNPs

FTIR analysis was used to characterize the TVE and TVE-CuNPs. As shown in Figure 4 and Table 1, the broad band at 3402 cm^{-1} is assigned to the O-H stretch of the polyphenol groups of *Thymus vulgaris* L extract. The peaks that appeared at 2934 cm^{-1} can be assigned to C–H in methyl and alkanes. The fewer absorption peaks are observed at regions of 1630 and 1421 cm^{-1}, which belong to the C=O stretching of amide in the carbonyl stretch in proteins and enzymes. These proteins can be combined with metal nanoparticles via the carboxylate of amino acid remnants or free amine groups.

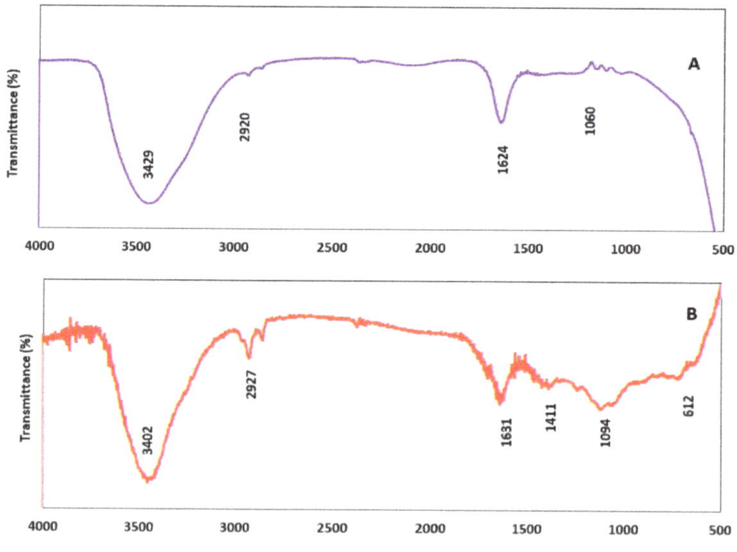

Figure 4. FTIR spectra of *Thymus vulgaris* extract (**A**) and the synthesized TVE-CuNPs (**B**).

3.3. Characterization of AgNPs

The AgNPs (1 mg/mL) utilized in this investigation were obtained commercially. In deionized water, the solution was diluted to 100 µg/mL (Figure 5A). TEM and UV–Vis spectroscopy were used to examine the size, shape, and homogeneity of the AgNPs. The absorbance spectra revealed a single high peak at 405 nm (Figure 5B), indicating the existence of spherical AgNPs. TEM verified that the particles were spherical in form (Figure 5C). The AgNPs had an average diameter of 20–50 nm (Figure 5D).

Table 1. Peaks were obtained by FTIR analysis and corresponding functional groups in *Thymus vulgaris* extract and TVE-CuNPs.

Thymus vulgaris Extract		TVE-CuNPs	
Peak (cm^{-1})	Functional Groups	Peak (cm^{-1})	Functional Groups
3429	Amine group (N–H) and the hydroxyl group (O–H)	3402	Amine group (N–H) and the hydroxyl group (O–H)
2920	C–H in methyl and alkanes	2927	C–H in methyl and alkanes
1624	C=O stretching of amide in carbonyl stretch in proteins/enzymes	1631	C=O stretching of amide in carbonyl stretch in proteins/enzymes
1060	C–O of proteins	1094	C–O of proteins

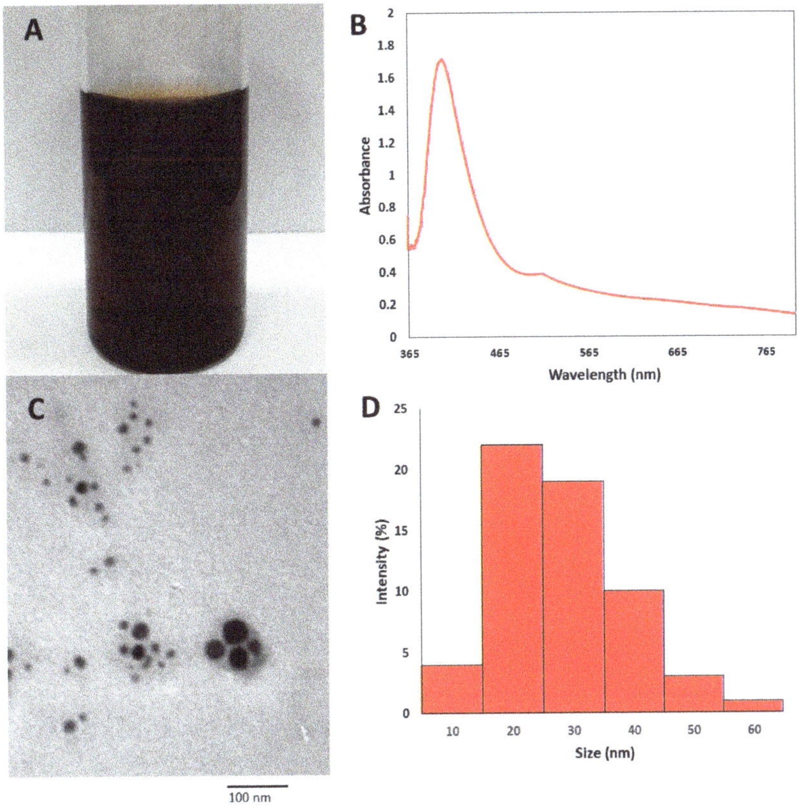

Figure 5. AgNPs characterizations: (**A**) AgNP solution; (**B**) UV–Vis spectrum absorption of the AgNPs; (**C**) AgNP TEM; and (**D**) AgNP size distribution.

3.4. Drug Release Determination

Figure 6 depicts the metronidazole release (%) from the modified GIC groups C and D, revealing a two-step release: The quick first-step release, which is the metronidazole release (%) from GIC, in groups 4 and 6, reached 58% and 64% after 24 h, respectively. After 240 h, the delayed second-step release reached about 84% and 87%, respectively.

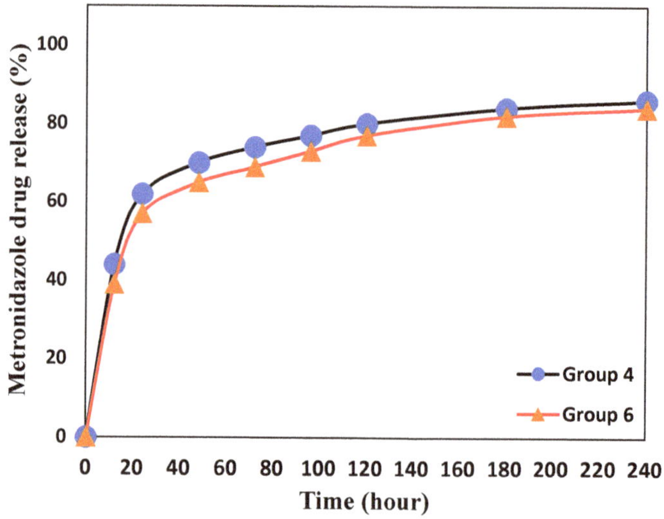

Figure 6. In vitro metronidazole release profile from GIC Group 4—1.5% metronidazole + 0.5% TVE-CuNPs + 98% GIC. Group 6—1.5% metronidazole + 0.5% AgNPs + 98% GIC.

3.5. Antimicrobial Activity of the Tested Biomaterials

The results of agar disc diffusion assays against *S. mutans* and *S. aureus* revealed that groups 3 and 4 of GIC had a significant effect on bacterial growth inhibition when compared to other groups, as shown in Figure 7A,B. Furthermore, as shown in Figure 7C,D, GIC groups 5 and 6 inhibited bacterial growth significantly more than GIC groups 1 and 2. This effect was more pronounced when metronidazole and TVE-CuNPs or AgNPs were added, as in groups 4 and 6, which showed the statistically highest inhibition zones compared to groups 3 and 5, which contained TVE-CuNPs or AgNPs, (Group 4 > Group 3 > Group 2 > Group 1), as well as with GIC groups (Group 6 > Group 5 > Group 2 > Group 1), as shown in Table 3. The efficacy of GIC combined with metronidazole and TVE-CuNPs in group 4 was slightly lower than that of GIC combined with metronidazole and AgNPs in group 6. The inhibition zone and significant effects of metronidazole and TVE-CuNPs or AgNPs were the greatest in groups 4 and 6.

The mean inhibitory zones of the modified groups of GIC in the case of *S. aureus* were significantly higher than the mean inhibitory zones of these groups in the case of *S. mutans*. However, when compared to the other tested groups, groups 4 and 6 had the highest inhibition zones (see Figure 7 and Table 3).

Figure 8 depicts the MDCT assay results of GIC groups against *S. mutans* and *S. aureus*. The modified GIC (Group 4 > Group 3 > Group 2 > Group 1) and modified GIC (Group 6 > Group 5 > Group 2 > Group 1) had a significant antimicrobial effect against *S. mutans* and *S. aureus* when the antimicrobial effects of the GIC groups were compared. These were modified after one hour. In comparison to other groups, the modified GIC groups with AgNPs or TVE-CuNPs with metronidazole were more effective. At the end of the month, groups 4 and 6 had the highest antimicrobial efficacy against *S. aureus* and *S. mutans* when compared to the other groups (Figure 8), though the antimicrobial efficacy of groups 3 and 4 was slightly lower than that of groups 5 and 6, respectively, against *S. aureus* (as shown in Figure 8A,C) and *S. mutans* (as shown in Figure 8B,D). As a result, when compared to the other groups, GIC combined with antimicrobial agents (AgNPs or TVE-CuNPs and 1.5% metronidazole) demonstrated a superior antibacterial activity against *S. mutans* and *S. aureus*.

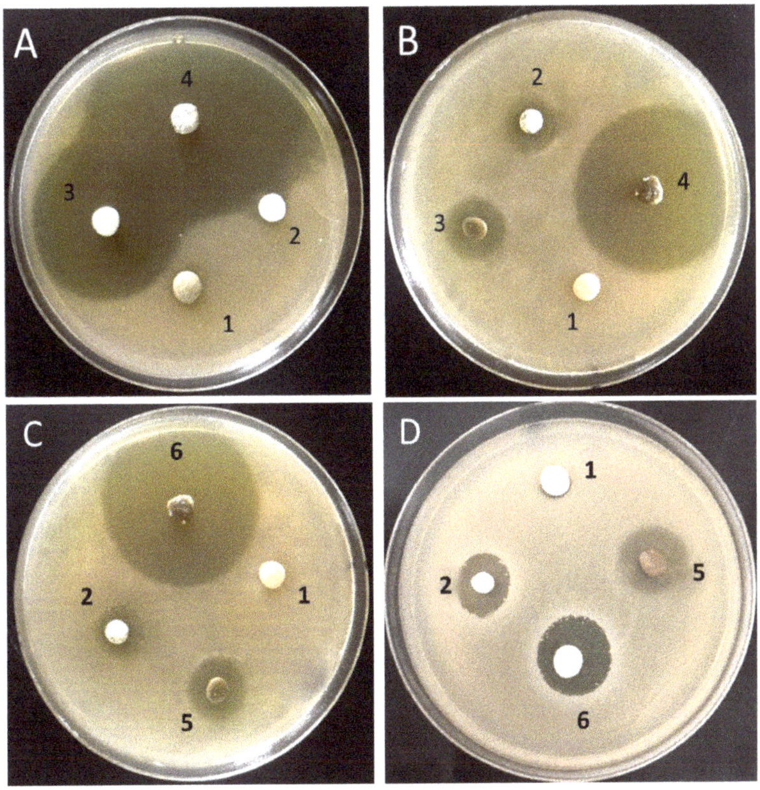

Figure 7. Antimicrobial activity of the biomaterials against *Staphylococcus aureus* (**A**,**C**) and *Streptococcus mutans* (**B**,**D**). 1—(Group 1) GIC; 2—(Group 2) 98.5% GIC with 1.5% metronidazole; 3—(Group 3) 99.5% GIC with 0.5% TVE-CuNPs; 4—(Group 4) 98% GIC with 0.5% TVE-CuNPs + 1.5% metronidazole; 5—(Group 5) 99.5% GIC with 0.5% AgNPs; 6—(Group 6) 98% GIC with 0.5% TVE-CuNPs + 1.5% metronidazole.

3.6. Compressive Strength Measurement

Table 2 shows the compressive strength of GIC in the following order: Group 3 ≥ Group 4 > Group 2 > Group 1 and Group 5 ≥ Group 6 > Group 2 > Group 1. Table 2 shows that, when GIC groups 4 and 6 were coupled with metronidazole and nano-copper or Ag-NPs, the compressive strength was slightly increased compared to GIC alone. Compressive strength measurements were not significantly different, indicating that the incorporation of metronidazole and/or nanoparticles did not reduce the strength of GIC for dental applications. The difference between all GIC groups was statistically insignificant ($p > 0.05$).

Table 2. Tested GIC samples' compressive strength (MPa).

Tested Groups	n	Mean ± SD	ANOVA F Value	ANOVA p Value
Group 1	8	41.2 ± 3.4	2.74	0.165
Group 2	8	42.8 ± 2.8		
Group 3	8	44.2 ± 3.8		
Group 4	8	43.9 ± 3.7		
Group 5	8	45.9 ± 4.1		
Group 6	8	45.0 ± 4.5		

1—(Group 1) GIC; 2—(Group 2) 98.5% GIC with 1.5% metronidazole; 3—(Group 3) 99.5% GIC with 0.5% TVE-CuNPs; 4—(Group 4) 98% GIC with 0.5% TVE-CuNPs +1.5% metronidazole; 5—(Group 5) 99.5% GIC with 0.5% AgNPs; 6—(Group 6)-98% GIC with 0.5% TVE-CuNPs +1.5% metronidazole.

Table 3. Antimicrobial efficiency of modified dental luting cements against the tested bacteria.

Bacteria	Samples (Groups)	Inhibition Zone (mm)			ANOVA *F* Value	ANOVA *P* Value	Tukey's Post Hoc
		1 Day	2 Weeks	1 Month			
S. aureus	a. Group 1	NA	NA	NA	NA	NA	NA
	b. Group 2	14 ± 1.62	13 ± 1.52	11 ± 0.57	7.62	0.02	1 day > 2 week and 1 month
	c. Group 3	20 ± 1.53	17 ± 1.21	15 ± 1.35	11.81	0.01	1 day > 2 week and 1 month
	d. Group 4	29 ± 1.11	22 ± 1.27	19 ± 1.56	13.43	0.001	1 day > 2 week and 1 month
	e. Group 5	24 ± 1.41	19 ± 0.57	16 ± 1.00	8.9	0.01	1 day > 2 week and 1 month
	f. Group 6	30 ± 1.71	23 ± 1.52	18 ± 1.15	14.30	0.001	1 day > 2 week and 1 month
	ANOVA F Value	16.71	14.65	17.35			
	ANOVA *p*-Value	0.0001	0.001	0.001			
	Tukey's post Hoc	f > d > e > c > b	f > d > e > c > b	f > d > e > c > b			
S. mutans	a. Group 1	NA	NA	NA	NA	NA	NA
	b. Group 2	18 ± 1.93	16 ± 1.67	12 ± 1.58	9.62	0.03	1 day > 2 week and 1 month
	c. Group 3	19 ± 1.26	16 ± 1.31	13 ± 1.55	11.9	0.01	1 day > 2 week and 1 month
	d. Group 4	26 ± 1.34	21 ± 1.88	18 ± 1.17	16.30	0.001	1 day > 2 week and 1 month
	e. Group 5	20 ± 1.65	18 ± 1.53	14 ± 1.66	11.9	0.01	1 day > 2 week and 1 month
	f. Group 6	27 ± 1.35	23 ± 1.93	17 ± 1.69	16.30	0.001	1 day > 2 week and 1 month
	ANOVA F Value	16.26	15.52	14.36			
	ANOVA p-Value	0.001	0.001	0.001			
	Tukey's post Hoc	f > d > e > c > b	f > d > e > c > b	f > d > e > c > b			

1—(Group 1) GIC; **2**—(Group 2) 98.5% GIC with 1.5% metronidazole; **3**—(Group 3) 99.5% GIC with 0.5% TVE-CuNPs; **4**—(Group 4) 98% GIC with 0.5% TVE-CuNPs +1.5% metronidazole; **5**—(Group 5)- 99.5% GIC with 0.5% AgNPs; **6**—(Group 6)-98% GIC with 0.5% TVE-CuNPs +1.5% metronidazole. All samples were analyzed in triplicate.

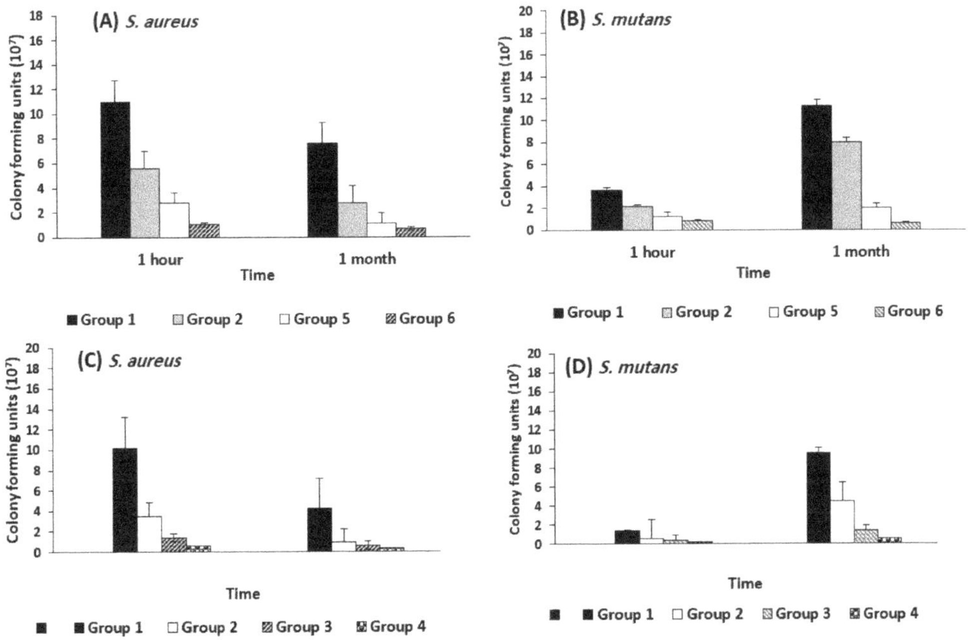

Figure 8. Number of CFU normalized to the control after one hour MDCT on the modified GIC cement and modified GIC cement after one month in suitable media (n = 3). 1—(Group 1) GIC; 2—(Group 2) 98.5% GIC with 1.5% metronidazole, 3—(Group 3) 99.5% GIC with 0.5% TVE-CuNPs; 4—(Group 4) 98% GIC with 0.5% TVE-CuNPs +1.5% metronidazole; 5—(Group 5) 99.5% GIC with 0.5% AgNPs; 6—(Group 6) 98% GIC with 0.5% TVE-CuNPs +1.5% metronidazole against *S. aureus* (**A**,**C**) and *S. mutans* (**B**,**D**).

Currently, metronidazole and TVE-CuNPs or AgNPs with GIC are successfully biofabricated and exhibit the highest antibacterial efficiency against resistant *S. aureus* and *S. mutans* when compared to other modified materials (GIC with TVE-CUNPs or AgNPs). Because of the AgNPs effect, the modified GIC with metronidazole and TVE-CuNPs had a slightly lower antimicrobial efficiency than the GIC with metronidazole and AgNPs. However, both groups 4 and 6 show superior antimicrobial efficacy when compared to the other tested forms, which adds to the study's novelty and significance in terms of controlling dental caries lesions.

4. Discussion

In this study, commercial AgNPs and TVE-CuNPs were synthesized using TVE and coupled with GIC with or without metronidazole to generate a new dental biomaterial. When GIC was mixed with the tested nanoparticles and metronidazole, its antibacterial effects were demonstrated when compared to GIC combined with AgNPs, TVE-CuNPs, or metronidazole alone. The current work validates the success of the biosynthesis of TVE-CuNPs from TVE based on our findings. The production of brown color in an aqueous solution of TVE-CuNPs was induced by surface plasmon resonance (SPR) excitation [21]. The production of TVE-CuNPs at the absorption band of around 573 nm due to SPR in TVE-CuNPs was verified by UV-Vis spectroscopy [22]. The TVE includes phytochemicals including phenol and protein amino acids, as well as enzymes that function as reducing and stabilizing agents for the greenly synthesized TVE-CuNPs and may also be responsible for the reduction of Cu+ to Cu0 [23]. TVE-CuNPs were spherical and monodispersed, with sizes ranging from 10 to 25 nm, and these results are in agreement with those of Suarez-

Cerda et al. [24]. The TVE-CuNPs XRD pattern revealed three main peaks at 2θ values. The XRD pattern's significant intensity and small breadth, and the sharpness of TVE-CuNP diffraction peaks indicated that the TVE-CuNPs were well crystalline. This study's XRD data revealed the purity of TVE-CuNPs, which is similar to prior research [25,26]. The presence of large absorption peaks in TVE's FTIR spectra proved the existence of reducing and stabilizing agents, such as proteins, amino acids, and phenolic chemicals (i.e., ellagic acid, thymol, -terpinene, -pinene, carvacrol) [23,25].

The AgNP characterization was undertaken. The spectra revealed a single high peak at 405 nm, indicating the presence of AgNPs, and TEM verified the particles' spherical form [27]. Smaller nanosilvers have a peak at 400 nm, but bigger nanosilvers have higher scattering and expanded peaks pushed towards longer wavelengths, as observed in earlier research [27,28].

The drug release from GIC groups 4 and 6 was carried out in two stages. The primary rapid release during the first 24 h is very dependent on the strength of the drug attachment and might be caused by weak electrostatic interactions between metronidazole molecules and the GIC surface, hydrogen bond breakdown, and non-covalent connections [29]. The surface of GIC is susceptible to chemical and hydrolytic deterioration. Water can infiltrate the resin matrix and enhance matrix solubility, which is accelerated by a low pH [30]. As a result, the inorganic component of the GIC on the surface can dissolve and metronidazole can be released. For the next 216 h, a protracted and progressive release profile was seen. The release during this extended stage might be attributable to medication molecules trapped in the cement's resin matrix. The two-step and prolonged release process is important in therapeutic treatments because the initial rapid release provides a therapeutic dosage rapidly and the subsequent long-term release maintains this dose over a long period of time. After 240 h, the released percentage was 86% of the total quantity.

The MDCT and agar disc diffusion assay results of GIC groups against *S. mutans* and *S. aureus* show that, when metronidazole and TVE-CuNPs or AgNPs were added, as in groups 4 and 6, there were statistically highest antimicrobial effect compared to groups 3 and 5, which contained TVE-CuNPs or AgNPs that were conjugated with SH groups in proteins and disrupted neuclic acids [6]. These findings are consistent with those of Mittal et al. [31] and Aguilar-Perez et al. [32], who found that the antimicrobial efficiency was affected by the concentration of metronidazole and nanoparticles used. Additionally, these findings are in agreement with the previous studies, which explain that the addition of nanoparticles to antibiotics enhances the antibacterial efficacy of clinically approved drugs [33] because metronidazole revives when mixed with TVE-CuNPs or AgNPs.

The findings of the GIC groups containing TVE-CuNPs were equivalent to those including commercial AgNPs against the tested bacteria, which exhibited somewhat decreased antibacterial activity, which was consistent with the findings of Zia et al. [34]. When compared to copper nanoparticles, silver nanoparticles were shown to be more reactive in inhibiting bacterial growth. According to Gad El-Rab et al. [35], *S. mutans* and *S. aureus* are very sensitive to nanoparticles containing antibiotics, which can be efficiently taken up by the bacteria because AgNPs and TVE-CuNPs regenerate the metronidazole to be active. Furthermore, AgNPs and TVE-CuNPs were conjugated with the SH group in proteins and damageed nucleic acids. In the current research, and for the first time, as shown in Table 4, GIC was combined with metronidazole and TVE-CuNPs or AgNPs, and the result demonstrates a significantly higher antimicrobial efficacy against the tested microbes when compared to the GIC with a single antimicrobial agent as well as the conventional GIC without any combinations [32,36,37]. Furthermore, the compressive strength difference across all GIC groups was statistically negligible ($p > 0.05$). These findings are consistent with those of Enan et al. [18], who discovered that adding nanoparticles to GIC did not result in a substantial improvement in compressive strength.

Table 4. Comparison of earlier research with restorative biomaterials for antibacterial activity against oral bacteria.

Materials	Concentration	Bacteria	Reference
• GIC + CuNPs	• 96% GIC + 4% CuNPs	• *S. mutans* and *S. sanguinis*	• Aguilar-Perez et al. [32]
• GIC + antibiotics	• 98% GIC + 2% antibiotic	• *S. mutans*	• Singer et al. [36]
• GIC + AgNPs	• GIC + 0.5% AgNPs	• *S. aureus*, *S. mutans* and *C. albicans*	• Ashour et al. [37]
• GIC + TVE-CuNPs + metronidazole	• 98% GIC + 0.5% TVE-CuNPs + 1.5% metronidazole	• *S. aureus*, and *S. mutans*	• This study
• GIC + AgNPs + metronidazole	• 98% GIC + 0.5% AgNPs + 1.5% metronidazole	• *S. aureus*, and *S. mutans*	• This study

Currently, the combination of GIC, metronidazole, and AgNPs or TVE-CuNPs is successfully produced and exhibits high antimicrobial activity against the tested strains because metronidazole revives when mixed with TVE-CuNPs or AgNPs. It appears to have superior antimicrobial efficacy when compared to other forms. The GIC groups containing TVE-CuNPs were equivalent to those containing commercial AgNPs against the tested bacteria, although they exhibited somewhat decreased antibacterial activity. These findings add to the novelty and significance of the study that can be applied in the dental field to prevent dental caries. In the future, we will study the antimicrobial activity of these novel materials in vivo.

5. Conclusions

To conclude the results of the present study, *Thymus vulgaris* extract is a potential natural product for the green biosynthesis of TVE-CuNPs by the reduction of copper salts using reducing agents, such as phenolics and proteins, in addition to the enzymes present in TVE and for forming TVE-CuNPs. When metronidazole was combined with AgNPs or TVE-CuNPs, it was revived, and this combination with GIC formed novel nanobiomaterials. The GIC groups containing AgNPs exhibited somewhat increased antibacterial activity when compared with the GIC groups containing TVE-CuNPs. So, this material has superior antimicrobial activity against the resistant tested strains. Finally, this novel bio-material has potential to increase antimicrobial efficacy against the resistant tested strains and will result in the development of cost-effective dental materials for controlling bacterial infections and dental caries in the future.

Author Contributions: Conceptualization, M.F.F., A.A.A., S.B., N.H.F., E.T.E., M.M.H. and S.M.F.G.E.-R.; methodology, S.B., A.A.A., N.H.F., E.T.E. and S.M.F.G.E.-R.; software, M.F.F., S.B. and S.M.F.G.E.-R.; validation, S.B., A.A.A., N.H.F., E.T.E., M.M.H. and S.M.F.G.E.-R.; formal analysis, M.F.F. and S.B.; investigation, M.F.F., S.B., A.A.A., N.H.F. and E.T.E.; resources, M.M.H. and S.M.F.G.E.-R.; data curation; writing—original draft preparation, M.F.F., S.B. and S.M.F.G.E.-R.; writing—review and editing, S.M.F.G.E.-R.; visualization, M.F.F. and S.M.F.G.E.-R.; supervision, S.B. and A.A.A.; project administration, S.B. and S.M.F.G.E.-R.; funding acquisition, S.B. and A.A.A. All authors have read and agreed to the published version of the manuscript.

Funding: This work was funded by Deanship of Scientific Research, Taif University (research group project No. 1/439/6084), Taif, Saudi Arabia. The authors declare that the funding bodies had no role in the design of the study, the collection, analyses, and interpretation of data, or in writing the manuscript.

Institutional Review Board Statement: The study was conducted according to the guidelines of the Declaration of Helsinki, and approved by the Institutional Review Board (or Ethics Committee) of Taif University (protocol code-41-1107-00152 and 1/12/2019).

Informed Consent Statement: Written informed consent was obtained from all subjects involved in the study.

Data Availability Statement: Data will be made available upon request from the corresponding author.

Acknowledgments: We would like to express our gratitude to Deanship of Scientific Research, Taif University, Taif, Saudi Arabia for financial support under the research project number (1/439/6084).

Conflicts of Interest: The authors declare that they have no conflict of interest.

Sample Availability: Samples of the compounds are not available from the authors.

References

1. Naik, R.G.; Dodamani, A.S.; Khairnar, M.R.; Jadhav, H.C.; Deshmukh, M.A. Comparative assessment of antibacterial activity of different glass ionomer cements on cariogenic bacteria. *Restor. Dent. Endod.* **2016**, *41*, 278–282. [CrossRef]
2. Wang, S.P.; Ge, Y.; Zhou, X.D.; Xu, H.H.; Weir, M.D.; Zhang, K.K.; Wang, H.H.; Hannig, M.; Rupf, S.; Li, Q.; et al. Effect of anti-biofilm glass-ionomer cement on Streptococcus mutans biofilms. *Int. J. Oral Sci.* **2016**, *8*, 76–83. [CrossRef] [PubMed]
3. Horowitz, A.M. Introduction to the symposium on minimal intervention techniques for caries. *J. Public Health Dent.* **1996**, *56*, 133–134; discussion 161–163. [CrossRef] [PubMed]
4. Liu, Y.; Ren, Y.; Li, Y.; Su, L.; Zhang, Y.; Huang, F.; Liu, J.; Liu, J.; van Kooten, T.G.; An, Y.; et al. Nanocarriers with conjugated antimicrobials to eradicate pathogenic biofilms evaluated in murine in vivo and human ex vivo infection models. *Acta Biomater.* **2018**, *79*, 331–343. [CrossRef] [PubMed]
5. Allaker, R.P.; Yuan, Z. Nanoparticles and the control of oral biofilms. *Nanobiomaterials Clin. Dent.* **2019**, 243–275.
6. Gutierrez, M.F.; Alegria-Acevedo, L.F.; Mendez-Bauer, L.; Bermudez, J.; Davila-Sanchez, A.; Buvinic, S.; Hernandez-Moya, N.; Reis, A.; Loguercio, A.D.; Farago, P.V.; et al. Biological, mechanical and adhesive properties of universal adhesives containing zinc and copper nanoparticles. *J. Dent.* **2019**, *82*, 45–55. [CrossRef]
7. Xu, V.W.; Nizami, M.Z.I.; Yin, I.X.; Yu, O.Y.; Lung, C.Y.K.; Chu, C.H. Application of copper nanoparticles in dentistry. *Nanomaterials* **2022**, *12*, 805. [CrossRef]
8. Almuhaiza, M. Glass-ionomer Cements in Restorative Dentistry: A Critical Appraisal. *J. Contemp. Dent.* **2016**, *17*, 331–336. [CrossRef]
9. Bollu, I.P.; Hari, A.; Thumu, J.; Velagula, L.D.; Bolla, N.; Varri, S.; Kasaraneni, S.; Nalli, S.V. Comparative Evaluation of Microleakage Between Nano-Ionomer, Giomer and Resin Modified Glass Ionomer Cement in Class V Cavities- CLSM Study. *J. Clin. Diagn. Res.* **2016**, *10*, ZC66–ZC70. [CrossRef]
10. Slavin, Y.N.; Asnis, J.; Häfeli, U.O.; Bach, H. Metal Nanoparticles: Understanding the Mechanisms behind Antibacterial Activity. *J. Nanobiotechnol.* **2017**, *15*, 65. [CrossRef]
11. Wang, Y.W.; Tang, H.; Wu, D.; Liu, D.; Liu, Y.; Cao, A.; Wang, H. Enhanced Bactericidal Toxicity of Silver Nanoparticles by the Antibiotic Gentamicin. *Environ. Sci. Nano* **2016**, *3*, 788–798. [CrossRef]
12. Castilho, A.R.F.; Duque, C.; Kreling, P.F.; Pereira, J.A.; Paula, A.B.; Sinhoreti, M.A.C.; Puppin-Rontani, R.M. Doxycycline-containing glass ionomer cement for arresting residual caries: An in vitro study and a pilot trial. *J. Appl. Oral Sci.* **2018**, *26*, e20170116. [CrossRef] [PubMed]
13. Mitchell, D.A. Metronidazole: Its use in clinical dentistry. *J. Clin. Periodontol.* **1984**, *11*, 145–158. [CrossRef]
14. Smith, A. Metronidazole resistance: A hidden epidemic? *Br. Dent. J.* **2018**, *224*, 403–404. [CrossRef] [PubMed]
15. Gad El-Rab, S.M.F.; Basha, S.; Ashour, A.A.; Enan, E.T.; Alyamani, A.A.; Felemban, N.H. Green Synthesis of Copper Nano-Drug and Its Dental Application upon Periodontal Disease-Causing Microorganisms. *J. Microbiol. Biotechnol.* **2021**, *31*, 1656–1666. [CrossRef]
16. Abolghasemi, R.; Haghighi, M.; Solgi, M.; Mobinikhaledi, A. Rapid synthesis of ZnO nanoparticles by waste thyme (*Thymus vulgaris* L.). *Int. J. Environ. Sci. Technol.* **2019**, *16*, 6985–6990. [CrossRef]
17. Felemban, N.H.; Ebrahim, M.I. Effects of adding silica particles on certain properties of resin-modified glass ionomer cement. *Eur. J. Dent.* **2016**, *10*, 225–229. [CrossRef]
18. Enan, E.T.; Ashour, A.A.; Basha, S.; Felemban, N.H.; Gad El-Rab, S.M.F. Antimicrobial activity of biosynthesized silver nanoparticles, amoxicillin, and glass-ionomer cement against Streptococcus mutans and Staphylococcus aureus. *Nanotechnology* **2021**, *32*, 215101. [CrossRef]
19. Hudzicki, J. *Kirby-Bauer. Disk Diffusion Susceptibility Test Protocol*; American Society of Microbiology: Washington, DC, USA, 2016.
20. Lewinstein, I.; Matalon, S.; Slutzkey, S.; Weiss, E.I. Antibacterial properties of aged dental cements evaluated by direct contact and agar diffusion tests. *J. Prosthet. Dent.* **2005**, *93*, 364–371. [CrossRef]
21. Chandraker, S.K.; Ghosh, M.K.; Lal, M.; Ghorai, T.K.; Shukla, R. Colorimetric sensing of Fe^{3+} and Hg^{2+} and photocatalytic activity of green synthesized silver nanoparticles from the leaf extract of *Sonchus arvensis* L. *New J. Chem.* **2019**, *43*, 18175–18183. [CrossRef]

22. Mallick, K.; Witcomb, M.J.; Scurrell, M.S. In situ synthesis of copper nanoparticles and poly(o-toluidine): A metal–polymer composite material. *Eur. Polym. J.* **2006**, *42*, 670–675. [CrossRef]
23. Adewale Akintelu, S.; Kolawole Oyebamiji, A.; Charles Olugbeko, S.; Felix Latona, D. Green chemistry approach towards the synthesis of copper nanoparticles and its potential applications as therapeutic agents and environmental control. *Curr. Res. Green Sustain. Chem.* **2021**, *4*, 100176. [CrossRef]
24. Dutta, D.; Phukan, A.; Dutta, D.K. Nanoporous montmorillonite clay stabilized copper nanoparticles: Efficient and reusable catalyst for oxidation of alcohols. *Mol. Catal.* **2018**, *451*, 178–185. [CrossRef]
25. Honarmand, M.; Golmohammadi, M.; Hafezi-bakhtiari, J. Synthesis and characterization of SnO2 NPs for photo-degradation of eriochrome black-T using response surface methodology. *Environ. Sci. Pollut. Res.* **2021**, *28*, 7123–7133. [CrossRef]
26. Agnihotri, S.; Mukherji, S.; Mukherji, S. Size-Controlled Silver Nanoparticles Synthesized over the Range 5–100 Nm Using the Same Protocol and Their Antibacterial Efficacy. *RSC Adv.* **2014**, *4*, 3974–3983. [CrossRef]
27. Paramelle, D.; Sadovoy, A.; Gorelik, S.; Free, P.; Hobley, J.; Fernig, D.G. A Rapid Method to Estimate the Concentration of Citrate Capped Silver Nanoparticles from UV-Visible Light Spectra. *Analyst* **2014**, *139*, 4855–4861. [CrossRef]
28. Ma, Y.-W.; Wu, Z.-W.; Zhang, L.-H.; Zhang, J.; Jian, G.-S.; Pan, S. Theoretical Study of the Local Surface Plasmon Resonance Properties of Silver Nanosphere Clusters. *Plasmonics* **2013**, *8*, 1351–1360. [CrossRef]
29. England, C.G.; Miller, M.C.; Kuttan, A.; Trent, J.O.; Frieboes, H.B. Release kinetics of paclitaxel and cisplatin from two and three layered gold nanoparticles. *Eur. J. Pharm. Biopharm.* **2015**, *92*, 120–129. [CrossRef] [PubMed]
30. De Paula, A.; De Fúcio, S.; Alonso, R.; Ambrosano, G.; Puppin-Rontani, R. Influence of chemical degradation on the surface properties of nano restorative materials. *Oper Dent.* **2014**, *39*, E109–E117. [CrossRef]
31. Mittal, S.; Soni, H.; Sharma, D.K.; Mittal, K.; Pathania, V.; Sharma, S. Comparative evaluation of the antibacterial and physical properties of conventional glass ionomer cement containing chlorhexidine and antibiotics. *J. Int Soc. Prev. Community Dent.* **2015**, *5*, 268–275. [CrossRef]
32. Aguilar-Perez, D.; Vargas-Coronado, R.; Cervantes-Uc, J.M.; Rodriguez-Fuentes, N.; Aparicio, C.; Covarrubias, C.; Alvarez-Perez, M.; Garcia-Perez, V.; Martinez-Hernandez, M.; Cauich-Rodriguez, J.V. Antibacterial activity of a glass ionomer cement doped with copper nanoparticles. *Dent. Mater. J.* **2020**, *39*, 389–396. [CrossRef] [PubMed]
33. Kaur, A.; Kumar, R. Enhanced bactericidal efficacy of polymer stabilized silver nanoparticles in conjugation with different classes of antibiotics. *RSC Adv.* **2019**, *9*, 1095–1105. [CrossRef]
34. Zia, R.; Riaz, M.; Farooq, N.; Qamar, A.; Anjum, S. Antibacterial activity of Ag and Cu nanoparticles synthesized by chemical reduction method: A comparative analysis. *MRE* **2018**, *5*, 075012. [CrossRef]
35. Gad El-Rab, S.M.F.; Ashour, A.A.; Basha, S.; Alyamani, A.A.; Felemban, N.H.; Enan, E.T. Well-Orientation Strategy Biosynthesis of Cefuroxime-Silver Nanoantibiotic for Reinforced Biodentine™ and Its Dental Application against Streptococcus mutans. *Molecules* **2021**, *26*, 6832. [CrossRef]
36. Rahman, S.A.; Umashankar, G.K.; Selvan, A.; Sharma, R.; Maniyar, R.; Kavya, M.J. To evaluate the antimicrobial efficacy of conventional glass ionomer cement incorporated with different antibiotics: An in-vitro study. *Int. J. Oral Health Med. Res.* **2016**, *3*, 30–34.
37. Ashour, A.A.; Basha, S.; Felemban, N.H.; Enan, E.T.; Alyamani, A.A.; Gad El-Rab, S.M.F. Antimicrobial Efficacy of Glass Ionomer Cement in Incorporation with Biogenic Zingiber officinale Capped Silver-Nanobiotic, Chlorhexidine Diacetate and Lyophilized Miswak. *Molecules* **2022**, *27*, 528. [CrossRef]

MDPI AG
Grosspeteranlage 5
4052 Basel
Switzerland
Tel.: +41 61 683 77 34

Antibiotics Editorial Office
E-mail: antibiotics@mdpi.com
www.mdpi.com/journal/antibiotics

Disclaimer/Publisher's Note: The title and front matter of this reprint are at the discretion of the Guest Editors. The publisher is not responsible for their content or any associated concerns. The statements, opinions and data contained in all individual articles are solely those of the individual Editors and contributors and not of MDPI. MDPI disclaims responsibility for any injury to people or property resulting from any ideas, methods, instructions or products referred to in the content.